TIME AND PROCESS:

Interdisciplinary Issues

TIME AND PROCESS:

Interdisciplinary Issues

The Study of Time VII

Edited by J. T. Fraser and Lewis Rowell

INTERNATIONAL UNIVERSITIES PRESS, INC.
Madison, Connecticut

Library of Congress Cataloging-in-Publication Data

Time and process: Interdisciplinary issues / edited by J T. Fraser and Lewis Rowell.
p. cm. — (The Study of time, ISSN 0170-9704:7)
Papers from a conference held at Glacier National Park in July 1989.
Includes bibliographical references and index.
ISBN 0-8236-6544-5
1. Time. 2. Process philosophy. I. Fraser, J. T. (Julius Thomas), 1923– . II. Rowell, Lewis Eugene, 1933– . III. Series
QB209.S85 vol. 7
[BD638]
115 S—dc20
[115] 92-973
CIP

Manufactured in the United States of America

Table of Contents

Foreword — *L. Rowell* — vii

1. Change, Permanence, and Human Values — *J. T. Fraser* — 1

Perspectives — 25

2. Cycles of Time: Calendrical and Astronomical Reckonings in Early Science — *W. M. Stevens* — 27
3. Process as a Primitive Physical Category — *C. J. S. Clarke* — 53

Perspectives — 71

4. Transformations of Kronos — *M. D. Akhundov* — 73
5. The Revolutionary Transition and the Modern Problem of Process — *M. Elchardus* — 87

Perspectives — 108

6. Central and Peripheral Mechanisms in Timing — *F. Macar* — 109

Perspectives — 125

7. Music as Process — *L. Rowell* — 127
8. Duration and Rhythmic Process in Music — *C. F. Hasty* — 147
9. Metaphor and Deep Time in Music — *G. L. Brodhead* — 167

Perspectives — 185

10. Time and Process in Marina Tsvetaeva's Definition of Lyric Verse — *O. P. Hasty* — 189

11. Order and Disorder: Creating Temporality in Literary Experience — *M. P. Soulsby* — 205

Perspectives — 217

12. Clarifying Concrescence in Whitehead's Process Philosophy — *L. F. Werth* — 219
13. The Process of Polity — *C. M. Sherover* — 243
14. Emerson and the Sacredness of Becoming — *B. P. Helm* — 263

Name Index — 283
Subject Index — 289

Foreword

To a musician a boulder is precisely what it seems to be—stable, enduring, perhaps even permanent; to a geologist it is a record of its history and the residue of a number of ancient and continuing processes. To a geologist Beethoven's *Eroica* Symphony may appear to be an enduring masterpiece, in the form of a stable, fixed text that can be actualized on demand—in effect a permanent musical object; to a musician the *Eroica* is heard as a series of dynamic, interwoven sound events and transitions—unless he or she has made the all too common mistake of confusing the printed score or vinyl disc with the essence of the work. Both are right, and both are wrong. A boulder and a symphony have little else in common, but each has been marked by the traces of its genesis and evolution, and in each case the illusion of permanence and stability is just that—an illusion.

We crave stability. Wishful thinking is a persistent habit of mankind, and we are easily seduced by the illusions of stability and permanence that encompass much of our ordinary experience. The present volume is a case in point: just as the towering geological structures of Glacier National Park may more accurately be seen as records of many ancient, ongoing processes, so is this volume the documentation of a complex process that has had multiple beginnings, events, stages, and transitions. Above all, it is the record of another forward step in the quarter-century evolution of the International Society for the Study of Time (ISST)—the Seventh Conference of the Society, held at Glacier National Park, Montana, U.S.A., in July 1989.

"Time and Process" was the theme of this conference, and is the theme of this book. Looking back, it seems inevitable that the idea of process would have emerged as a major concern for scholars engaged in the study of time from multiple perspectives, for the human experience of time is shaped mostly by time's contents, less so by what we may regard as time's "facts" (such as the temporal spectrum and how it can be measured). It is tempting to say that everything that occurs in time is a part of one process or another, but perhaps this too is wishful thinking and an attempt to find meaning in the contingencies of life by considering them to be parts of an ongoing process. So, although there may not be anything inherently processive in the idea of time (beyond the motions of the bodies by which we reckon it), it is clear that process is inherently temporal—in all its manifestations, real and supposed—and there is much we can learn about time from the study of process.

When we speak of process in general, we are immediately plunged into an uncomfortable level of abstraction, for which the best remedy is to make some concrete applications. We all have a reasonable notion of how certain processes work, from the gyrations of the stock market to the delicately brutal negotiations for peace between adversaries. But what is it that they have in common? Shall we adopt J. T. Fraser's definition of process as "change that follows a stable pattern long enough for us to recognize continuity, transient as the continuity itself may be"? Or, in the words of my address, as "a series of progressive and interdependent steps by which an end is attained"? Or is it "the interweaving of invariance and variance" (Elchardus), "a tension between linear succession and sequential recurrence" (Gould, as cited by Brodhead), "a developmental career" (in this case, the career of a social body, according to Sherover), or "a becoming of continuity" (Whitehead, as interpreted by Werth)? Each offers unique insights.

Perhaps an even simpler way of approaching the issue is to recognize that to label a course of events and changes as a process is one of the most basic and instinctive habits of human explanation. To examine this explanatory tactic in the thinking of various disciplines is to grasp more fully how humans have come to understand the nature of their world and have arranged their own affairs in meaningful sequences. For readers who are willing to suspend their search for a single, all-encompassing definition of process and follow the efforts of scholars in diverse fields of study as they seek to identify and understand the processive nature of their subjects, the ISST offers this volume of studies in aesthetics, calendar making, literature, music, mythology, philosophy, political theory, psychology, physics, and sociology.

One of the main tasks of this foreword is to acknowledge the substantial contributions of many of our members. First of all, the continuing efforts of J. T. Fraser, the founder and tutelary genius of the ISST, deserve the continuing gratitude of all who share our curiosity about time. He is also the coeditor of this volume. The theme of the 1989 conference sprang from the fertile brain of Jacob A. Arlow, M.D., a member of the ISST Council and the Conference Committee. I am grateful to the other members of the Conference Committee: former presidents George H. Ford and David Park, Richard A. Block, Richard Martin, and John G. Weihaupt.

One of the highlights of the meeting was a special presentation by Gert H. Müller, our fourth president; the session in which he presented his "Reflections about J. T. Fraser's Theory of Time," followed by Dr. Fraser's response, will long be remembered by those present. The editors regret that the lack of recording facilities made it impossible to recapture this spontaneous exchange of ideas for the present volume. A second highlight, similarly beyond our power to preserve in its original form, was Jonathan D. Kramer's lecture-demonstration on "Time and the Composer."

Part of the business of the conference was the election of a new slate of officers. Samuel L. Macey, a distinguished scholar of eighteenth-century English literature and Professor Emeritus at the University of Victoria, was elected President, and I offer him my congratulations. Professor Richard A. Block, of Montana State University, takes over as our new Treasurer after a spectacularly successful six-year term of service by Mark H. Aultman, to whom the Society owes an enthusiastic vote of thanks. I am also grateful to outgoing Council members Anindita Balslev, George H. Ford, Jonathan D. Kramer, John A. Michon, and Helga Nowotny. New members of the ISST Council are as follows: Richard A. Block, A. I. Rabin, Charles Sherover, Marlene P. Soulsby, Wesley M. Stevens, and Lord Young of Dartington. It should also be noted that Professor Rabin has succeeded President Macey as Membership Chairman, and Marlene Soulsby has taken over the editorship of the ISST newsletter *(Time's News)* from the capable hands of Richard Martin, who has served nobly for a number of years.

Finally, we remember with sadness members of the Society who have passed away: Gilbert Cantor, F. C. Haber, Hans Kalmus, Frederick T. Melges, and Howard Trivers. Their contributions live on in the process by which the present volume was generated.

Lewis Rowell
President, 1986–1989

Contributors

Murad D. Akhundov, Ph.D. Professor, Institute of Philosophy, Moscow, Russia; Visiting Scholar (1991–1993), Russian Research Center, Harvard University, Cambridge; his special fields are philosophy and history of science; author of *Conceptions of Space and Time: Sources, Evolution, Directions* and other books.

Garry L. Brodhead, Ph.D. Professor of Music Theory, Ithaca College, Ithaca, New York.

Christopher J. S. Clarke Professor of Applied Mathematics, University of Southampton, U.K.; works in relativity theory, biomagnetism, and consciousness physics; author (with F. de Felice) of *Relativity on Curved Manifolds*.

Mark Elchardus Professor of Sociology: *Vrije Universiteit* (Free University) *Brussels;* Associate Editor of *Time and Society;* member of the board of the *Programme in European Leisure Studies* (Pels); his special areas of interest are the temporal dimensions of social life, and the analysis of culture, especially with regard to meanings, motivations, and politics.

J.T. Fraser Founder of the International Society for the Study of Time; author of *Time, the Familiar Stranger, Of Time, Passion, and Knowledge,* and other time-related books, and editor of the eight volumes of *The Study of Time* series.

Christopher F. Hasty, Ph.D. Associate Professor, The University of Pennsylvania.

Olga Peters Hasty, Ph.D. Lecturer, Department of Slavic Languages, The University of Pennsylvania; specialist in Russian literature of the early twentieth century; coauthor of *America Through Russian Eyes* and numerous articles on Russian poetry.

Bertrand P. Helm, Ph.D. Professor of Philosophy and Distinguished Scholar, Southwest Missouri State University, Springfield, Missouri; author of *Time and Reality in American Philosophy*.

Françoise Macar Director of Research, Centre National de la Recherche Scientifique, Laboratoire de Neurosciences Cognitives, Marseille, France; author of *Le Temps: Perspectives Psychophysiologiques;* editor with Viviane Pouthas and William J. Friedman of *Time, Action and Cognition: Towards Bridging the Gap*.

Lewis Rowell Professor of Music, Indiana University; past President, Treasurer, and Member of the Council, The International Society for the Study of Time; author of books and articles on Indian music and the philosophy of music and editor of *Music Theory Spectrum* (1982–1985).

Charles M. Sherover, Ph.D. Professor of Philosophy, Hunter College, City University of New York; author of *Time, Freedom and the Common Good: An Essay in Public Philosophy, Heidegger, Kant, and Time, The Human Experience of Time,* and of numerous articles, as well as editor of an anthology in democratic theory, and translator of Rousseau.

Marlene Pilarcik Soulsby Associate Professor of German and Comparative Literature, Penn State University, Worthington Scranton Campus; author of articles on time and literature, East–West literary relations, and foreign language pedagogy.

Wesley M. Stevens, Ph.D. Professor of History, University of Winnepeg; author of "Bede's Scientific Achievement" (The Jarrow Lecture for 1985), a critical edition of the Latin *Computus of Hrabanus Maurus* (1979), and many articles about mathematics, astronomy, and developement of computistical studies in early medieval schools; awarded a Humboldt-Forschungspreis 1992 by the Alexander von Huboldt-Stiftung (Germany).

Lee F. Werth, Ph.D. Associate Professor of Philosophy, Cleveland State University, Cleveland, Ohio; publications focus on the philosophy of time, metaphysics, and foundations of science.

I

Change, Permanence, and Human Values

*J. T. Fraser**

Abstract The nature of the true, the good, and the beautiful have been of continued interest to reflective thought at least since the birth of dialectical reasoning in ancient Greece. In contemporary scholarship, the study of what constitutes truth is of concern mainly to epistemology, the scope and nature of the good is dealt with in ethics, the ramifications of what is judged beautiful is a subject of aesthetics. Although these three domains of critical thought embrace a substantial portion of the world of the intellect, it is difficult to find within their boundaries a recognition of the significance of time awareness in the generation of human values.

This paper examines the role that the experience and idea of time have played in the creation, maintenance, and revision of values. The framework for the inquiry is the hierarchical theory of time.

The need for human values is seen as arising from the necessity to control the unbounded imaginative powers of the mind. That control takes the form of collectively approved guidelines for beliefs, formulated in terms of truths and untruths; of guidelines for conduct, stated in terms of right and wrong; and guidelines for the management of emotions, in terms of the beautiful and the ugly.

To relate human values to time and remain consistent with the general usage of the terms, truth is defined as a predicate which asserts that a belief is judged permanent; good is defined as a predicate which asserts that certain conduct, intent, or character promotes stability in the affairs of man and society; beautiful is defined as a qualitative judgment upon feelings which we would like to remain lasting. When so understood, human values are seen to join the perennial attempts of our species to oppose the passage of time.

Normative teachings about values do so in many ways, among which their bid for permanence is the most obvious. They generally maintain that believing in what they say is true, doing what they claim is right, and appreciating what they judge beautiful will help achieve stability in the life of the individual and society here on earth, or at least for the person post mortem, elsewhere. But the witness of history suggests that value judgments as guides to thought and action—such as the pursuit of truth through

*The Founder's Lecture, July 10, 1989.

science, justice through law, or beauty through the arts—have never secured lasting balance among the passions of mind, body, and society.

The purpose of the paper is to defend the thesis that the role of human values in the development of civilizations has not been one of leveling. On the contrary, human values have helped generate and keep alive those unresolvable conflicts of the mind and of society which have been responsible for the stunning creativity and frightening destructiveness of our species.

This paper interprets the nature of human values in terms of the hierarchical theory of time. It concludes that although value systems often claim that abiding by their judgments would secure stability in the individual and communal affairs of people, the evolutionary role of value judgments has been one of creating and maintaining instability and thereby promoting social change and individual development.

The Need for Value Judgments

The need to select responses appropriate to the immediate challenge of food, friend, or foe must be coeval with life itself. Such responses, useful for the perpetuation of the life of the species or the survival of one of its members, may be said to be of value for that species or for its member. This meaning of value also applies to men and women insofar as they are biological organisms. Protecting a child, for instance, involves patterns of behavior that humans share with all other species that protect their offspring.

But making sacrifices for an ideal, a belief, or in behalf of a person long dead is a human trait not shared with any other species. Protecting an idea, for instance, and trying to assure the continuity of its life requires patterns of conduct that could not have arisen before the human brain crossed a certain threshold of complexity, beyond which it could function in ways we call mental. The set of peculiarly human motivations that are said to justify such conduct are usually associated with the ideas of the true, the good, and the beautiful. As a group, these ideas constitute the family of human values.

To understand the origins of human values in the mind and, through that understanding, gain insight into the roles played by them in the lives of persons and civilizations, it is useful to take a historical and developmental approach. Accordingly, we begin by inquiring into the evolutionary origins of mental functions. By mental functions I mean an ability to perform symbolic transformations of experience and, with those symbols, draw a map of a world richer and freer in important ways than the material and living world from which it arose.

An example of such a transformation would be the changing of sense impressions about tigers into the line, "Tiger! Tiger! burning bright," writing down the line, and by that two-step process—a creative and a communicative one—adding one's self-expression to the myriad variety of cultural goods held by a civilization.

The evolutionary origins of mental functions may be closely identified with the coming about of our species. Guided by what is known about the emergence of our species from its hominid ancestors, I would describe that process as one which involved several simultaneous and mutually reinforcing changes. I would also imagine that in their collective effects they resembled positive feedback in that they amplified change, rather than negative feedback which would have maintained stability.

Those steps surely included an increasing division of labor together with refinements in cognitive powers that evolved toward the recognition of a highly differentiated environment. One may also safely assume the appearance of increasingly articulated methods of communication that helped enlarge a store of signals that stood for themselves, by adding a store of signs that stood for events and things other than themselves.

I imagine that these changes conspired to expand the temporal horizons of the early hominids. At one time, the future-related behavior of these hominids must have been almost entirely instinctive, that is, genetically based. Yet, at some later date a nongenetic type of conduct came to allow planning as a complement to instinctual action. During the eons that witnessed the coming about of nongenetic, goal-directed behavior the rudiments of the human sense of time, with its horizons expanding beyond the immediate future and past, must have been born.

Sooner or later, each member of the species must also have noted the existence of a curious object. It shared the external world with sticks and stones and creatures that moved, but it could also feel hungry, experience anger, and desire a mate. This object was the distant forerunner of the self. Somewhere along our ancestral history, the inevitable demise of that object must have come within the scope of a thinkable future. That discovery, I suspect, extended mortal terror—which is an episodic response to immediate threats to life—by a type of concern unknown in the animal world. It was a lingering anxiety related to the necessary passing and the eventual demise of the self. Nonpassing permanence—expressed or expressible in structures, whether physical or mental, then came to acquire a special appeal to the creature who knew that sooner or later, he had to die.

Besides suggesting to the mortal self that death may only be a mirage, being able to perceive the world in terms of structures and processes that appeared unchanging had its advantages in the strategy of survival. For that reason, so I speculate, it was favored by selection. Specifically, images—thoughts—of permanent objects and patterns of behavior made it possible to play out scenarios of future behavior and, with their assistance, place present conduct in the service of long-term planning.

Let me now make the deliberate mistake of representing human time by a straight line, with one of its ends standing for the conception of a person, the other for his death. On this geometrical metaphor the beginning and ending of life are symmetrical events. But in our minds the relation between the beginnings and endings of our lives are anything but symmetrical because the self, which will die with the body, did not originate with conception but was mentally constructed during the first years of life. It is "I" who will die, but the fertilized egg in my mother's womb was never "I the egg" but "the egg from which I came."

Where a person was before he or she was conceived is a question of limited interest, at least to the Western mind, whereas the question of where a person will be after death, if anywhere, mobilizes the whole spectrum of human emotions as well as the intellect. A longing for immortality in the form of life after death, made possible by symbols which stood for the self and could be thought of as surviving the body, has informed the development of all civilizations.

When our distant ancestors began burying food and weapons with their dead, they attested to their belief that death was not final. To borrow a line from Dylan Thomas, they expressed their "rage against the passing of the light." To be able to live with this rage, people created a variety of opiates—objects, acts, and ideas—that could make the passage of time appear unreal or, at least, insignificant. In his novel *Two on a Tower,* Thomas Hardy (1905, p. 36) makes his protagonist remark that if "you are restless and anxious about the future, study astronomy at once. Your troubles will be reduced amazingly. But your study will reduce them in a singular way, by reducing the importance of everything" of ordinary human concern.

These opiates of all people, these bids for conquering time, when taken together are the stuff of human knowledge. Among the many forms of theoretical and practical knowledge, we have to include the ability to construct scales of values pertaining to beliefs, to conduct, and to the management of emotions. Human values then join the other efforts of our species directed toward lessening the threat of passage. They disclose that role in their self-descriptions: I know of no assertion of truth, or teaching about right and wrong, or the claim of beauty for a piece of art or a poem that would not make a claim for a measure of permanence.

But, if human time and human values are part and parcel of the same constellation of feelings and thoughts, then a natural philosophy of time should be an ideal location for an epistemology of human values. For, through the many fields of knowing that contribute to the interdisciplinary study of time, the architect of such an epistemology would have direct access to insights relevant to the origins, scope, and nature of values.

In what follows, I shall attempt to sketch the outlines of an epistemology built on a particular natural philosophy of time.

The Hierarchical Theory of Time

The hierarchical theory of time, also called the theory of time as conflict, is a conceptual system in the natural philosophy of time. As is often the case with theories that are built on assumptions alien to established views, it is difficult to understand some of the details of the theory before it is understood as a whole, and it is difficult to understand the theory as a whole before its details are understood.

Yet, I cannot elaborate here the content of that theory in any detail but only sketch those of its teachings that are relevant to the theme of change, permanence, and human values. Unfortunately, such a sketch tends to leave the impression that the ideas outlined were born ready and complete, as was Athene from the head of

Zeus. Of course, they were not so born. Those who would want to learn more about the hierarchical theory of time may wish to consult some of my other writings.[1]

The theory adopted, and extended, the idea of stable integrative levels of nature, argued in the works of Joseph Needham and others, who speak about those levels as being arranged along a scale of increasing complexity. Let three works represent the extensive literature: Needham (1944, pp. 233–272), Needham (1968, pp. 106–169), and Pattee (1970). Specifically, the theory recognizes five distinct integrative levels, each possessing a different complexity, each governed by different kinds of laws and logics, and each characterized by a different temporality. The integrative levels form a nested hierarchy of increasing richness of content as we proceed from the world of electromagnetic radiation to that of particles, to massive physical systems, to that of life, and of the mind of man. As we rise along the scale of complexities, each temporality is seen as subsuming in it that or those beneath it, and each allows for qualitatively new forms of creative freedom.

Let me define and describe these temporalities, starting with the immensely rich yet most familiar kind of time: that of the human experience of time's passage.[2]

Nootemporality or noetic time. This is the temporal reality of the mature human mind. It is characterized by a clear distinction among future, past, and present, by unlimited horizons of futurity and pastness, and by the *mental present* with its vast temporal horizons.

Biotemporality, or biological time. This is the temporal reality of living organisms, including man as far as his biological functions are concerned. It is characterized by a distinction among future, past, and present. But the horizons of futurity and pastness are very limited when compared with those of noetic time. The boundaries of the *organic present* are species-specific and appear to be fixed.

It is with respect to the organic and mental presents that future and past, and with them the metaphor of a flow of time, acquire meaning.

Eotemporality is the time of the physicist's *t*. So named after Eos, the Greek goddess of dawn, this is the temporal reality of the astronomical universe of massive matter. It is continuous but is without a definable now and therefore necessarily undirected. It is a kind of time to which our ideas of a present, future, or past cannot be applied and hence it cannot be said to "flow."

The time of the world of elementary objects bears the name *prototemporality* for proto-, the first in a series. It is an undirected time and also one in which events have only probabilistic locations.

The world of electromagnetic radiation is termed *atemporal*. Atemporality does not signify nothingness but a condition to which none of our ordinary notions

[1]*Of Time, Passion, and Knowledge* (Fraser, 1990) remains the main reference. The dialectical nature of the theory was elaborated in *Time as Conflict* (1978a). The theory's implications for physics were the subject of *The Genesis and Evolution of Time* (1982). A good first entry is *Time, the Familiar Stranger* (1987).

[2]The idea of different temporalities is closely tied to the question of what is to be meant by their reality and by reality in general (see the definitions and explications given, with slightly different foci, in Fraser [1982, pp. 19–23; 1987, pp. 108–110; 1990, pp. 75–76]).

associated with time apply. It is akin to Plato's idea of complete chaos (quite unlike the chaos of chaos theory) and to eternity, if thought of as a state without time.

From the atemporal, the most primitive level of time, let me pass back "up" to a very complex kind of time. *Sociotemporality* is the name given to time on the organizational level of societies. Although much has been written about it, it is extremely difficult to identify its hallmarks with the same authority with which we can describe those of the lower order temporalities. The reason for the difficulty is, I believe, that we are both the makers and the subjects of sociotemporality and hence an observational platform, a stable perspective, is not, or not yet, available.

The five, or perhaps six temporalities constitute the *canonical forms of time*.

The stable integrative levels of nature, as already mentioned, are distinct in their complexities. Complexity is defined with the assistance of the relationship between minimal programs and complexity, an extension of Gödel's incompleteness theorem proposed by Chaitin (1975). This relationship is then so generalized that it could be applied to the sets of members that comprise an integrative level of nature. This generalization made it possible to obtain numerical values for level-specific complexities. The numbers so obtained fall into five distinct ranges in apparent support of the qualitative differences among the integrative levels, a fact which is intuitively obvious.[3]

The distinction among integrative levels in terms of their complexities recommended that level-specific languages, causations, and with them increasing regions of undeterminacy be also recognized.

At this point the theory made one of those leaps of imagination that are not at all arbitrary, yet may be shown as having been justified only ex post facto. For it is not possible to arrive at them through deductions from propositions admitted as correct by earlier reasoning.

That leap of imagination was the postulate that the canonical forms of time are not different aspects of time which become noticeable as we rise to systems of increasing complexity, but are stations along the open-ended evolutionary development of nature.

Although speaking about the evolution of time seems to be a contradiction in terms, there should be no difficulty—on the basis of logic—in admitting the possibility of a genesis and evolution of time. It is generally held that the expanding universe does not expand to fill preexisting space but expands space itself in a manner that might be called creative. All that the hierarchical theory of time claims in this respect is that time does likewise. It does not evolve within a preexistent expanse of time but creates new kinds of temporalities. But whereas the evolution of space is one of quantitative increase, the evolution of time involves quantitative as well as qualitative steps both in the physical world and upon the integrative levels of life, mind, and society. Such qualitative changes include increasing degrees of freedom or unpredictability.

[3]The details are rather technical, but the rounded out numbers that can stand as measures of complexity on a logarithmic scale are as follows: 0 for the atemporal, 2 for the prototemporal, 6–7 for the eotemporal, 30–40 for the biotemporal, 10^9 for the nootemporal (see Fraser [1982, pp. 154–156, 167]).

Before turning to the issue of human values, it is necessary to speak about one other teaching of the theory: that of the nature and origins of existential or constitutive tension. The best way to introduce that idea is by describing what the theory holds about the origins of life.

Imagine that in the Precambrian Era certain molecular aggregates began to display a collective rhythm, such as for instance, being able to absorb or emit energy in synchrony with the tides. If such an internalized and expressed rhythm offered any advantages for survival then it was selected for, and organic evolution was on its way.[4]

From one point of view, having internalized and expressed the tidal rhythm was a step in adaptation: it made the behavior of these molecular clock shops fit the appropriate niches of the cyclic spectrum of their environment. But the same improvements in this speculative scenario also opened the way for finer tuning to cycles that were unrecognizable earlier, such as for instance to the spring and neap tides or the semidiurnal inequality of the tides. After these and other periodicities were internalized, they had to be coordinated with each other and with yet other cycles, such as with the daily and yearly motion of the sun.

Each step of improved adaptation refined and enlarged those domains of organic functions where the expected and prepared-for conditions could differ from the conditions actually encountered. These ever-present differences amounted to a form of stress, to a chronic condition of life, to needs whose satisfaction had to be regularly delayed. I called this need existential tension. (In Fraser [1990] see index entries under "Existential tension.")

With the appearance of man 3½ billion years after the first forms of life, the scope of existential tension became greatly enlarged. For, to the biological needs that had to be satisfied there came to be added new needs for possible, as well as impossible things, events, and conditions. Existential tension in humans became manifest as an all-informing hunger for food, for mate, for knowledge and power, for life, and even for death.

The idea of existential tension, when interpreted in terms appropriate to the different integrative levels, is the basis of the metaphysics of the hierarchical theory of time, giving it its alternate name: the theory of time as conflict.

That metaphysics sees life, mind, and society as processes defined by their unresolvable, creative conflicts.

Here process means change that follows a stable pattern for a long enough time for us to recognize continuity, transient as the continuity itself may be.

Conflict means the coexistence of two opposing trends, regularities, or groups of laws in terms of which those processes may become intelligible.

Unresolvable means that if those conflicts cease, so does the integrity of the process.

[4]On this see Fraser ([1987, pp. 136–142]; index entries under "Biogenesis" [1990]; the chapter on "Biogenesis and Organic Evolution" [1982, pp. 144–175]).

Creative means that under certain conditions the conflicts may give rise to the processes and laws of a higher integrative level—itself subject to the principles of unresolvable conflicts.

Let me now illustrate the meaning of what was just said.

The life process is often thought of as one of self-organization or growth. This view is valid but it is only half of the story. To make it the complete story, one must think of life as a conflict between growth and decay, maintained and coordinated from instant to instant in the organic present. This conflict is unresolvable because if it ceases, if there is no more growth that may be opposed by decay, then the animal or man is either dying or dead. The existential tensions of life may thus be seen as functional corollaries of biotemporality.

Likewise, the process of minding—the use of one's mind—certainly involves the brain's labor of self-organization. Again, this view is only half of the story. To make it the whole story, it is necessary to think of minding as a conflict between those processes of organization and decay that go on in the brain when we think. This conflict is also unresolvable because if it ceases, such as in senility, if there is no more selfhood that must continuously struggle against the loss of identity, then there may still be a living body but not a mind. The existential tensions of the mind may thus be seen as the functional corollaries of nootemporality.

Analogous arguments hold for society. It is true enough that society is a group of people organized for a common purpose, but once again it is only one half of the story. It is better to think of society as a process, a conflict between the forces of growth and decay. If that conflict vanishes, such as when social organization collapses, there may still be persons around, but not a society.[5]

Now, with the hierarchical theory of time as an aid in the inquiry, we may turn to an epistemology of human values.

Time and Human Values

Let human values be represented by the ideas of the true, the good, and the beautiful. I do not mean Platonic Ideas lodged in eternity, but constellations of thoughts and feelings ordinarily designated by those notions.

I intend to argue that truth and untruth are socially approved guidelines to the selection of beliefs; good and evil are guidelines to conduct; the beautiful and the ugly are judgments upon the quality of feelings, useful in the management of

[5]The principle of unresolvable conflicts may be extended down into the domain of undirected temporalities, where it takes the form of opposition between entropy decreasing and increasing processes. Tracing these processes through physical cosmology to their cosmic origins suggests that what arose from the primeval chaos was not order but conflict between orderability and disorderability. This new perspective upon the standard model of cosmic creation—when traced forward in time—permits the recognition of continuity between inorganic and organic evolution and leads to an interpretation of biogenesis in the temporal domain (see Fraser [1978a] index entries under "Conflicts/Resolutions/Fiascos" and Fraser [1990] under "Conflicts").

emotions. I will consider each one of these categories of values in terms of the antinomies of change and permanence, in the context of the hierarchical theory of time.

Guidelines to Beliefs: Truths and Untruths

As a working definition, I propose that we think of truth as an assertion that a belief is judged permanent.

How may one test whether a belief deserves to be called permanent and therefore represents truth? Perhaps by making predictions based on the belief and test their correctness in the course of time. Examples may be: "Ten years from today Mars will be at its closest point to earth because its orbit is elliptical"; "by the turn of the twentieth century AIDS will have killed over 60 million people"; "Christ will come to judge the living and the dead."

Many readers will recognize the proposed prescription for testing for truth as having been borrowed from the practice of the exact sciences, where predictability is judged a necessary test for demonstrating verity. But the kind of testing for permanence that works well for physics cannot be uncritically extended to the testing for truth in the complex processes of life, human behavior, and society, for reasons that need to be explained.

The notion of permanence or changelessness, implicit in physical theory and experiment, first entered Greek thought some twenty-four centuries ago. It was then, as it is today, an idealization based on a knowledge of nature limited by whatever is available to sense impressions. The limitation of the received meaning of permanence is discovered, however, as soon as those domains of nature that can only be explored through scientific instruments and abstract, mathematical reasoning are included in the survey. Specifically, one can find only permanence–analogues; that is, aspects of the world that remind one of the ordinary notion of permanence but do not correspond to it. The character of the permanence-analogues are seen to depend on the integrative level one considers. As they vary, so does the relation of truth to time.

To illustrate the meaning of these remarks, I will first discuss what corresponds to the idea of permanence in the integrative levels of directed temporalities (those of life, mind, and society), then consider the physical world of undirected temporalities, then return again to the domains of directed temporalities so as to gain a perspective upon the distinctions among physical, biological, mental, and social forms of truth.

In the biological domain, organic evolution separates those aspects of the umwelt of a species that appear permanent from those that do not.[6] Behavioral traits

[6]"Umwelt" is a German word first used by the theoretical biologist Jakob von Uexküll to signify a concept that has been variously translated as "self-world," "phenomenal world," and "species-specific universe." Because of the awkwardness of all these translations, the word became naturalized in English. The most satisfactory definition of "umwelt" is given in English and English (1958, s.v. "umwelt") as "the circumscribed portion of the environment which is meaningful and effective for a given animal

of food, friend, and foe which regularly repeat themselves are retained by natural selection and are deposited as cumulative residues in the programs of biological inheritance. Genetic material constitutes biological memory of what has been learned about permanence through the evolutionary history of a phylum.

This biological memory (the permanence-analogue of biology) is quite unlike computer memory (which is an example of physical permanence) because of the vastly different complexities of organic and inorganic systems. Whereas a computer printout is a one-to-one mapping of what has been stored in the computer and is, therefore, predictable, genetic memory retains only regions of possibilities. This is the reason why phenotypic development and behavior—unlike planetary motion—is predictable only in very broad terms. The structures and functions of the phenotype are the results of responses, permitted by the genotype, to the unpredictable challenges of the developmental environment.

In man the slow biological learning process of sorting out the permanent (in the sense explained) and the contingent is enlarged by mental learning. Also, the horizons of time for which predictions could be made were opened up through a recognition of increasingly more refined patterns in the environment. While the ability of the mind to make predictions based on recognizable stable patterns of the environment has thus reached its present level of great sophistication, human conduct acquired an equally impressive level of unpredictability. The freedom so gained has been sufficiently broad to have permitted the rise of civilizations. The myriad details of those civilizations, though governed by the lower order regularities of physics and biology, are unpredictable from those lower order principles.

As I already mentioned, permanence, expressed in structures and processes that seem to be unaltered by time, has a special appeal for humans. There is even support for the belief that we instinctively search for whatever appears permanent (see the section on "Resolution of Perceptual Conlicts" in Fraser [1990, pp. 91–93, 312 and 374]). The search begins in infancy and is developed under social control. The universe of the infant is a chaos of sensory impressions out of which he constructs an increasingly patterned world. Those aspects of the world that change very slowly are judged permanent and are reified in symbols, such as in the nouns of natural languages. Processes, that is, aspects of the sensible world which involve changes that follow enduring patterns, are described by words called verbs. Those aspects of the world which involve change without a recognizable pattern cannot and therefore do not have names.

How does the child learn to identify processes? He learns to separate himself from his mother, from other objects, and eventually establish his identity. Selfhood (the "I") so established is a mental construct. It is precisely because it is a mental construct that it may be assigned a status of unchanging continuity. Although "I change," there is always an imaginary object represented by "I" that does not. Once established, the feeling of identity serves as a pardigm for permanence or oneness, and oneness is the basis of counting. In its turn, counting is the basis of

---

species. . . ." Note that the "environment" of which the animal's umwelt is a portion, is our own human unwelt (see Fraser [1988, pp. 108–110]).

quantifying experience and of studying the laws of quantities, such as those of mathematics, universally judged as permanent.

Let us, therefore, consider number—and quantities in general—so that by interpreting quantitative relationships revealed through physics, we may identify the permanence-analogues of the physical world. Our findings will then become a part of the inquiry about truth-as-permanence.

If one is unwilling to believe with Plato that mathematics is a direct link to a preexisting store of truths lodged in eternity, then one must ask this: Through what kind of evolutionary process did mathematical knowledge arise?

A case can be made for the belief that mathematical thinking arose through a selection process that favored quantitative thought. Mathematics may then be seen as a mental ability for creating symbolic transformations of experience, a skill that became programmed into our genetic makeup as was, for instance, three-dimensional vision. And, like three-dimensional vision and its innumerable expressions in art, mathematical thinking remains subject to individual talent and social conditioning.

Our animal ancestors might have counted their offspring, our distant human ancestors the bodies of their dead enemies; none tried to count electrons or light years. When we learned how to use mathematical science for the exploration of worlds not directly available to the senses, we began to receive counterintuitive reports about time, space, and causation in the physical world.

This, then, is what can be said about permanence analogues appropriate to the different integrative levels.

In the atemporal world nothing can correspond to predictability, to permanence, to causation, or to rest, which is the spatial analogue of permanence.

In the prototemporal world, events in time and locations in space are of a probabilistic, statistical nature. Causation is also probabilistic and with it, so is truth itself. Since probabilistic truth means only likelihood, the truth of the quantum world does not correspond to our ordinary ideas of truth.[7]

As elementary particles combine into the solid matter of the eotemporal world, the probabilistic causation of the prototemporal world gets smoothed out to yield the deterministic causation of the eotemporal world. It is in the Newtonian (eotemporal) universe of massive matter that we first find permanence and predictability of the kind that is close to the received, idealized versions of these concepts.

This is not surprising, for the whole structure of classical philosophy as well as early modern science is built on those features of the Newtonian universe that are within the range of human senses, unaided by instruments.

As we enter the higher integrative levels, the nature of the way in which

[7]The strong human desire for identifying all things permanent is beautifully illustrated by a view held by Albert Einstein. He believed that the absence of deterministic causation in the quantum world, expressed in the probabilistic laws of particle behavior, is not a statement about the true behavior of particles but a sign of our ignorance of their true laws which, he felt, should permit precise predictability. Contemporary view firmly holds that the probabilistic nature of predictability in the quantum world is correct.

events are connected once again changes. Thus, the characteristic connectivity, or causation, of the bio-, noo- and sociotemporal worlds relates to final goals, defined by the self-interests of living organisms, persons, and societies. The deterministic predictability of the macroscopic physical world remains applicable, but in the case of life, mind, and society it is complemented by increasing degrees of freedom.

What emerges from these reflections is the following. Change and motion are primeval states of the universe and for that reason they must be taken as ontologically prior to permanence and rest.[8] Out of the totally impermanent world arose in the course of evolution different ways whereby events may be connected, to wit, in their hierarchically nested order: not at all, probabilistically, deterministically, and by final causation.

Western ideas of what constitutes truth comprise a large set of propositions that have varied with the philosophical, religious, and scientific fashions of the centuries.

Two millennia after Plato, the notion of truth as discoverable permanence reached its apotheosis in the magnificent structure of the natural sciences. The ideas of scientific law and scientific truth were born when the Platonic ideal of timeless forms assumed the shape of mathematical images, known as formulas. In creating formulas, the early purpose of Western science had more to do with demonstrating the wisdom of God than with serving the practical needs of passing life. Bit by bit, however, the religious trimmings gave way to a metaphysics of science and technology. This metaphysics endowed scientific pronouncements with the same kind of uncritical admiration that, earlier, was reserved for religious pronouncements.

The idea of truth in exact science remained the image of a world divided into time and the timeless, a division to which nothing in nature corresponds. No wonder that this dichotomy has given rise to certain problems.

For instance, the many ways of knowing came to be separated into those which demanded predictability and precision as a test for truths and those that could not and, therefore, did not so demand. These two groups may be associated with C. P. Snow's two cultures. But if the hierarchical nature of time is taken into account and with it, we recognize the evolutionary character of predictability, and hence the dependence of the quality of truth upon the complexity of the systems considered, then we find that knowledge does not gather into two distinct domains. Instead, it forms a continuous spectrum of interpretations along a scale of permanence-analogues.

This reasoning places even mathematical truth into the perspective of an evolving world. It suggests that mathematics, appropriate to the handling of level-specific issues of society, of the mind, and of life is likely to develop in a direction that would make some of its branches approach the kind of qualitative reasoning which

[8]Recognizing the ontological priority of motion to rest removes the reasons that have made Zeno's flying arrow paradoxical. Namely, contrary to all prior formulations that I know of, the question one must ask is not how to construct motion from elements of rest but rather, how to construct rest from elements of motion—and that is an easy task (see Fraser [1987, pp. 27, 254–255]).

biologists, psychologists and social scientists have already found necessary for their metier.[9]

In the spectrum of different kinds of truths the human paradigm of truth remains "this above all: to thine own self be true," which says nothing about precision or predictability. Truth, appropriate to nootemporality and sociotemporality, is closer to the beatitudes than it is to any claim based on quantitatively measurable comparisons, no matter how significant the statistical handling of biological, psychological, and social data may be.

Thomas Hardy has sensed the mischief resident in mistaking mathematical truth for truth as a human value. His *Two on a Tower,* from which I already quoted, is the story of a young astronomer who returns to claim his lady love after a lengthy and unfortunate separation. She sends him away with words that say, but do not mean, "Go!" But the young man leaves just the same because, as Hardy put it, "He was a scientist and took words literally. There is something in the inexorably simple logic of such men which partakes of the cruelty of natural laws that are their study" (Hardy, 1905, p. 331).

There is something inexorably simpleminded in the mad rush of our age for gathering raw data, thought of as knowledge, a kind of cruelty that comes from mistaking quantitative truth, uninformed of human passion, for truth as a human value. The confusion between the two helped make our age minute wise and century foolish.

Guidelines to Conduct: The Good and the Evil

As far back as it can be traced, people of all cultures have demonstrated their readiness to formulate rules of preferred conduct in terms of what is right and wrong, an equal readiness to uphold those rules against challenge, and again an equal readiness to challenge them. The cruelty of our species in meting out punishments to those who break the rules, and the suffering that people have been ready to endure to do just that, is nothing short of incredible.

Opinions as to what specifically constitutes right and wrong are tremendously varied. But there is a fair agreement among contending teachings that by doing what is right, a state of harmony and perfection may be approached. Descriptions of those states of fulfillment are also quite varied but, once again, they do have an element in common. They tend to claim that the state one ought to strive for should be a permanent one.

The goal of a permanent heavenly abode is clear in Christian salvation history. Plato, working within the Greek presumption of the cyclic rise and fall of all arrangements on earth, had hoped that after his soul had separated from his body upon death, it would gain eternal wisdom through the purification process it followed while still in his body. The great Mid-Eastern and Eastern religions tend to

[9]The central theme of Penrose (1989), the inadequacy of contemporary mathematical physics for the handling of mental functions, resembles but is not identical with what is being said here.

see the destiny of man in reaching a final state where in the words of Brandon's conclusion (1962, p. 384) "he will be secure from the everlasting menace of time's destructive logic." Confucianism and Taoism, philosophies and ways of life rather than codified beliefs, seek lasting harmony within the person, among individuals, and between society and nature.

The shared views of these many great traditions about the desirability of a lasting, final, and balanced state suggest a working definition of the good in terms of permanence.

Let good be defined, therefore, as a predicate which asserts that a certain conduct, intent, or character will promote permanent stability in the affairs of man and society, whether here on earth or elsewhere. The variations among moral teachings then shift from a shared desire for final, permanent fulfillment to the different specifications on (1) what form such permanent states would take; (2) the reasons why one should get there; and (3) the map of the road to be taken.

But neither moral courage in defense of approved conduct or intent, nor beastliness of attacks upon conduct or character judged wrong, have ever succeeded in securing anything but transient stability in human affairs. Whether in the face of unfavorable socioeconomic conditions or with the aid of favorable ones, human life has always been one of armed and unarmed conflicts.

I see history as a ceaseless turmoil of unresolvable creative conflicts, with an occasional decrease in the intensity of these conflicts, and not as an ascent of man with frequent troubles from the bad guys.[10] One may well object that this is too pessimistic a view, it is the seeing of a glass half empty when it should be seen as half full. My answer to this objection is that perceiving man as a creature in continuous conflict helps make sense of history, whereas an image of men and women as good by nature does not. This is an old debate, relevant to the theme of this section, but not one that can be discussed here in detail.

Is it, perhaps, that moral instructions of all kinds have failed to provide a steady balance among the passions and needs of man because they were poorly informed or their enforcement was incompetent? Would it be desirable to have across the globe only a single set of ideas about right and wrong? Is it the case that, given time, money, and more research, we should be able to identify the kind of moral tenets and modes of enforcement that would, finally, secure permanent social balance? Or is it that we misinterpreted the role of ethics and morality in the development of civilizations? The case, I believe, is this last alternative: a misjudgment of what ethical rules do, in fact, accomplish.

Let me defend this thesis by first noting an analogy between the history of life and that of morals. Organic evolution is made possible by the existence of genetic variations, for it is the availability of choice that allows for selection pressures to favor some and disfavor other phenotypes and through them, select genotypes.

[10]The phrase, "the ascent of man" is a reference to Bronowski's popular television series and book by that title (1973). The phrase reflects an Enlightenment view which, in its current use, is associated with the belief that in the scientific method humanity has finally found a way for conquering all ills that plague mankind.

In the case of man, the selective forces of biology are joined by forces resident in symbolic causes that are shared or opposed by members of a society. Considering that symbolic causes themselves change, it is justifiable to assume that there has existed all along a degree of choice in conduct between following or rejecting collectively held ideas of right and wrong. The choice I am postulating has been called free will or human freedom. With the assistance of the hierarchical theory of time, let me explore the dimensions of human freedom as a degree of unpredictability, specific to the nootemporal integrative level.

In the physical world causation is probabilistic or deterministic but in either case, unfree. One cannot speak of choice in behavior. There exist no autonomous systems in terms of whose needs choices could be made, nor is there a present, an anchorage in time to which the choices could be referred.

It is only in the world of living organisms that one may begin to speak of choices in behavior. The spectrum of life's freedom stretches from the meager alternatives available to prokaryotes to choices in rudimental planning available for the most advanced nonhuman species. Biological freedom, as far as it goes, also applies to man: the bodies of people in coma keep on making their choices in support of life.

The degree of freedom of choice takes a large step when one considers humans, because the mind is able to imagine both possible and impossible ways of behavior, and generate a practically unlimited store of recommendations for conduct. The mind can play with symbols on its inner stage, move them around in imagined space, and back and forth in imagined time. These gifts of imagination have been the most characteristically human driving forces in the story of man. An animal will fight if abused, but only humans can maintain that an injustice has been done and direct their long-term conduct accordingly. It is here that private conscience has been fighting with public authority, private needs with public demands, where social guidance for what is right and wrong is most clearly called for, and is very often uncalled for.

Socially endorsed values concerning conduct are built into language, art, artifact, and even into the rules of acceptable reasoning. These modes of communication tie individual and collective visions together and provide for unity as well as conflict among the symbolic continuities known as individual and social identities. The subtle ubiquity of ethical teachings make people, submerged in a single tradition, take the moral norms of that tradition as obvious and, therefore, undebatable truths. This pattern of behavior makes all societies conservative. As the Declaration of Independence observes, "all experiences hath shown that mankind are more disposed to suffer, while evils are sufferable, than to right themselves by abolishing the forms of which they are accustomed."

Still, symbolic continuities may be changed much more rapidly than genetic inheritance, a fact that leads to the paradoxical role of moral values in the development of societies. On the one hand, as I just maintained, they make all societies conservative. Yet on the other hand, it is the existence of moral values that provide the usual and most efficient means for changing societies.

But, effecting substantial changes in moral values is never easy. Socrates had to violate the laws of Athens, then drink the hemlock to make a point about freedom of speech. Christ had to violate the laws of Rome and drive himself to crucifixion to replace a thundering Jehova by the Son of Man. It seems that any substantial change in prevailing ideas of good and evil demands sacrifice. A nation, being a symbolic entity in the minds of its citizens, cannot experience joy nor can it make sacrifices. Nations may perish but only men and women can suffer and die. To examine the nature of sacrifice that is necessary for changing a prevailing set of moral tenets and relate that effort to time, let me turn to the form of drama known as tragedy.

The unfolding of tragedy demands uninterrupted reflections upon past and future and the continuous making of choices among alternative courses of conduct. The struggle of the hero involves decisions that pertain to a hierarchy of obligations (Fraser, 1990, p. 394). The tragic hero constantly weighs his duties and responsibilities by the standards of prevailing judgments, as well as in terms of new values which by the process of his struggle he, in fact, creates. Because of his decisions, the domain of possible alternative choices of conduct available to him systematically narrows, until he has no further choices left but must suffer the denouement of death.

The story of the tragic hero is a literary projection of the predicament of the human knowledge of time. On the biological level the body continuously makes its own decisions, manifest in a process called aging. Because of those decisions the domain of alternative biological choices available to men and women keeps on narrowing until there are no further choices left. Against the biological decisions that necessarily lead to death, the mind holds up the image of the self as a player on a stage of eternity, both in the sense of endless time and a world without time.

What for the tragic hero are external forces challenging his ideas of right and wrong is, in the life of every person, an awareness of time's passage. It is an internal challenge, a selection pressure by an idea called time, a power more implacable than any external one, more absolute than any social convention or political tyranny. The unavoidable conflict between the fantasy or image of eternity on the one hand and, on the other hand, the biological facts of passing lends human life its irreducible aspects of the tragic. This is a cross-cultural trait, independent of whether or not the literary form of tragedy is known to a culture, because all men are dreamers and all men are mortals.

Speaking about conflicts between imagination and reality in the context of conduct is a way of describing the existential tensions appropriate to the nootemporal integrative level. These tensions, expressed in the ideas of good and evil and in the guidelines of right and wrong in moral obligations, may then be understood as aids that help people manage to live with those conflicts.

If moral laws create that much tension, if we must force ourselves to live by them, would a society without moral laws be more stable and hence more desirable? I would think that in the absence of moral tenets there would be no human society at all. Communities are created when their peculiar conflicts are created.

With the emergence of the time-compact globe, individual memories are being increasingly replaced by institutional memory; individual desires overwhelmed by communal needs. As a new global sociotemporality is being fashioned, the nature of obligations is also changing and with it the specifications of what is right and wrong, good and evil.

Organic evolution by natural selection could not, and never did, lead to lasting balance but rather to the creation of new structures and functions. Likewise, so I conclude, the role of morality has been one of maintaining those unresolvable conflicts of the mind that nurture the tragic dimensions of human life and, thereby, promote changes in the rules of preferred conduct.

Guidelines to the Management of Emotions: The Beautiful and the Ugly

The purpose of this section is to place aesthetic values, represented by the idea of the beautiful, into the perspectives of the hierarchical theory of time.

It is assumed that anything and everything that humans can perceive, experience, or think may become aesthetically relevant, provided it succeeds in mobilizing human feelings. Let beauty and ugliness be regarded as predicates upon the quality of feelings so mobilized. Beauty may be thought of as a feeling one would like to remain lasting, ugliness a quality of emotion one would prefer to be transient.

Debates about emotions in respect to ideas of beauty span the centuries. In the *Republic* (X-606d) Plato took poets to task for watering and fostering such feelings as pain and pleasure whereas, he wrote, "what we ought to do is to dry them up." In our own epoch philosophers, Gestalt psychologists, information theorists, semioticians, behaviorists, psychoanalysts, existential psychologists, sociobiologists, and theorists of art assign broadly different importance to feelings in the creation and appreciation of the beautiful. The reasoning of this section appeals to an understanding of the origins and evolutionary development of aesthetic judgments.

King Lear, as he approaches the end of his tragic life, wishes for a world wherein he could pray and sing and "laugh at gilded butterflies" (V.iii.12). Do butterflies like pretty patterns? Biologists tell us that those designs are mostly for the birds.

Butterflies, writes E. O. Wilson (1975), "tended to develop poisonous and distasteful substances to repel vertebrates [their predators] while simultaneously developing audacious color patterns to provide warnings about their unpalatable conditions" (p. 241).

In the world of insects whatever is beautiful (for us) cannot be separated from its utilitarian use (for them). Nor would my definition of beauty as a predicate upon the quality of feelings be valid because I would not know what to mean by Lepidopteric emotions. Let us take leave, therefore, of our beautiful winged cousins and turn to higher animals that give ample demonstration of having feelings.

Darwin maintained that expressions of emotions both in animals and man were remnants of previously useful adaptive behavior. Writing about hatred and anger, he spoke about uncovering the teeth as an expression of defiance. Sneering in humans, for instance, involves the retraction of the lips and the exposing of the teeth on one side, possibly the remnant of the behavior of a toothy ancestor, ready to bite. "The expression of half-playful sneer graduates into one of great ferocity when, together with a heavily frowning brow and fierce eye, the canine tooth is exposed" (Darwin, 1872, p. 248). He added that a trace of the same expression is present in a derisive or sardonic smile.

From the utilitarian designs on butterfly wings to the sneer of a dog or a person, to a sardonic smile, to such a smile acted out, written up, or drawn on paper one may trace a path to illustrate the continuity of behavioral patterns from animals to man and, perhaps, draw some conclusions regarding evolutionary path of emotions. But when it comes to trying to understand expressions of those emotions in the making of objects, or in the creation of conditions that might be called beautiful, our outsider status with respect to other species poses an unsurmountable difficulty.

Namely, familiarity with the inner experience associated with aesthetic judgment is an essential ingredient in understanding the ideas of the beautiful and the ugly. I do not listen to music in the privacy of my home or in a bistro because others—dancing men and women or artistic monkeys—do or would behave in certain ways. I listen because music fills *me* with joy, or with a sense of sadness that dissolves the feeler in the felt, or because it speaks about *my* memories and hopes. Sooner or later, all inquiries about the nature of the beautiful come upon the issue of personal identity. For it is in the selfhood of a person's mind and body that the feelings we associate with the beautiful are born and manifest.

In animals, even in the more advanced species, we can do no more than observe and ask questions by means of experiments. When it comes to humans the situation radically changes. There I can insert myself in the study as an expert witness in matters of emotional experience known from within, serve as an object of my own observation partly from without, and use the immense flexibility of human language to help explore the role that feelings play in aesthetic experience.

An aesthetic judgment made by someone attending to his feelings—those of a person different from all other persons—is a very intricate process because identity is an intricate symbolic construct (Fraser, 1987, pp. 157–159). Understanding the nature of such judgments becomes even more intricate when one considers the social dimensions of the exercise. Although I am the one who registers his feelings and attributes their causes to the outside world (by calling this object or happening beautiful and that ugly), what I so register is conditioned by collective standards. And, I am both the cocreator and the subject of those standards.

People express their emotions in a myriad of different ways. By those very expressions and by giving the feelings names (this is love, this is anger, this is sorrow) they standardize and socialize the qualities of affects. The decisions made by a writer or an artist, or the artist in all of us, concerning which thoughts and emotions ought to be recorded and for what aesthetic reasons, involves a bargaining within each person. It is between his or her judgment as an individual, and his or her judgment as a representative of the standardized and socialized qualities of affects.

Socially preferred views on the beautiful and the ugly serve him or her as guides to the management of those very emotions which help formulate the collective judgments in the first place. Aesthetic values, no less than those of truth and morality, are messages passed back and forth between a person and a group. They define certain feedback circuits as it were, conditions that are continuously close to resonance: they selectively amplify and mute but are never passive or quiescent.

Opinions as to what specific objects, conditions, or thoughts are beautiful and which ones are ugly differ from age to age and place to place, and what is ugly for one person may be beautiful for another. What humankind shares are not specific tastes but the desire to reach for whatever is judged beautiful and reject whatever is thought of as ugly. The readiness to make such distinctions joins the readiness to offer judgments upon beliefs and conduct, as the third great cross-cultural universal of human values.

Whenever there is even a modicum of agreement among members of a society about the nature of feelings engendered by aesthetically relevant aspects of the world, judgments of beauty or ugliness are promptly attributed to those objects or events. These attributes are then maintained through the tendency of language—and the total human communication network including art and artifact—to resist changes in the established direction of the cultural process (Fraser, 1978b, p. 350). Because of this inertia, certain images, actions, thoughts, will be judged intrinsically beautiful or ugly. These judgments tend to remain valid for the group until its collective ideas on beauty once again change through the dialectic of aesthetic evaluation.

What is known about the evolutionary origins of human feelings, the primary sources of aesthetic value judgments?

Jean-Jacques Rousseau, a keen observer of the inner landscapes of men and women, remarks at the very beginning of his *Confessions* that we suffer before we think. Be it suffering or pleasure, we certainly feel before we think. And there are sound biological and psychological reasons supporting Rousseau's example of common wisdom.

Emotions, such as fear, rage, sexual arousal, aggressiveness, and motivation, are regulated by the limbic system of the brain. The limbic system includes the hypothalamus, a group of cells that governs the internal environment of the body. Control of the internal milieu, no less than the control of the primitive emotions, is an archaic function of the nervous system. The neural center for phylogenetically more recent functions—speech, memory, thought—is located in the cerebrum, the part of the brain that surrounds the limbic system and evolved later.

Feelings are not only older than is reasoned thought, they also appear manifest in the development of the individual long before reasoning does. A child can tell what to him or her appears beautiful or ugly and express those feelings in words, drawing, music, or acting, long before he or she is able to offer reasoning regarded as mature. In the course of growth and development, these early judgments become integrated with other value judgments, on their way to becoming a part of the mental world of the grown person.

Although judgments based on emotions are more deeply seated functions of the mind than are those of reasoning, in a mature person they are not independent, neither are they on an equal footing. They are hierarchically related. It is possible to experience intense and rich feelings without the presence of reasoned thought, but it is not possible to reason without a world of underlying emotions. Scientific theories and even individual mathematical formulas are often described as beautiful. Calling them beautiful lends to reasoning the added dimension of emotions. But although qualities of feelings expressed in the arts, letters, or in daily discourse may profit from reasoned justification, they do not demand it.

It will be recalled that for scientific judgments of truth to be valid, agreement by others is essential; for aesthetic judgments, although the dialectic between the person and the group is always present, it may be irrelevant. What a person selects from among his feelings as those worth perpetuating and even sharing with others, is necessarily a privileged disclosure. The labor of the selection process is a part of the instinctive quest of all people for permanence. In its most reassuring version, beauty is an absolute and eternal Platonic form. In the theory of time as conflict it is neither eternal nor absolute. It is seen, instead, as both a guide and a strategy for creating order and permanence in the turmoil of the ill-defined and ever-changing shapes of feelings.

That strategy is formalized—expressed and revised and reexpressed—in the arts and letters. Each and every literary or artistic object is a record of what the writer or artist judged worth retaining. Every sculpture, painting, or flower arrangement represents whatever its maker judged as worth preserving. Each and every composition, choreography, or narrative writing expresses what its composer or writer regarded as an unfolding process which should not by that unfolding, pass.

Is longevity in public appreciation, so sought after by artists and artisans alike, a test of beauty? From a collective point of view it is a part of the test. Clocks and calendars are more or less the socially accepted scales used to demonstrate aesthetic value. The Venus de Milo, so the argument might run, is still admired after more than two millennia: it must be beautiful.

Individuals, however, have access to a method more subtle than measuring time by a clock. It is one that recommends itself as a check on whether or not a created work has been successful in offering a haven against the flux of time. What I have in mind is the experience of timelessness occasionally engendered by something judged beautiful, whether or not man made, whether actually present or existing only in expectation or memory.

Imagine that while reading a poem you were overcome by a feeling of timelessness. Was it due to a passing glimpse of the inexpressible absolute script of Creation, of which the lines you read were but poor, earthly copies? This could be a poetic, Platonic, or religious interpretation.

Experiences of timelessness have been recognized in all civilizations. They have often been regarded as privileged states, direct links to an almighty and everlasting divinity or to the Evil One. The explanation in contemporary terms is much more earthy and more intriguing. It maintains that the feeling of timelessness is the result of what may be described as a mental journey to the archaic regions of one's own mind, with their temporal unwelts different from the nootemporal one. I have dealt with this phenomenon of "time travel" in various contexts, elsewhere (Fraser, 1981, pp. 5–7; 1987, pp. 290–309; 1990, pp. 74–75, 305–311). Here are the details relevant to the current argument.

Consider first that reports on feelings of timelessness never mean the complete absence of time. They mean the phrase, "an odd kind of time," one that lacks some of the features of noetic time. It may lack a sense of continuity, a sense of temporal ordering with respect to a present, it may lack a present itself, or a directedness, or open horizons. Such adventures, true time travels, are made possible by the fact that the human brain retains some of its history. Structurally, it comprises three evolutionary forms, with the later ones surrounding the earlier ones, as already mentioned in connection with the cerebrum that surrounds the limbic system. The innermost part of the limbic system has sometimes been called the reptilian brain.

The three layers—the reptilian, the paleomammalian (or limbic system), and the neomammalian (cerebrum)—share the cranial cavity and are in constant communication. Still, each does for the human body what it would have done for its archaic owner before the later layers evolved. Each retains its peculiar way of assessing reality, including the nature of time. The different realities, each with its peculiar temporality, are continuously present among our mental functions and interpret sense impressions simultaneously. In the normal, waking state the most recent, the noetic control system predominates, although, with Freud, not without the influence of the older systems. But one or another of the older control systems can become dominant while the evolutionarily younger systems are short-circuited as in sleep, illness, danger, or feelings of elation.

Let us recall the hallmarks of the canonical forms of time. Having severely limited temporal horizons is the hallmark of biotemporality; the absence of a preferred direction is the hallmark of eotemporality; fragmentation of a directionless time describes prototemporality, complete temporal chaos characterizes atemporality.

The affective dimensions of these temporalities may be described as their "moods" (Fraser, 1978, pp. 284–293; 1987, pp. 291–309; 1989, pp. 11–13).

Thus, the atemporal mood is the suffering of the schizophrenic who feels the pull of chaos and panics because of his or her loss of nootemporal reality.

The prototemporal mood is a feeling of disjointedness with respect to time. A visual representation of this mood may be a Jackson Pollack painting with its locally coherent islands of form and color which, when taken together, have no coherence.

The eotemporal mood is a sense of directionless time, it is the "oceanic feeling" described by Freud.[11]

[11]The term is due to the French man of letters, Romain Rolland. It is described by Freud, who paraphrased a part of a letter he received from the author. It is "a peculiar feeling, which he himself [Romain Rolland] is never without, which he finds confirmed by many others . . . which he would like to call a sensation of 'eternity,' a feeling of something limitless, unbounded—as it were 'oceanic' " (Freud, 1930, p. 64).

The biotemporal mood is the happiness or fright of a moment that has no distant future or past, while the nootemporal mood is the reasoned feeling of human reality with its vast temporal horizons and sense of possibilities.[12]

There are an infinite number of ways in which the various moods of the temporalities may mix and enter conscious awareness. It is not surprising, therefore, that people often report about strange experiences related to time and, having no other words available, describe them as those of timelessness.

The Greek word for such an experience was *ekstasis,* which meant "to cause to stand outside one's self." Do experiences described as those of timelessness help us stand outside ourselves in a world that is somehow superior to that of the mature, waking mind? Not at all. Instead, they take us to the perceptive worlds of the archaic portions of the brain, less evolved than the noetic.

May we regard experiences of timelessness, related to feelings of beauty (or ugliness), as tests of some kind? Yes, but not the kind of tests that the exact sciences could admit as valid, because they are unrepeatable. Each and every one of them is unique, different, and almost as fleeting as the lives of certain elementary particles. Yet they are tests, nevertheless. They are measurements.

It is a fundamental principle of measurement theory that one must compare like with like. One measures length with length, variation in biological structures with other, similar variations. One can compare feeling only with feeling.[13]

For instance, a report of "oceanic feeling" may indicate that the temporal mood of a present feeling, as compared with the mood of noetic time experience, lacks something, such as directionality or a well-defined present. That such an inner comparison was made, in regard to the effects of an object or condition, suggests that the experience has mobilized the emotions to a sufficient degree to make the experience in question aesthetically relevant.

Decisions as to whether a particular aesthetic experience should be described as beautiful remains a socially conditioned personal one. According to the definition of the beautiful given earlier, each decision will depend on whether or not a person would wish that the feelings generated by the experience remain lasting.

Since peoples' states of mind brought to any experience are infinitely varied and also change with time, it is impossible to develop a universal, normative theory of artistic and literary taste. One has to remain satisfied with interpreting the nature of aesthetic judgments, as I have been attempting to do in this section.

One of the central claims of this essay has been that all humans seek to escape from the insecurity conferred upon them by the awareness of passage and reach for what appears to be permanent. The most complete philosophy that asserts the superiority of the permanent and the inferiority of process is that of Plato. He maintained that reality consisted of eternal, timeless forms or ideas, while temporal

[12]For a superb essay on the recognition of these moods in the literary arts see Soulsby (chapter 11, this volume). For a related analysis from the point of view of Derridian deconstruction, see Argyros (1991).

[13]If convincing reasons could be marshaled, one could measure anger by the distance that a person could throw a standard hand ball when in that state of mind. But the measurement would still be a comparison of like with like: distance with distance.

objects were only appearances or copies of those eternal, unchanging forms. He saw temporality itself as an inferior condition of man the creature, compared to the permanence of his creator (*Timaeus* 37c-d).

In Platonic philosophy the theory of values is closely tied to this view of time. We read in the *Republic* (509d–e) that human values may be arranged along a scale, a line which may be imagined as connecting heaven with earth. The good, the true, and the beautiful are placed close to the heavenly terminal of the line. Plato would say that these qualities always are (remain permanent), they never become (change). They are timeless and for that reason of the highest value. In contrast, all things that pass are placed close to the lower terminal of the line, and because they are temporal, they have the least value.

May we conclude, with Plato, that our aesthetic faculties open our minds to the vistas of eternal, unchanging, and therefore, most valued forms? I do not believe so. I believe that Plato's image of the divided line is upside down. What matters to people the most are those things that pass, such as life, love, or beauty. People seek truths and hold on to them for dear life because they know how vulnerable all verities are. The good is valuable because we know that moral judgments are subject to reckless and sudden changes, making what is judged good today evil tomorrow. Beauty is precious because a whiff of the wind can collapse the delicate configuration of the conditions—external and internal—by which those feelings have been generated.

The feeling of timelessness induced by the contemplation of man-made or natural beauty may bestow solace to a person under the relentless burden of individuation. But it does not open to his admiring view the vistas of eternal, unchanging forms and secure, through them, a lasting haven in the mind against the certainty of time's passage. The theory of time as conflict sees the role of our aesthetic faculties as lying in the opposite direction.

Aesthetic exploration is the foremost and freest method available to our species for the creation of new images and desires, including many to which no object or event in the external world does or could correspond. By this means, the perennial human search for lasting beauty generates not satisfaction stably fulfilled, but an intensifying conflict between the world as we find it to be and that other imagined, timeless world to which art gives a transient form and a changing name.

A Backward Glance, to Serve as a Summary

To the extent that any extensive and intricate argument may be summed up, what this essay proposes is that human values are self-generated selective forces which work through symbols. As modes of knowing, they may sometimes help stabilize human affairs in the short run. But in the long run they refine and amplify human conflicts. Their factual role in the development of cultures has been one of generating and maintaining, through conflicts, the stunning creativity and frightening destructiveness that characterizes our race.

References

Argyros, A. (1991), *A Blessed Rage for Order: Deconstruction, Evolution, and Chaos*. Ann Arbor: University of Michigan Press.

Brandon, S. G. F. (1962), *Man and His Destiny in the Great Religions*. Manchester, U.K.: Manchester University Press.

Bronowski, J. (1973), *The Ascent of Man*. Boston: Little, Brown.

Chaitin, G. J. (1975), Randomness and mathematical proof. *Sci. Amer.*, May, pp. 47–52.

Darwin, C. (1872), *Expressions of the Emotions in Man and Animals*. Chicago: University of Chicago Press, 1965.

English, H. B., & English, A. C. (1958), *A Comprehensive Dictionary of Psychological and Psychoanalytic Terms*. New York: McKay.

Fraser, J. T. (1978a), *Time as Conflict*. Basel: Birkhäuser Verlag.

——— (1978b), Temporal levels: Sociobiological aspects of a fundamental synthesis. *J. Social. Biol. Struct.*, 1:339–355.

——— (1981), Temporal levels and reality testing. *Internat. J. Psycho-Anal.*, 62:3–26.

——— (1982), *The Genesis and Evolution of Time*. Amherst: University of Massachusetts Press.

——— (1987), *Time, the Familiar Stranger*. Amherst: University of Massachusetts Press.

——— (1989), The many dimensions of time and mind. In: *Time and Mind*, ed. J. T. Fraser. Madison, CT: International Universities Press, pp. 1–14.

——— (1990), *Of Time, Passion, and Knowledge*, 2nd ed. Princeton, NJ: Princeton University Press.

Freud, S. (1930), Civilization and its discontents. *Standard Edition*, 21:57–145. London: Hogarth Press, 1961.

Hardy, T. (1905), *Two on a Tower*. New York: Harper & Brothers.

Needham, J. (1944), Integrative levels: A revaluation of the idea of progress. In: *Time, the Refreshing River*, ed. J. Needham. London: Allen & Unwin, pp. 233–272.

——— (1968), Hierarchical continuity of biological order. In: *Order and Life*, ed. J. Needham. Cambridge, MA: M.I.T. Press, pp. 106–169.

Pattee, H. H. (1970), The problem of biological hierarchy. In: *Towards a Theoretical Biology*, Vol. 3, ed. C. H. Waddington. Chicago: Aldine, p. 117.

Penrose, R. (1989), *The Emperor's New Mind*. Oxford: Oxford University Press.

Plato, *Timaeus*, trans. B. Jowett, in *The Collected Dialogues of Plato*, ed. E. Hamilton & H. Cairns. Princeton: Princeton University Press, 1973.

Wilson, E. O. (1975), *Sociobiology*. Cambridge, MA: Harvard University Press.

Perspectives

The two papers that follow deal, respectively, with medieval time reckoning (Wesley M. Stevens) and with process as a fundamental category in contemporary physics (C. J. S. Clarke). The epochs of the two themes bracket more than fifteen centuries in the history of Western science. Each in its own way is a record of attempts to relate the processes of nature to such permanent, unchanging principles as may be taken to manifest the immutable order beneath the phenomenal world.

Calendars in common use today may be found hanging on walls or displayed on wrist watches. If one wishes to know more than what day it is today, such as the precise times of the rising and setting of the sun and the planets or the phases of the moon, one may consult the tables of the Astronomical Almanac, *based on the calculations of the Jet Propulsion Laboratory in California.*

Calendars are also often parts of word processing programs. For instance, the software used for writing this page offers a calendrical display of months, with the possibility of tracing the months forward (press key F7) and backward in time (press key F8). If traced backward, the program does not extend to dates before February 1582. That was the month when Ugo Buoncompagni, better known as Pope Gregory XII, promulgated a new calendar in his bull, Inter gravissimas. *The opening words, "Among the most serious," refer to the duties with which the Apostolic See was charged by the 1563 Council of Trent. The Council decree says nothing about calendar reform: the necessity for that reform emerged from the task of revising the missal and the breviary.*

What is the ancestry of the knowledge upon which the computations of the Jet Propulsion Laboratory and the equations, programmed into my word processor are based? Their roots go back along the story of calendar making all the way to prehistoric times. Stevens's article deals with a single but important period of that history, with the art of computus, an intellectual skill, and the first extensive elaboration of medieval science.

Computus *is a term for time reckoning and chronology in general and the work of the computists (calendar makers) in particular. As a science, it was constructed from three elements, one of which was astronomy. Another one was a knowledge of mathematics, sufficient for the numerical handling of the incommensurable lengths of astronomical and ecclesiastical cycles. The third one was a strong religious motivation to reconcile cyclic natural processes with the Christian understanding of the position of humans in the linear process of salvation history.*

With Clarke's paper we leave the Middle Ages and turn to the end of the twentieth century. In physics, motion is described in terms of velocity. In its turn, velocity is measured by the ratio of lengths and times. But in human development velocity (or speed) is known before space and time are. The world of an infant is one of motion and change, of process, in which he learns to identify stable continuities. Much later he learns to distinguish between the ideas of distance and time interval. Eventually he constructs his notions of space and time. When he enters his first physics class, he is taught to reconstruct his early experiential world by combining distances and times to yield measures of motion.

Relativity Theory goes beyond combining distance and time measurements for the purpose of comparing velocities. It merges space and time themselves and represents their union through geometrical metaphors. For various reasons, this geometrization *of the combined space–time arena of all events and things has tended to make the temporal component of motion spacelike rather than the other way round. Yet, within this puritanical framework, a particular kind of motion—that of light—was retained as a fundamental as well as absolute reference. Process was thus smuggled back into the spatial–geometrical metaphors.*

Clarke tells us that space–time, if assumed to be the ultimate ground of physics, leads to inconsistencies. The early conceptual simplicity of relativistic space–time, first greeted with great enthusiasm, melted away as physicists had to accommodate processes to which our experiential notions of space and time—the elements from which space–time was constructed—could not be applied. As a replacement for the space–time framework, Clarke describes the mathematical principles of a process-based physics.

In his concluding remarks—as seen from the point of view of these Perspectives—he returns to the task of the computists: specifically, to a consideration of linking the coherence among natural cycles to the passage of historical (i.e., experiential) time. The formal means employed in this task are immensely more sophisticated than those that were employed by the computists. But the scientists of the two epochs do share a belief. Although the Ordainer of Order, so significant for the computists, has become irrelevant in defining man's relation to the cosmos, the metaphysical belief that the way to explore that relation must be mathematical has remained.

J.T.F.

2

Cycles of Time: Calendrical and Astronomical Reckonings in Early Science

Wesley M. Stevens

Abstract Calendar reckonings in early medieval schools were a result of the fruitful meeting of Hellenistic science and Christian piety. The Christian worship of God demanded a common festival for celebrating the life and death of Jesus. But the historical information for setting a date for celebrating either Passion or Resurrection was fragmentary and assumed either Jewish or Roman calendars. Predicting the dates of future celebrations thus required calculations of lunar and solar cycles and of their coordination in terms of calendar usages.

This paper reviews Hellenistic resources for calendar cycles, in so far as they were available to Hippolytus of Ostia, Christian scholars of Alexandria and Milano, Victurius calculator of Aquitaine, the Schythian Dionysius Exiguus, the venerable Bede and others. It emphasizes that their production of various sequences and cycles of time were never satisfactory but always stimulated further calculations of the regularities of the sun and moon (whose cycles are not commensurable), further observations of the irregularities of the wandering stars (planets), the use of Euclidean geometry, and diverse conceptions of the globe of the earth and its continents.

The discipline enveloping such studies was known as computus. Evidence from numerous manuscripts from the second to the eleventh centuries is cited to show that calendrical reckoning or computus was the context for continuing and intensive studies in the mathematical sciences of early Christian schools during the so-called Dark Ages.

Abbreviations Used in the Text

ed	edidit	rpr	reprinted
ep	epistola	s	saeculum, siècle
fl	floruit	seq	et sequentia
f	folio recto	tr	translatus
ms	codex manuscriptus	v	verso

Memorable events of our lives may be recalled in past sequences or anticipated in terms of future sequences. This recall may be in terms of stories recounted by experts in each community, but another way to account for past events of significance has been with written annals and chronicles. All of this organizing of social memory by sequences implies some sort of calendar, and all known peoples seem to have created calendars. Once a form of calendar has been found helpful within a culture, it seems to be accepted and applied without very much further thought about its structure and procedures. Before that form becomes common, however, there has usually been a great deal of serious thinking about how to make the calendar and how to use it.

I

The European calendar, which now seems so certain and accurate, has a long history of uncertainty and inaccuracy before it reached the common form in the hands of a Northumbrian monk of the eighth century, the venerable Bede, who lived from A.D. 672 until 735. In the several stories of earlier attempts to develop the European calendar, one may identify many Greek, Latin, African, and Scythian scholars at work during times when studies of this kind were much more difficult to pursue than now. The written product of such scientific and technical work is called *computus,* and it is in computistical tracts and tables that one finds how early mathematicians and astronomers were understanding cyles of time and were doing something useful with them.

The ancient word, computus, is first of all a Latin term denoting reckoning, the practice of arithmetic for any purpose; but it soon became applied especially for reckoning a series of dates or the cycles of time. Thus computus[1] often refers to the use of tables of dates, or to the methods for calculating the dates, or to explanations of the significance of such data as they refer to the circular courses of sun and moon. The earliest Christian literature of computus includes a fifth century *Calculus* by Victurius of Aquitaine (fl. A.D. 450–460)[2] and his *Cursus paschalis* to

[1]Computistical literature from the second to the eighth centuries was surveyed by Jones (1943, pp. 6–122), with bibliographical notes to the chapters of Bedae *De temporibus ratione liber*. A useful guide to the diverse works which bear the name of Bede mistakenly in the volumes of *PL* XC and XCIV has also been provided by Jones (1939). Computistical and other scientific labors by Scots, Britons, Saxons, and Angles have been discussed by Stevens (1981). See also an excellent survey of computi from the ninth to fourteenth centuries by Borst (1988). All Latin computistical works written prior to A.D. 1200 which have been printed, and those which have thus far only been identified in manuscript but not yet printed, will be included in Stevens, *A Catalogue of Computistical Tracts,* A.D. *200–1200* (in preparation).

An interesting survey of the social motivations for making calendars, teminologies, and achievements is Fraser (1987), chapter 2, "For Ever and a Day. The Reckoning of Time."

[2]Various selections from the basic arithmetic of Victurius have been published by Christ (1983, pp. 132–136); Friedlein (1871); and *idem* in *Boncompagni* (1871, p. 443). A partial edition from a poor manuscript was printed by Herwagen (1563, p. 147 seq.), rpr *PL* XC (1850), coll. 667–680. The entire prologue and text of this *Calculus* should be published in a critical edition.

be discussed below, the sixth century *Cyclus* and nine *Argumenta* by Dionysius Exiguus (A.D. 525–531) working in Italy, the ten items of computistical literature named in the early seventh century by Cummian (A.D. 632–633) in southern Ireland, and the many documents assembled at Wearmouth and Jarrow in Northumbria during the early eighth century, which were transcribed in ms Oxford Bodleian Library Bodley 309 (s.XI 2, from Vendôme) and several other similar collections of texts identified by Jones (1937, 1943, pp. 105–113). In the search for understanding during the medieval centuries, however, computus became something more; for it tended also to attract other mathematical and scientific materials far beyond the requirements of a calendar. Time-reckoning required recognition of the Zodiac and its twelve divisions and 48 or 60 or 360 parts as a basic scale for quantified observations of solar and lunar motions. Computus is also a term which could denote a whole essay about how the calendar works, or an entire codex in which such materials formed a large portion of the contents, though mixed with other subjects such as a catalogue of stars or studies of poetic meter or excerpts of liturgy with musical annotation, all of which assumed or were offered in terms of sequences or cycles of time. Computus may even include brief discussions of the whole order of the cosmos (Stevens, 1979).

Turning the leaves of computistical codices prepared for study in the eighth, ninth, or tenth centuries of our era, one finds even more diverse works of scientific interest; for example, there will be *mappaemundi* displaying the globe of earth with four zones or with three continents or with five parallels of klimata which we might call terrestial latitudes.[3] There will be *planisphaeria* with stereographic projections of stars and constellations for both the northern and the southern hemispheres; some will be projected from an infinite distance above the earth, looking down as Hipparchos imagined in the second century B.C.; and some will be projected from the south celestial pole, looking up as Ptolemy explained in the second century A.D.[4] Some of these books of computus describe and display alternate systems of planetary motion in epicyclic or in eccentric systems. Of course this means that sometimes the earth was shown as the exact center of all planetary motions, which were perfectly circular, consistent with Aristotle. But other diagrams in the early ninth century show planetary motions and their extremes or apses which were irregular and not concentric with the earth.[5] By the end of the ninth and during the

[3]*Monumenta cartographica* (1964), especially chapter 2 for which earlier manuscript examples could be cited: Aujac (1974), Stevens (1980).

[4]Six planispheres used in ninth and tenth century monastic schools have been listed by McGurk (1973, esp. pp. 200–201); other manuscript examples are known in each successive century. The earliest example is found in the Regensburg manuscript, München Bayerische Staatsbibliothek CLM 210 (A.D. 810–816) f.113v, briefly described by North (1975, pp. 386–388); a color reproduction may be seen in Bullough (1965, p. 189, plate 50). The two perspectives have been explained by Stevens (1980).

[5]Ms Paris Bibliothèque National Nouvelles acquisitions lat.1615 (s.IX 1) f.128–193 was part of a large computistical and astronomical collection dating from about A.D. 809 before it was copied at Auxerre about 830. It shows three diagrams: f.159v planets in circles whose distance is set by tonal proportions; f.160v planets on circles with different centers (earth is also shown), and the circles broken with zigzags representing retrograde motion; f.161 planets on other circles with diverse centers and blind areas when a planet is too near the sun to be seen. The drawings were used during ninth and tenth centuries in the school of St. Benoît-sur-Loire at Fleury.

tenth century, such planetary motions were also plotted on linear scales of 30 × 12 parts.[6] These computistical collections often contain multiple *argumenta* on the same topic, that is, not only alternate explanations of how to solve a problem but also irreconcilable data to be analyzed[7]

All of these documents have their origins in centuries which have been called "The Dark Ages," a period during which all learning was supposed to have been lost to the barbarian hordes in the West! For several generations of historians therefore, such materials were not supposed to exist in the pious schools of monks who only prayed and read Holy Scriptures. But these works were indeed written in those times, they were taught in schools for piety, and they do still exist in abundance. Between the decline of Roman control in the West and the flowering of Carolingian schools, there was scientific activity of importance, a small portion of which will be presented here. It is found in the meeting of Hellenistic science and Christian piety. In order to understand the development of computistical science, it is necessary to recall an essential element of all religion and particularly of the Christian religion: the cult or worship of God.

II

Christians should worship together. That is an obvious implication of the apostolic teachings about the historic Jesus and of the Hebrew Scriptures which they inherited and shared with the Jews. But it would be almost impossible to find a time when all or even most Christians have actually worshipped together for their chief festival of the year, *Pesach, Pascha, Easter*. Celebration of the resurrection of Jesus Christ occurred on different dates according to various calendars and according to divergent systems of reckoning within the same calendars (Richardson, 1973). This variety existed already in the Hellenistic society within which Jews and Christians tried to worship, and it is a product of intelligent reflections and decisions, not only

---

Other types of diagrams for studies of planetary motions have been published by Eastwood (1983a, 1983b, 1987).

[6]Ms Bern Burgerliche Bibliothek 347 (s.IX ex) f.22v–24 display the planets in figures similar to the Auxerre/Fleury ms Paris BN N.a.lat.1615, cited in the previous note. But ms Bern f.24v shows a rectilinear coordinate schema for displaying the patterns of latitudes for apogees and perigees of planetary orbits. Illustrations from the Bern, Paris, and other manuscripts were published by Eastwood. Late examples of the rectilinear schema were printed by Pedersen and Pihl (1974, p. 246).

[7]Good examples of this practice may be seen in the *Compilatio compvutistica et astronomica* A.D. *DCCCVIIII* which was probably compiled at the palace school of Charlemagne in the form of three books; earliest copies are ms München Staatsbibliothek CLM 210 (A.D. 810–818) f.4–163v and Wien Nationalbibliothek Lat.387 (A.D. 812–816) f.4–165, each of which gave the date 810 in argumenta from their exemplar, rather than the expected year 809. An expanded seven-book version of this *Compilatio* in ms Monza Biblioteca Capitolare f-9/176 (A.D. 869) f.7–92v, and other manuscripts. Outline of contents and a survey of the two groups of manuscripts were made by King (1969). It was probably from such a collection of texts for study that Walahfrid Strabo took his series of dates for the Incarnation in ms Sankt Gallen Stiftsbibliothek 878 (s.IX second quarter), pp. 284–292; see Stevens (1972).

about historical events but also about the observed and calculated cycles of lunar and solar motions.

In the Hellenistic tradition the earliest attempt at systematic coordination of lunar and solar cycles in tabular form appears to be an intercalation cycle of eight years, reported by Pliny (Plinius, *NH* II 97, 215). This 8-year cycle or octaëteris may have been received from the ancient Babylonians and used by Chaldeans and Greeks. It was based upon a solar year of 365 days and upon lunar months of either 29 or 30 days. If you observe the moon over a season or two, you will find that it reappears above the horizon sometimes after 29 days and sometimes after 30 days, and you may recognize a mean lunar month of 29½ days. Eight years will include 99 cycles of the moon; but if you multiply 29½ days by 99 cycles, the result will be half a day more than if you count the daily course of the sun about the earth during the same period:

99 months	× 29½ days	= 2920½	(1)
8 years	× 365 days	= 2920	

If one doubled the sequence in order to obtain a 16-year cycle, the surplus would be a full lunar day which could be skipped in order to begin again.

In the fifth century before Christ, an alternative system was created by two Athenian scholars. Euctemon and Meton, who recommended the use of a 19-year cycle of alternating full lunar months of 30 days, each with short lunar months of 29 days, to which they added seven full intercalary months of 30 days, resulting in a total of 6940 lunar days.[8] We do not know by what pattern the full, hollow, and intercalary months were alternated in the Metonic cycle in order to maintain accord with solar phenomena; but the overall numbers will be as follows:

118 full months	× 30 days	= 3540	(2)
110 hollow months	× 29	= 3190	
7 intercalary months	× 30	= 210	
235 lunar months	total days	= 6940	

Meton is justly famous for this advance in reckoning the times. But the Metonic lunar cycle is about 2 minutes too long, and the supposed solar cycle would be about 30 minutes too long; continued use will create problems. Other Greek scholars were not at all satisfied with it.

In order to reduce the problems of the Metonic cycle, another fourth century astronomer and pupil of Aristotle, Kallippos of Cyzicus (ca. 370–300 B.C.), reckoned a longer period.[9] He calculated that 912 lunar months would accumulate

[8]*PW* VI/1 (1907, pp. 1060–1061) and XV (1932, pp. 4458–1466); Ginzel II (1911, pp. 391–392); Duhem I (1914, p. 108); *DSB* IV (1971, pp. 459–460).

[9]Geminos, *Elementa astronomiae* VIII, pp. 120–122: the 76-year period of Kallippos; *PW* X (1919, pp. 1662–1664); Heath (1913, pp. 212–213); Duhem, I (1914, pp. 123–128; *DSB* III (1971, pp. 21–22).

26,918 days, with alternating 29- and 30-day periods for which we have not received an explanation. To this he intercalated 29 full lunar months of 30 days each to add 840, for a total of 27,758 days. According to later reports this result was one day short of four Metonic cycles or 76 solar years. That information requires the solar cycle to be 365¼ days. We can only guess at Kallippos' method of calculation and sequence for 940 lunar months, 76 solar years, and 27,758 days whose relationships are not reported in the evidence.

The variety of such Hellenistic calculations was very great, and evidence for them survives only in fragments. But from the time that Julius Caesar went courting Cleopatra on a barge in the Nile River and returned to Rome with an astronomer, Sosigenes, the 19-year cycle with fourth year embolism, more commonly called the extra day of leap year, was accepted in the solar calendar of the Roman Republic and thus came to be used in capital cities of the expanding Roman Empire.[10] Furthermore, the Metonic cycle had been developed into its classic form with mean lunar months of 29½ days each[11]:

12 mean lunar months × 29½	=	354 days	(3)
19 years × 354	=	6726	
7 intercalary months × 30	=	210	
Embolisms during 19 years	=	4¾	
total days	=	6940¾	
19 solar years × 365¼ days	=	6939¾	

There was a remainder of one lunar day which they called *Saltus lunae;* that is, if you simply skip over that one extra lunar day, you may repeat the 19-year cycle and remain rather well in accord with the course of the sun. It was in this form that the 19-year cycle was adapted to Christian usage at Alexandria and Milan, later at Carthage, and eventually at Rome.

III

The evidence for the Christian development of computus or calendar reckoning is very complex but quite abundant because it was necessary. A bishop for example must teach the sequence of prayer and praise as well as the meaning of Christian festivals throughout the year. But so much of the meaning and so much of the timing depends upon the Easter festival, that any bishop must certainly know when to do it. As with so many other obvious practical matters, the letters of bishops which survive from early centuries imply that such questions are being dealt with

[10]Suetonius, *Divus Julius,* p. 40: on calendar reform, without mention of Cleopatra or her astrologer; but see Plinii *NH* II 8.39 and XVIII 211; Ginzel II (1911, pp. 274–281); *PW,* Zweite Reihe III/2 (1927, pp. 1153–1157); *DSB* XII (1975, p. 547): Sosigenes.

[11]The classic form of the Metonic cycle is explained by Rühl (1897, pp. 114–116); Schwartz (1905, pp. 15–17); Ginzel II (1911, pp. 91–392); Jones (1943, pp. 20–21).

but do not repeat the details which they tell priests, deacons, or other leaders of their scattered flocks. One of them was quite verbose, however, the famous St. Augustine, and many of his letters and works have survived.

Augustine grew up in Roman Africa. He taught as a grammaticus in Carthage and Rome, and then with the influence of his Manichaean friends he obtained the prominent and lucrative post of imperial rhetorican in Milan. After ten years of a successful career in the capital cities of Rome and Milan, he changed his religion and returned to Africa where he served as a Christian priest in Carthage and Thagaste from A.D. 389/390 to 395 and was then dragooned into becoming bishop of Hippo Regis (A.D. 395–430).[12] Augustine wrote several great books which set forth his thoughts on grace, free will, consciousness and memory, the nature of man, and the nature of God. These writings include many philosophical reflections upon history in which the concept of "time" is often mentioned and sometimes discussed.[13] There are aspects of history and time which have apparently seemed to be too mundane for doctrinal history or philosophical theology but which the man himself could not have avoided: when did this bishop observe Easter during the course of a Julian year? What system did he use to set the day of the month for Easter Sunday each year within his diocese in order that all churches could celebrate together?

A letter had come to Augustine from Ianuarius about Easter, and we have the bishop's answer—a long answer: twenty-five Latin pages in the *CSEL* edition and thirty-three pages of English translation.[14] Ianuarius had asked: Why does the annual commemoration of the Lord's Passion not come around on the same day of the year, as does the day on which he is said to be born? If Easter is irregular because of the sabbath and the moon, what does it mean to observe the flux of sabbath and moon?

In his response Augustine emphasized the Easter festival as the great sacrament of grace through the historical Incarnation, and not any Christmas celebration or date. That is to say, he emphasized the Passion of Jesus and not his birth. Christ died and rose again, and thereby something was passed over to us: *Pasch, transitus, Passover*—something to be received. Incidentally he also affirmed the value of

[12]There are many good studies of his life, such as Bonner (1963) and Brown (1967), and of his political concepts, such as Deane (1963) and Adams (1971). But the best introduction may still be H.-I. Marrou, *Saint Augustine and His Influence Through the Ages* (1957). A reliable and often enlightening group of surveys of his many works is *A Companion to the Study of St. Augustine* (1954).

[13]In most essays concerning "time and history" in the thought of Augustine, there is not a word about real time in daily life. Only the juxtaposition of large concepts and wide philosophical meanings are found in the oft-cited essay by Marrou (1950). Another example is O'Daly (1981) which has nothing to do with any sort of measurement, much less of time. "Meaning of History" appears not to lead authors to discuss actual time-sequential experience and its structural necessities.

[14]Epistola LV ad Ianuarium, ed. A. Goldbacher, *CSEL* XXXIV/2 (1898, pp. 169–213), Eng. transl. W. Parsons in *Fathers of the Church,* Vol. 12 (1951, pp. 260–293). A large part of this letter was available for medieval schools in the *Excerpta ex operibus s. Augustini,* prepared in the sixth century by Eugippius, Fulgentius of Ruspe, and Paschasius diaconus, ed. P. Knoell, *CSEL* IX/1 (1885, pp. 425–444). Discussion may be found in Jones (1943, pp. 35–37, 87–89, 400 et passim); and Van der Meer (1961, pp. 285–293).

astronomy with its calculations and eclipses of sun and moon, and he warned against Manichaean fables about such things. Furthermore he twice reiterated the regularity of the "councils and fathers" who had found the date of Easter in terms of:

1. The month of new corn.
2. Luna XIV to XXI.
3. Sunday, the eighth and first day.
4. Forty days for Lent; fifty days for Pentecost.

It will help to recall that the Hebrew calendar was lunar, not solar, and it called its first month Nisan or the month of new corn. Commencing with the first sliver of reflected light as the new moon or luna I appeared somewhere in the sky, the moons were counted to the fourteenth or full moon which corresponded with the vernal equinox. Within the eight days and nights from luna XIV to luna XXI—from the fourteenth to the twenty-first day of the new moon—therefore, the day of Jesus' death would be observed on a Friday, and the third day thereafter (counted inclusively) would be Sunday: the time for celebrating his Resurrection (Richardson, 1940; Jones, 1943, pp. 6–11, 18–20). The lunar cycle itself, however, introduced uncertainties as to the fourteenth day, in that the observed first moon may rise during the light of day or at any time during the night. And some people customarily reckoned the beginning of a 24-hour day at dawn, noon, sundown, or midnight. Such regional and cultural variances thus usually required any set of guidelines to be adapted by local scholars to local practices in order to fulfill the intended goals.

Observance of Christian Sunday rather than Jewish Saturday was repeatedly justified by the bishop without speaking ill of anyone. Christians had not always reorganized weekdays to commence with Sunday, and it may have been adopted first in Alexandria by the third century if not earlier. Although it would seem to be an easy adjustment of the Jewish pattern for communities to make, especially if Christians wished to give priority to the day of Resurrection, such social changes do not always find ready acceptance. Thus celebration of the Resurrection of Christ on a Sunday does not appear to have found general usage within all of the Christian communities before the fifth century. In terms of Roman culture, the first day of the week could equally be counted the eighth, beginning and completion in one, by which Augustine gave rhetorical emphasis to celebrating on Sunday. Quadragesima or Lent was reckoned back from Easter Sunday, but intervening Sundays were not counted for this purpose; and Quintagesima was counted forward beyond Easter Sunday to Pentecost in the same manner.

The system which Augustine explained therefore was dominated by liturgical concerns, as one should expect of a bishop, and it is consistent with the fragmentary reports from the Council of Nicaea. It is also quite simple and is similar—perhaps identical—to what I learned from my mother as we walked to church on Sunday mornings:

Easter is the first Sunday
after the first full moon
after the vernal equinox.

Augustine was well known to be attentive to his episcopal duties, careful and diplomatic in fulfilling them. He was responsible for setting the date each year for Easter to be celebrated in every parish within his diocese. He sat in many church councils of the province of Roman Africa, and he consulted with other bishops in more distant provinces. What he explained to Ianuarius, therefore, ought to have been in accord with his actual practice if not complete in detail. We may fairly ask of him: Are his rules sufficient for his diocese, for the province, for the whole church?

That is a difficult question to answer. Many scholars attempted this task in the early Christian centuries and have left considerable evidence for the dates of Christian Easters from the early third through the ninth centuries and various explanations for the sophisticated systems which they used to determine those dates. From their fine works it will be possible now to describe eight sets of evidence for the chronological work of early Christian scholars which bear directly upon the context in which Augustine worked and the rules which he provided; but they will not be easy for us to explain.

1. The octaëteris or eight-year calendar cycle of Babylonians, Chaldeans, and Greeks seems to have been doubled and applied by Hippolytus to guide the choice of Sundays to celebrate Easter for 16 years.[15] He was a presbyter at Ostia in the Roman civitas who lived about A.D. 170–236 and was exceedingly hostile toward Calixtus, contemporary overseer or episcopus of Rome (217–222), earlier convicted of embezzlement and whom he also accused of heresy. Two books by Hippolytus were written in Greek, including a *Synagoge* or chronicle which placed the birth of Christ in the *annus mundi* 5500. The *Cyclus Hippolyti* which began with luna XIV on Saturday, April 13, 222, survives on two marble tables together with his third century bust at the public entrance to the Lateran Museum of the Città del Vaticano, Rome. His parallel series of dates were only for the full moon or luna XIV and for the following Sunday of Easter; the latter evidently could range between limits of luna XVI to XXII. Later the cycle was applied for the years 241–256 by adjusting the dates (Krusch, 1880, pp. 189–192; Richard, 1950, p. 243). It was partially explained by an *Expositio bissexti* for use during 64 years, and then it was extended for application during 7 × 16 or a total of 112 years.[16] The evidence

[15]The Easter tables and other works of Hippolytus have been edited and discussed by Isaac Argyrus [= Jacob Christmann], *De correctione paschatis* (1595), rpr by J. J. Scaliger, *Hippolyti episcopi Canon paschalis cum commentario* (Leiden, 1595, pp. 2–24), with long excerpts from Argyrus appended in this and other of Scaliger's books; the comments were collected and translated into Latin by Denis Petau, *Uranologion* (Paris, 1630, pp. 359–392). The cycle of Hippolytus was printed and its author discussed also by Aegidius Bucherius, *De doctrina temporum commentarius in Victorium Aquitanum* (Antwerp, 1634, pp. 289–312), rpr *PG* X 1857, coll. 877–884); *Monumenta ecclesiae liturgica,* I (1904, coll. 25*–27*); *DACL* IV/2 (1924, coll. 2419–2434); Jones (1943, pp. 11–13); Richard (1950).

[16]Ms London British Library Cotton Caligula A.XV (s.VIII 2) f.97v–105v; and a manuscript once in the Bibliothèque Saint-Remy of Reims, transcribed by Jean Mabillon into ms Vat. Regin.lat.324 (s.XVII 2) and edited by J. Wallis in *Sancti Caecilii Cypriani opera,* ed J. Fell (Oxford, 1682), rpr *PL* IV (1844,

deriving from Hippolytus shows that third-century Roman practice allowed Easter Sunday to be observed as early as March 18th or even the vigil of that day which would have been March 17th, a full week or more earlier than the Julian equinox of March 25th, and as late as April 21st. But the Passion of Jesus was primary for these Christians: the *Cyclus Hippolyti* avoided celebrating Easter Sunday on either luna XIV or luna XV which it considered to be the moons of Crucifixion and improper for days of fasting.

2. Bishops Demetrios (d. A.D. 232) and Dionysios (fl. 247) of Alexandria also used an 8-year cycle to set limits of luna XIV to XXI for observance of the Easter full moon, from which they then counted forward to the first day of their seven-day week in order to determine Easter Sunday. But in Alexandria they corrected the vernal equinox from the Julian date of March 25th to March 21st, and they would not celebrate during the hours of light on that day but only in the evening which they called luna XV or March 22nd. Their rules allowed Easter Sunday to range as late as April 26th.[17]

3. Anatolios was a teacher in Alexandria before he became bishop of Laodicaea about A.D. 269 to 280. Some of the dates for Easter Sunday used by him have been preserved by Eusebios, and they show that Anatolios set limits of luna XIV to XX for the Easter full moon and presumably accepted the Alexandrian vernal equinox of March 21st (*HE* VII, pp. 14–19, 32).[18] Surviving evidence unfortunately does not make it clear whether he also may have used the 16-year cycle or the Metonic 19-year cycle, but modern attempts to fit his data to the latter have not been convincing.[19]

4. Many changes took place in setting the dates of Easter for the Roman civitas on the Italian peninsula. Presidents, presbyters, and eventually bishops of its dominant basilica tried at various times to use luna XIV to XX, or at other times

---

coll. pp. 939–967); ed, G. Hartel, *CSEL* III/3 (1871, pp. 248–271); *Monumenta ecclesiae liturgica,* II (1913, coll. 82–87); Richard (1950, pp. 237–247); Ogg (1950); the lost Reims manuscript apparently attributed its text of *Expositio bissexti* to Cyprianus, and the editors have thought that this meant the bishop of Carthage (s.III med), to whom a *Chronicon* was later attributed by Paul the Deacon in the late eighth century. The *Expositio bissexti* in ms Ivrea Biblioteca Capitolare XLII (A.D. 813) f.50v–51v is also accompanied by a cycle of 16 × 4 = 64 years which may represent the Hippolytan tradition, and it should be published.

[17]Demetrios wrote paschal letters announcing the date of Easter to the bishops of Antioch, Jerusalem, and Rome according to the annals of Eutyches, ed *PG* CXI (1863, col. p. 989); but doubtless others also received these letters. Dionysios did the same, at least for the churches of Egypt, and one of his paschal letters for A.D. 247 included an octaëteris mentioned by Eusebii *HE* VII, pp. 20–22.

Within the same area of influence Epiphanios, episcopos of Constantia or Salamis on Cyprus (367–403), was still using such an eight-year cycle at the end of the fourth century; Krusch (1880, p. 23); Jones (1943, p. 14).

[18]The reference is VIII 28 in the Latin translation edited by T. Mommsen, "Die lateinische Ubersetzung des Rufinus," in *Eusebius Werke* II/2 (1908, pp. 723–725). Ms Paris Bibliothèque Nationale Lat. 10318 (s.VIII 2), p. 189: Cyclus Paschalis Grecorum seu Macedonum Post Annos XCV (extending from 258 to 352), ed Van de Vijver (1957, p. 11).

[19]Krusch (1880, pp. 311–316); Anscombe (1895); Turner (1895, pp. 699–710); Rühl (1897, pp. 114–116); MacCarthy (1901, pp. lxv–lxvi); Schwartz (1905, pp. 15–17, 138–139); Jones (1943, pp. 20–21).

luna XVI to XXII, as limits for determining the moon which Easter Sunday would follow. Once they even celebrated Easter as late as May 15th (*idus Maii*).[20] But eventually the Roman bishops gave up the attempt to limit the lunar sequence within which to determine the Easter moon. At least by the years A.D. 412 to 414, they began trying to regulate Easter Sunday itself without reference to the equinox and instead set arbitrary limits of XI *Kalends Aprilis* to XI *Kalends Maii* (that is, March 22nd to April 21st) in order to meet a peculiarly local need. They considered it important to avoid the carnival of *natalis urbis dies Romae,* the wild times around an annual commemoration of the founding of the city.[21]

5. Augustine's conversion and his transition, from an ambitious imperial rhetor and aspiring neo-Platonic philosopher to the more moderate life of a Christian theologian, occurred in Milan. There he probably learned the technical terms for Christian festivals from Ambrose, bishop of Milan (A.D. 374–397). Those terms were the Alexandrian limits of luna XIV to XXI, which by the end of the fourth century were found not only in northern Italy and in Roman Gaul (Provence) but also in most of southern Italy (including the heavily populated Neapolitan civitas) and Sicily. In Roman Africa to which Augustine returned, however, the Julian calendar still gave the vernal equinox as March 25th, not the Alexandrian correction to March 21st. Use of Alexandrian tables there with a different equinox would require new calculations and would doubtless cause considerable confusion. Fortunately there is evidence of how various scholars faced their computistical problems in Roman Africa.

A conflict over the calendar had arisen earlier in Carthage. It seems that Augustalis, a third-century bishop, used an 84-year table which extended from A.D. 213 to 297 and that he extended it to A.D. 312 for the convenience of rounding it off to an even 100 years. This Easter table has been reconstituted by Krusch (1880, pp. 17–19). In an attempt to improve upon the Hippolytan tradition by local changes in criteria, it applied the *Saltus lunae* every fourteenth year so that a successive cycle of 84 years could begin again with precisely the same data in all columns. According to this computist the data should include an *epact* of one on the first of January, and this may be the first computus which specified the age of the moon from one to thirty on the first of January each year as a number of *epactae*.

The *Laterculus Augustalis* was continued in Africa by Agriustia of Thimidia Regis who created more tables in A.D. 412, by which he supposed that he could

[20]Tertullian (ca. A.D. 160–240) argued that the Romans were wrong to allow luna XIV to fall as early as March 18th, or the evening of March 17th, as had the Hippolytan tables; and he also thought that they had erred by allowing Easter Sunday to be delayed as late as May 15th. His occasional remarks on these matters were conveniently given in *The Calendar of St Willibrord* (1918, pp. 24–25), by H. A. Wilson. Discussions of Tertullian and the most detailed analyses of his writings (Barnes, 1971) have ignored all reference to the date of Easter and to Easter tables, even when citing Hippolytus.

[21]The earliest indication of this practice is the letter of Innocentius, bishop of Rome (A.D. 401/2–417), to Aurelius of Carthage advising him to observe Easter Sunday on luna XVI in the year 414, rather than on luna XXIII. Thus he would set aside the vernal equinox as an early limit for the Easter moon, rather than let it wander into the confusions of carnival. This letter is found in all the collections of early canons in the Roman area of influence; for example, *PL* LXXXIV (1850, pp. 657–658).

maintain the older practices of the Roman civitas and correct the errors of others. He seems to have believed that Easter practices long since abandoned in Rome would be better than the new practices from Alexandria. Agriustia wanted to be more Roman than the Romans (Krusch, 1880, p. 29; Jones, 1943, pp. 15–17).

Such difficulties were furthered also by a mid-fifth century *Ratio paschae* which has also been called *Computus Carthaginensis*.[22] This table commenced with A.D. 439, near the end of an 84-year cycle which culminated 420 years (84 × 5) after the crucifixion in *annus Passionis* XXIX, thus A.D. 449; and all data were presumed to recur thereafter. It attributed many faults to both Augustalis and Agriustia, especially rejecting their use of *Saltus lunae* after the fourteenth year (e.g., 449) and favoring the twelfth in an attempt to make the 84-year sequence truly cyclical; but its 12-year *Saltus* actually appeared following A.D. 449 as well, with potential confusion. It also rejected attempts to reconcile Alexandrian and Milanese limits of luna XIV to XXI or XV to XXI for the paschal full moon, as well as new Roman limits of March 22nd to April 21st for Easter Sunday. The purpose of the Carthagenian *Ratio paschae* was to reassert earlier Roman practices, including those which allowed Easters on or prior to the vernal equinox, as in the Hippolytan tables, or which delayed Easter Sundays as late as the Ides of May. This new computus, however, would show how those rules were valid and meet the difficulties anticipated in the year 455 by recognizing luna XIV on March 22nd and setting Easter Sunday on April 17th.

This controversy in the Roman Africa of Augustine apparently provided occasion also for a peace-keeping effort called the *Acta Synodi Caesareae*. According to Eusebios (*HE* V, p. 24) there was a council at Caesarea presided over by Theophilos in A.D. 180 which was concerned with the tradition of the Passover and with the sending of letters "so that in the same manner and at the same time we keep the sacred day" with Alexandria. These extant *Acta* cite that place and name the bishop but do not address the same questions.[23] The document responded to a different local need during the early fifth century, which was to reconcile those Christian communities who used Alexandrian limits of luna XIV to XXI and those who used one or another of the more recent Roman limits for Easter Sunday. Thus the *Acta* were written after Romanists of Carthage who had rejected both of the current Roman usages and had tried to go back to the earliest and impossible Roman dates. As we shall see, Roman bishops did not fall back on any of these confusing and contradictory ideas of the various "Romanists" in the fifth century.

6. The Romans themselves had objected to using the Alexandrian Easter in A.D. 382 when it would have required that the Resurrection be celebrated on April

[22]Fragments of this computus survive in ms Lucca Biblioteca Capitolare Felinana 490 (A.D. 796–816), f.282–286v; Schiaparelli (1924, pp. 17–19). It seems to have been noticed first by Etienne Baluze (1630–1718), librarian for Colbert during A.D. 1667–1700, and was published in *PL* LIX (1847, pp. 545–560). A later edition is by Krusch (1880, pp. 279–297), and see further pp. 28, 138–140, 170; Jones (1943, pp. 44, 46).

[23]Four versions of the *Acta* will be sorted out from thirty-two manuscripts and twelve editions in Stevens' *Catalogue* (in preparation). The best modern editions have mixed them by collation of two or more versions in the attempt to create a single critical text: Krusch (1880, pp. 306–310), and Wilmart (1933); Jones (1937; 1939, pp. 44–45; 1943, pp. 87–89).

22nd. About A.D. 395 another Theophilos, bishop of Alexandria (A.D. 385–412), sent an Easter Table of 100 years to Rome, so that those bishops would know the correct dates for Easter Sunday.[24] But Easter Sunday fell on April 22nd in the year 417, again at the beginning of the carnival. Either Cyril, bishop of Alexandria (436–444), or his successor, Dioscores (444–451), gave assistance to Rome by sending their paschal tables for years 437–531 with rubrics and explanations in order to make certain that they could be understood there by bishops and curial officials as they already were understood in Milan. A serious effort was made in the Roman churches to follow this Alexandrian table; but it soon faced the same problems of local practice.

In the year 444 Easter Sunday was expected to fall on April 23rd according to the Alexandrian cycle, but at Rome that was two days into carnival. Therefore Leo, bishop of Rome (440–461), enquired of Pascasinus, bishop of Lilybaeum (Sicily), whether this could be avoided. Pascasinus' response[25] noted that the Romans had difficulties in the past partly because they were still using the outdated Julian equinox of March 25th for luna XIV and that they seemed ignorant of the better tables from the bishops of Alexandria. He courteously but firmly advised acceptance of Easter Sunday on April 23rd, 444. A few years later Leo wrote again to Pascasinus concerning an even later Easter Sunday anticipated for April 24th, 455. This time he also approached the emperor Marcianus (450–457), Julianus, bishop of Kios (Bithynia Pontica, ca. 454/5), and finally Proterios, bishop of Alexandria (451–457), to whom Marcianus had referred the question.[26] By that time the Alexandrians had absorbed their octaëteris into the first eight years of a nineteen-year cycle. They still insisted upon the lunar limits of XIV to XXI but, assuming that a full day extended from noon until noon, they excluded the daylight hours for luna XIV. However, people in the Western regions had a different notion of what constituted a day (midnight to midnight), so that Alexandrian luna XIV to XXI would usually have been expressed in Latin as luna XV to XXI if the moon did not rise before night (N.B.: lacking a rule, if the moon did rise between nightfall and midnight, the problem could still exist). Proterios' answer to Leo in A.D. 454

[24]Theophilos prepared a table of Easter data for 100 years from A.D. 380 to 479, commencing with the first consular year of Emperor Theodosius I (378–395). It no longer exists, but there are manuscripts and editions for an *Epistola ad Theodosium* and a *Prologus,* including *questiones* which provided the classic rules in both Greek and Latin for Christian use of the 19-year cycle. The prologue is known in two Latin versions, ed. Krusch (1880, pp. 220–226), and Jones (1943, pp. 29–33); see discussion by Krusch, pp. 84–88. Jones (1943, pp. 26–27) also mentioned several early compromises for the sake of common celebrations of Easter.

[25]Leo's first letter has not survived, but Pascasinus' reply is found in eighteen manuscripts. The earliest is ms London British Museum Cotton Caligula A.XV (s.VIII 2, f.93v–95), *Brittanniae ecclesiarium antiquitates* (Ussher, 1639), p. 926 and rpr (London, 1687), p. 480. The Ballerini edition from ms Vat. Regin.lat.586 (s.X 2) f.115v–117, was reprinted in *PL* LIV (1846, pp. 606–610); Krusch (1880, pp. 245–250); Stevens, *Catalogue*.

[26]The Ballerini edition of these letters (1753) has been reprinted in *PL* LIV (1846): ep.88, 121, 122, 127, 133, 137, 142. Some of them have been edited from other manuscripts by Krusch (1880, pp. 212–217, 255–265, 269–278); they are also discussed by Jones (1943, pp. 56–61). Details of the texts, manuscripts, editions, and bibliography will be found in Stevens, *Catalogue* (in preparation).

upheld Alexandrian practice of course, but he went on to demonstrate the validity of its extremes: that is, Easter Sunday had been celebrated in Egypt as late as April 26th, 387, and this would recur in 482 and again in 550—salt in the wounds of the Romans!

Ultimately, Leo always yielded to the more learned Alexandrians, and in Epistle 138 he announced their date for Easter Sunday on April 24th, 455, within his provinces of central Italy, Gaul, and Spain for the sake of unity with the apostolic sees. It appears from Leo's letters and answers to them that at this time the Roman bishop's influence did not extend to Magna Graecia (southern Italy and Sicily) or to Roman Africa and other African provinces.

7. One result of Leo's concerns was that his archdeacon Hilarus commissioned the *calculator scrupulosus,* Victurius of Aquitaine, to create new Easter tables which would fulfill the goal common to both Alexandria and Rome, that is, to celebrate the principal Christian festival in unity. Victurius worked from Alexandrian tables but did so in a new way and stretched the classic rules for a 19-year cycle quite a bit.[27] He allowed the Easter moon to be recognized as early as March 18th, and he approved of Easter Sundays as late as April 24th. But his primary point of reference was the Passion of Christ on luna XIV of annus mundi 5229 which he reckoned to have been the consular year of the Gemini; this he described as *annus Passionis* I.[28] With reference to the 19-year cycle his *a.P.*I was to be the fourth year, but this had the result that *Saltus lunae* followed *a.P.*XVI. With these shifts of lunar and solar cycles Victurius' year sixteen fell unhappily upon year six of the standard Alexandrian cycle whose *Saltus* followed the ninteenth year. In consequence the Easter terms would be applied differently in the two cycles during 13 years (VII–XVIIII) of the Alexandrian cycle. This made it quite difficult for the churches to celebrate Easter on the same Sundays unless Alexandria agreed to change its usage—a most unlikely prospect which even Victurius did not have the temerity to suggest. In order to obviate this problem he supplied alternate dates, *Graeci* and *Latini,* so that each bishop could use his own cycle; apparently he did not realize that some of the particular Greek dates which he proposed would not be tolerated in Alexandria.

The *Cyclus Paschalis* of Victurius of Aquitaine provided a mechanism for Romans to adapt their local practices to the Alexandrian 19-year cycle. But on many dates the apostolic sees would not be celebrating the Resurrection on the same Sundays. It may be for this reason that Leo and other bishops of Rome, Ravenna, Milan, and Aquileia did not accept them. Some bishops gained approval for them in parts of Gaul, and they seemed to have found favor temporarily in

[27]The prologue and tables of Victurius were published by Denis Petau (1627), A. Bucherius (1634), Th. Mommsen (1892), and E. Schwartz (1905); all were superseded by the edition by Krusch (1938, pp. 17–25), rpr *PL Supplementum* III (1963, pp. 381–426). Thus far, fourteen manuscripts of the prologue and five manuscripts of the tables have been verified for Stevens, *Catalogue* (in preparation). Other manuscripts of authentic Victurian tables may yet be found.

[28]Later, the Victurian *annus Passionis* I might have corresponded with *annus Domini* XXVIII in the Dionysian cycle but for the fact that in his actual reckoning Victurius used *annus mundi* 5230 [= *annus Passionis* I].

Ireland. Occasionally the Romans celebrated Easter on one of the Sundays which appears in the Victurian options, but there seems to be no direct evidence that this *cyclus* was actually used in Rome or other parts of Italy and North Africa.

8. The Roman ecclesiastical leaders did not abandon the project of finding some way for observing the Easter celebration together with other Christians according to some predictive system. Yet another attempt to solve the problems was made about A.D. 523 when Bonifatius, *primicarius notariorum* or (as one might say) papal chancellor, asked Dionysius Exiguus to translate Alexandrian rules into Roman practice. Dionysius was a Scythian scholar whose work in the Roman curia was primarily with codification of conciliar decisions. When his Easter tables were ready Bonifatius and his colleague Bonus recommended them to bishop Iohannes (A.D. 523–526). This was a very important development and is well documented.[29] The tables and explanations created by Dionysius Exiguus provided the pattern which eventually prevailed in the Latin churches. He applied the classic rules for the Alexandrian 19-year cycle without significant alteration, and he extended the earlier Alexandrian data for 95 years from DXXXII to DCXXVI. Although he did not suppose that a perpetual calendar had thereby been achieved, Dionysius nevertheless supplied later scholars with the mechanisms by which a nearly perpetual 532-year cycle could be used until the changes of A.D. 1582 brought further adjustments. Nevertheless there is no clear evidence to show that churches of the Roman *civitas* or of the other *civitates* in Italy did use Dionysian Easter tables until the tenth century. By that time they prevailed in schools and cathedrals throughout the Carolingian and Germanic kingdoms, Hispanic regions, Ireland, Cornwall, Wales, Scotland, and England. But apparently Rome and its provinces were the last Western areas to accept them (Krusch, 1938, pp. 4–15; Jones, 1943, pp. 61–67; Stevens, 1981, pp. 92–93).[30]

These eight sets of evidence show that from the second to fifth centuries much scholarly labor was dedicated to discovering the relation of the crucifixion, death, and resurrection of Christ Jesus to the sabbath and to the changeable moon. But

[29]Ms Würzburg Univ.-Bibliothek M.p.th.f.46 (ca. A.D. 800) f.lv–6 and Berlin Deutsche Staatsbibliothek Phillipps 1830 (s.IX 2) f.4–10v are the earliest manuscripts containing unaltered Dionysian tables; an earlier fragment is ms Paris Bibliothèque Nationale Lat.9527 (s.VIII) f.20.

An *Exemplum Bonifatii* recommended the tables as meeting both the criteria of the Council of Nicaea and the need to avoid carnival in Rome; it is extant in five manuscripts. Apparently Iohannes died before he had accepted them. There are also twenty-one manuscripts of Dionysius' *Epistola ad Petronium*, twenty-six manuscripts of his *Epistola ad Bonifatium et Bonum*, and at least six manuscripts of his *Argumenta paschalia* explaining the rules for his table. Many other argumenta for Easter tables were modeled on those of Dionysius. There are many editions of these works by Dionysius; the best are by Krusch (1938, pp. 69–86).

Interesting lectures on the contributions of Dionysius by B. Hoffman were printed only in *PL* LXVII (1848, pp. 453–492). See also Jones (1934; 1938, pp. 204–205; 1943, pp. 68–75).

[30]The presence of Alexandrian indictions in a Latin document have often been taken to mean that Dionysian tables were then being used, as for example Rühl (1897, p. 171): Roman documents of A.D. 584. But the Roman civitas and diocese often did business with the imperial chancery, for which indictions provided official dating. Dating by indiction therefore does not imply the use of Dionysian tables.

for a bishop the results were still uncertain. What was Augustine to say to Ianuarius about these questions? His own concern for the date of Easter was quite limited, and he was far less precise than were many Christian scholars during his era. He only proposed that one should observe it on the Sunday which falls within luna XIV to XXI of the month of new corn. That is, everyone should celebrate Easter on the first Sunday after the first full moon after the vernal equinox. Beyond that: not to worry!

IV

These examples of different sets of evidence generated through the fourth, fifth, and sixth centuries can be traced by their uses and elaborations in the many Christian schools of Spain, Gaul, Italy, Germany, Ireland, and England through the sixth, seventh, eighth, and ninth centuries. They display very intense and usually competent use of arithmetic and astronomy for the purpose of determining the proper date of Easter and predicting it for the future. And this would not exhaust the evidence by any means. Many more complications would underline the recognition that calendar and computus engaged the minds of experts at technical levels of arithmetic and astronomy and that the results interacted significantly with the lives of early Christians at worship.

It was a period during which Roman controls, courts, and taxes were disappearing in the West. During the late Roman Empire the well-to-do leaders of Roman society were worried about invaders, potential or actual. The movement to build high and sturdy walls around cities in many parts of the Empire testify to a sense of threat.[31] Furthermore from the third century onwards, Italian and Gallo-Roman patricians practiced economic restraint in education, gymnasium, and the baths. Not only did they enclose their cities with larger walls, but often they simply withdrew their produce from the markets, leaving lesser castes to bear the brunt of the economic chaos which such actions induced. Except for military employment, aristocrats usually refused to associate with the new Asiatic and many Germanic groups which moved in a few at a time with different notions of raising children and perhaps a fear of emasculation through literacy (Musset, Bachrach).[32] One could therefore suppose a break in the tradition of every school with masters and students which had existed up to that period (Roger).[33]

[31]Not only military and political problems but also the sense of crisis as it increased from A.D. 161 to 518 is discussed by Remondon (1964), with emphasis on the third century as a period of reaction which had major consequences especially for cities in the Western regions of the Empire. See also Bayless (1972).

[32]Bachrach (1972, 1973) identified several Asiatic as well as Germanic peoples who were invited and settled within the Western Empire by Roman generals specifically for the sake of using their skills. They did not mix with local leaders or peoples for several centuries.

[33]For this point of view, the fundamental work was M. Roger, *L'enseigement des lettres classiques d'Ausone à Alcuin* (1905). Marrou (1948) also completed his great study of Hellenic and Hellenistic education with insistence that the tradition was lost in the subsequent period: chapter 9 emphasized the diappearance of classical schools after the third century; chapter 10 outlined the appearance of Christian

Doubtless there was great loss, yet the evidence shows also considerable continuity. When did aristocrats fail to tutor their own children at home, even as patrons of small enclaves of students on their estates? Some of the Roman aristocrats were also Christian bishops who looked after the training of children and of priests in their dioceses for pastoral duties (Riché).[34] The Latin classical studies were continued in the households of bishops, encouraged by the positive example of Gregory the Great as bishop of Rome (590–604) and the influence of his highly literate writings, as well as by the Cassiodorean *artes et scientiae vel disciplinae* which extended the tradition of the seven liberal arts.[35] Thus there was no time when schools failed to be found in various regions of the late Roman Empire.

On the other hand if we look ahead several centuries, we find that by the late eighth century endowments and personnel once more had become available for schools on a larger scale, both East and West of the Rhine river, both North and South of the Alps, in Ireland and the British isles. Costly books were supplied again; instructors were trained. But it was not the same. These later schools were not Roman and rhetorical in a Hellenistic sense but were Christian and pietistic, serving quite different functions in monasteries and cathedrals. The disciplines of advanced learning were taking new forms.[36]

Because historians have found an abundance of piety in the new centers of sacred study, and little taste for Horace's nuances of language in the bulk of surviving Latin manuscripts, they have been too sure that there was also no science in the early Christian schools. But they have simply overlooked the demands which the new development of Christian worship and thus the Christian calendar placed upon masters and students for complex calculations of dates and all the exactitude possible in astronomy.

It was not the trivium and quadrivium that made up the new curriculum of monastic schools; it was *Grammatica, Computistica,* and *Cantica:* language, reckoning, and singing—without which neither a cathedral nor a monastery could meet its obligations to God and to man (Jones, 1947; Stevens, 1979).

Thus it is necessary to turn again to the evidence of *Computistica.* What did the masters do with the computus in monastic schools? As illustrated above, it is

schools ''of the medieval type''; and his Epilogue affirmed ''The end of the school of antiquity.'' Beyond the ''Silver Age'' of Latin literature however, Marrou relied too much upon Roger, and he did not apply the same careful analysis to the evidence for education in the later periods that he had given to the earlier.

Nevertheless by 1972 Marrou had accepted and emphasized continuity in his ''L'école de l'antiquité tardive''; and see also the discussions of his paper on pp. 203–211, especially remarks of F. Prinz.

[34]In section one, ''Survivance de l'école antique et organisations des écoles chrêtiennes dans les royaumes barbares méditerranéens (480–533),'' Riché provided abundant evidence for continuity of classical forms of education well into the sixth century.

[35]This development was also analyzed by Riché in the first part of his section two, ''La fin de l'éducation antique et le développement des écoles chrêtiennes en Italie, Gaule, Espagne (533—premier tiers du VIIe siècle).''

[36]This is the principal theme of Riché's third section, ''Les débuts de l'éducation médiévale,'' challenging the view that the new Irish, Anglo-Saxon, or Carolingian schools were recovering a lost classical curriculum.

not at all a simple matter to use a 19-year cycle, much less to repeat it in successive periods of nineteen years. But it is only from the Christian literature of the Latin middle ages that one may learn the complex details both of its theory and of its practice. The following examples of calculations are summarized from Bedae *De temporum ratione liber* (A.D. 725) and have been explained more fully elsewhere by Stevens (1992).

Although lunar months may be reckoned as alternately 29 or 30 days, twelve such months account for only 354 days, leaving more than 11 days to deal with in terms of a solar year. Three years like that and you have lost more than a month. One keeps track of such things in two ways: first of all, some years were designated common and other embolismic in order to coordinate the lunar with the solar cycles. Thus an embolismic month of 30 days was added to the lunar cycle following years 3, 6, 8, 11, 14, 17, and 19. The eighth year in this series was chosen because of the early tradition of octaëteris, the 8-year cycle. (It is remarkable that the ninth year might have served better than the eighth but was never used.) An extra adjustment was therefore required to maintain the 8-year cycle, so that an embolismic month of 30 days is counted at the end of year 8 in order to keep track of the *epactae*. But in order to do that, the computist must look ahead and borrow two days, one from year 9 and another from year 10.

Second, after twelve lunar months of 354 days there still should be eleven days to make up the 365 days of a solar year, three times out of four. The eleven days which are lacking are reckoned as lunar *epactae,* and a running account is kept of them: after two years one has counted 22 lunar epacts; after three years 33 lunar epacts. At this point the count of epacts has surpassed the days of one full month; therefore an embolismic month had to be noted and its 30 days subtracted from 33, leaving three epacts to carry forward. After four years therefore one will count 3 + 11 = 14 lunar epacts, after five years 14 + 11 = 25 lunar epacts, after six years 25 + 11 = 36 when another embolismic month must be noted and its 30 days subtracted from the count of 36 lunar epacts, leaving six epacts to carry forward; and so on for 19 years—keeping in mind that after eight years the running count of epacts will be rounded up to 30 by borrowing. In summary it looks like this:

8	years × 11 lunar epacts	= 88 + 2 = 90	(4)
	less 30 days × 3 embolisms	= 90	
11	years × 11 lunar epacts	= 121 − 2 = 119	
	less 30 days × 4 embolisms	= 120	
	Remainder	= − 1	
or, 19	years × 11 lunar epacts	= 209	
	less 30 days × 7 embolisms	= 210 − 1 = 209	

As a remainder, one day of the embolismic months has been subtracted more than the running count of *epactae*. In order to begin a new 19-year cycle properly, that extra day is skipped over and is called *Saltus lunae*.

Nevertheless the overall count of lunar and solar days does not come out right:

19 years	$\times$ 354 days	=	6726	(5)
7 embolisms	$\times$ 30 days	=	210	
	Total	=	6936 lunar days	
19 years	$\times$ $365^1/_4$ days	=	$6939^3/_4$ solar days	

From equation 3 (page 32 above) it was evident that the 19-year cycle should have $6940^3/_4$ lunar days, from which the *Saltus lunae* could be subtracted in order to bring sun and moon together again; it should not have 6936. Where is the error?

Imagine for yourselves the situation of masters, students, and scribes redoing their figures, comparing their tables, looking up each running count of epacts to balance their figures for the 19-year Easter tables. The manuscripts show that this was in fact the single most repeated problem in monastic schoolwork; and its solution led to the greatest number of *argumenta* to be found in computistical handbooks for fourteen centuries. The error stems from that fraction $^1/_4$. The day count of $365^1/_4$ is not applied in the computus by quarters but only by integers: counting 365 every year and adding one more day every fourth year, or leap year. It is a simple procedure, but one in which the use of a fraction will lead to error! Use of a fraction in that way allowed the computisti to forget that for each day there is not only a sun to reckon but also a moon. Therefore if one adds a day to the solar cycle, one must add the same day to the lunar cycle; thus

Solar Cycle	19 $\times$ 365 days	=	6935	(6)
	4 $\times$ 1 extra day	=	4	
	Total	=	6939 days	
Lunar Cycle	19 $\times$ 354 days	=	6726	
	7 $\times$ 30 days	=	210	
	4 $\times$ 1 extra day	=	4	
	Total	=	6940 days	

This count of days (6939, 6940) will occur only one time out of four successive 19-year cycles, one in which there would be four leap years. But for the other three 19-year cycles in a sequence of four, there would be five leap years in each cycle; then in each of those three 19-year cycles the solar and lunar totals would both increase by one day (6940, 6941). In each case there will be an excess moon, and that remainder represents the *Saltus lunae* which is skipped in order to begin the next 19-year cycle with sun and moon in accord. By use of the fraction the computist could easily lose track of a full lunar day in counting $19 \times 4 = 76$ years; but despite their awkwardness at some points, by use of integers the reckonings of lunar and solar cycles will correlate without fail, and the date of Easter Sunday may be predicted reliably.

V

By no means do these reckonings exhaust the problems or the hidden complications to be met in applying the nineteen-year cycle itself, and they only begin to touch the technical and scientific riches of the broader computistical literature. When Frankish scholars in ninth- and tenth-century monastic and cathedral schools noticed a slippage in the position of the equinoxes—the calculations undertaken became increasingly complex and the concepts became steadily more sophisticated. On this basis schoolbooks of arithmetic, of astronomy, and even of geometry were generated—with contents both theoretical and practical (Stevens, 1979, 1981, 1992). It is also on this basis that there was an audience for the more nearly complete books of Euclid and Ptolemy when these later became available in Latin, an audience eager and informed.[37]

The modern philologists' search for the classical quadrivium faltered and partly failed when it was made to depend upon literature that cited and discussed only the classic four fields. Those discussions in late Roman and early medieval letters and encyclopedias were significant and had some influence upon schoolmasters. But during hard times the real need and pressure for studies was stimulated in other terms and by other needs which were not Hellenistic. Oddly enough, it was in the schools of Christian piety that masters and students thought it necessary to drill, to learn, to understand the computus and thus not only to use but also to comprehend the calendar and improve it. In order to do that, basic arithmetic and astronomy were required at a modest level. It has been found that masters and students in monastic and cathedral schools went well beyond those basic elements of computistical science and added planetary theory and plane geometry which were not at all necessary. The best and thus far the least known evidence for studies in early medieval schools may still be found in those 9000 or more Latin manuscripts which include the tracts, tables, and argumenta of computus wherein are calculated repeatedly and diversely the cycles of time.

References

Adams, J. (1971), *The Populus of Augustine and Jerome*. New Haven: Yale University Press.

Anscombe, A. (1895), The paschal canon attributed to Anatolius of Laodicea. *Eng. Hist. Rev.*, 10:515–535.

Aujac, G. (1974), "L'image du globe terrestre dans la Grèce ancienne." *Revue d'hist. sci.*, 27:193–210.

Bachrach, B. (1972), *Merovingian Military Organization, 481–751*. Minneapolis: University of Minnesota Press.

——— (1973), *The History of the Alans in the West*. Minneapolis: University of Minnesota Press.

[37]M. Folkerts, *Euclid in Medieval Europe* (Questio II de rerum natura, 1989); and F. Benjamin, "Campanus of Novara," in Benjamin and Toomer (1971, pp. 3–24). Campanus was a computist and calculator whose commentary on Euclid's *Elements* (1255–1259) and *Theorica planetarum* (1261–1264) were based upon Ptolemaic solar, lunar, and planetary models; both of his works became widely known from the thirteenth to the seventeenth centuries.

Barnes, T. (1971), *Tertullian, a Historical and Literary Study*. Cambridge, U.K.: Clarendon Press.

Bayless, W. N. (1972), *The Political Unity of the Roman Empire During the Disintegration of the West, A.D. 395–457*. Unpublished doctoral dissertation. Providence, RI: Brown University.

Benjamin, F., & Toomer, G. J. (1971), *Campanus of Novara and Medieval Planetary Theory*. Madison: University of Wisconsin Press.

Boncompagni, B. (1868–1887), *Bulletino de Bibliographica e di Storia delle Scienze matematiche e fisiche*, 20 volumes. Rome.

Bonner, G. (1963), *St. Augustine of Hippo: Life and Controversies*. London: SPCK.

Borst, A. (1988), Computus: Zeit und Zahl im Mittelalter. *Deutsches Archiv*, 44/1:1–82.

Brown, P. (1967), *Augustine of Hippo*. Berkeley: University of California Press.

Bucherius, A. *De doctina temporum commentarius in Victorium Aquitanum*. Antwerp: 1964.

Bullough, D. A. (1965), *The Age of Charlemagne*. London: Elek.

Cabrol, F., & Leclercq, H., eds. (1900–1904), *Reliquae Liturgicae Vetustissimae*, Vols. 1–2. In: *Monumenta ecclesiae liturgica*. Paris: Didot.

The Calendar of St Willibrord, H. A. Wilson, ed. (1918), Henry Bradshaw Society Publications, LV. London.

Christ, W. (1983), "Victorii Calculi," Sitzungsberichte der königliche Akademie. *Philos.-hist.-Klasse*, 1:132–136 (München).

A Companion to the Study of St Augustine (1954), ed. R. W. Battenhouse. New York: Oxford University Press.

CSEL (1866 seq), *Corpus scriptorum ecclesiasticorum latinorum*, multivolume series. Vienna: C. Geroldus.

Cummian's letter De controversia paschali and the De ratione conputandi, ed. M. Walsh, & D. ó Cróinín. Toronto: Pontifical Institute of Mediaeval Studies, 1988.

Deane, H. A. (1963), *The Political and Social Ideas of St. Augustine*. New York: Columbia University Press.

DACL (1907–1953), *Dictionnaire d'archéologie chrétienne et de liturgie*, 15 vols., ed. F. Cabrol, H. Leclercq et H. I. Marrou. Paris: Librairie Letouzey et Ané.

DSB (1970–1980), *Dictionary for Scientific Biography*, 16 vols., ed. C. C. Gillispie. New York: Charles Scribner's Sons.

Duhem, P. (1914–1959), *Le Système du monde*, 10 vols. Paris: Hermann.

Eastwood, B. S. (1983a), Mss Madrid 9605, Munich 6364, and the evolution of two Pinian astronomical diagrams in the tenth century. *Dynamis: Acta Hispanica*, 3:272–276.

——— (1983b), Origins and contents of the Leiden planetary configuration (ms Voss.Q.79, f.93v). *Viator*, 14:1–23.

——— (1987), Plinian astronomical diagrams in the early middle ages. In: *Mathematics and Its Applications to Science and Natural Philosophy in the Middle Ages: Essays in Honor of Marshall Clagett*, ed. E. Grant & J. Murdoch. Cambridge, U.K.: Cambridge University Press.

Eusebius Werke. In: *Griechische christliche Schriftsteller*, ed. E. Schwartz. Vol. 9. Leipzig: J. C. Hinrichs, 1903–1904.

Folkerts, M. (1989), *Euclid in Medieval Europe*. Questio II de rerum natura, ed. W. M. Stevens. Winnipeg: The Benjamin Catalogue for History of Science.

Fraser, J. T., & Lawrence, N., eds. (1975), *The Study of Time*. Berlin: Springer-Verlag.

——— (1987), *Time, the Familiar Stranger*. Amherst: University of Massachusetts Press.

Friedlein, G. (1871), Der Calculus des Victorius. *Zeitschr. für Mathemat. und Physik*, 16:42–79.

Geminos, *Elementa astronomiae*, Vol. 8, ed. C. Manitius. Leipzig: Teubner, 1898.

Ginzel, F. K. (1906–1914), *Handbuch der mathematischen und technischen Chronologie*, 3 vols. Leipzig: J. C. Hinrichs.

Heath, T. C. (1913), *Aristarchos*. Cambridge, U.K.: Oxford at the Clarendon Press.

Herwagen, J. (1563–1564), *Opera Bedae venerabilis omnia*, 8 vols. Basel: Verlag Herwagen.

Jones, C. W. (1934), The Victurian and Dionysiac paschal tables in the West. *Speculum*, 9:408–421.

——— (1937), The lost Sirmond manuscript of Bede's computus. *Eng. Hist. Rev.*, 52:204–219.

——— (1939), *Bedae pseudepigrapha: Scientific Writings Falsely Attributed to Bede*. Ithaca, NY: Cornell University Press.

——— (1943), *Bedae opera de temporibus*. Cambridge, MA: Medieval Academy of America.

——— (1947), *Saints' Lives and Chronicles in Early England*. Ithaca, NY: Cornell University Press.

King, M. H., & Stevens, W. M., eds. (1979), *Saints, Scholars & Heroes: Studies in Medieval Culture in Honour of Charles W. Jones*, 2 vols. Collegeville, MN: Hill Monastic Manuscript Library, Saint John's Abbey and University.

King, V. H. (1969), *An Investigation of Some Astronomical Excerpts from Pliny's Natural History Found in Manuscripts of the Earlier Middle Ages*. University of Oxford Bachelor of Letters Thesis, unpublished.

Krusch, B. (1880), *Studien zur christlich-mittelalterlichen Chronologie: Der 84 jährige Ostercyclus und seine Quellen*. Leipzig: Verlag Von Veit. Abbreviated *Studien* I.

——— (1938), *Studien zur christlich-mittelalterlichen Chronologie: Der Entstehung unserer heutigen Zeitrechung*, Abhandlungen der Preussische Akademie zu Berlin, Jahrgang 1937, Philol.-hist.-Klasse Nr. 8. Berlin. Abbreviated *Studien* II.

MacCarthy, B. (1901), Introduction. *The Annals of Ulster*, Vol. 4. Dublin: Irish Record Publications, 3.

McGurk, P. (1973), Germanici Caesaris Aratea cum scholiis, a new illustrated witness from Wales. *Nat. Lib. Wales J.*, 18:197–216.

Marrou, H.-I. (1948), *A History of Education in Antiquity*. New York: Sheed and Ward, 1956.

——— (1950), *L'ambivalence du temps de l'historie chez S. Augustine*. Montréal: Institut des Etudes médiévales.

——— (1957), *Saint Augustine and His Influence Through the Ages*. London: Longmans.

——— (1972), L'école de l'antiquité tardive. *Settimane di Studio* (Spoleto), 19/1:127–143.

Mommsen, Th. *Chronica minora*, 3 vol. in *Monumenta Germaniae Historica*, Vols. IX, XI, XIII. Berlin: 1892–1898.

Monumenta cartographica vetustioris aevi, A.D. *1200–1500*, ed. R. Almagia & M. Destombes. Amsterdam: N. Israel, 1964.

Musset, L. (1965), *The Germanic Invasions: The Making of Europe*, A.D. *400–600*, trans. E. and C. James. London: Elek, 1975.

North, J. D. (1975), Monasticism and the first mechanical clocks. In: *The Study of Time*, vol. 2, ed. J. T. Fraser & N. Lawrence. Berlin: Springer-Verlag, pp. 381–398.

O'Daly, G. J. P. (1981), Augustine on the measurement of time: Some comparisons with Aristotelian and Stoic texts. In: *Neoplatonism and Early Christian Thought*, ed. H. J. Blumenthal & R. A. Markus. London: Variorum, pp. 171–179.

Ogg, M. (1950), *The Pseudo Cyprianic De pascha Computus*. London: SPCK.

Parsons, W. (1951), English translation of Augustini *Epistola LV ad Ianuarium*. In: *Fathers of the Church XII*, ed. A. Goldbacher. Washington: Catholic University of America Press, pp. 260–293.

PG (1857–1886), *Patrologiae cursus completus, Series graeca*, 161 vols., ed. J.-P. Migne. Paris: J.-P. Migne.

PL (1844–1882), *Patrologiae cursus completus, Series latina*, 221 vols., ed. J.-P. Migne. Paris: J.-P. Migne.

Pedersen, O., & Pihl, M. (1974), *Early Physics and Astronomy: A Historical Introduction*. New York: Neale Watson.

Petau, D. *Opus de doctrina temporum*, 2 vols. Paris: 1627.

Plinius Secundus, Gaius (A.D. 73–77), *Naturalis historia*, 37 vols., ed. J. Jan & C. Mayhoff. Leipzig: Teubner, 1906.

PW (1893–1968), *Real-Encyklopädie der klassischen Altertumswissenschaft*, multivol. series, ed. A. Pauly, G. Wissowa et al. Stuttgart: A. Druckenmüller.

Remondon, R. (1964), *La Crise de l'empire romain de Marc-Aurele à Anastase*. Paris: Presses Universitaires de France.

Richard, M. (1950), Comput et chronographie chez Saint Hippolyte. *Mélanges de science religieuse*, 7:237–257.

Richardson, C. C. (1940), The quartodecimans and the synoptic chronology. *Harvard Theolog. Rev.*, 33:177–190.

——— (1973), A new solution to the quartodeciman riddle. *J. Theolog. Studies*, 24/1:74–84.

Riché, P. (1962), *Education et culture dans l'Occident barbare, VIe-VIIIe siècle*. Paris: Editions du Seuil. Second edition 1967; third edition 1972. [*Education and Culture in the Barbarian West, Sixth Through Eighth Centuries*, trans. J. Contreni. Columbia: University of South Carolina Press, 1976.]

Roger, M. (1905), *L'enseigement des lettres classiques d'Ausone à Alcuin*. Paris: A. Picard et Fils.

Rühl, F. (1897), *Chronologie des Mittelalters und der Neuzeit*. Berlin: Reuther & Reichard.

Schiaparelli, L. (1924), *Il Codice 490 della Biblioteca Capitolare de Lucca*. Rome: Pressa La Biblioteca Vaticana.

Schwartz, E. (1905), *Christliche und jüdische Ostertafeln*. Abhandlungen der Königlichen Gesellschaft der Wissenschaften zu Göttingen, *Philol.-hist.-Klasse*, Neue Folge 8/6. Berlin.

Stevens, W. M. (1972), Walahfrid Strabo—a student at Fulda. In: *Historical Papers 1971*, ed. J. Atherton. Ottawa: Canadian Historical Association, pp. 13–20.

——— (1979), Compotistica et astronomica in the Fulda school. In: *Saints, Scholars & Heroes*, vol. 2, ed. M. H. King & W. M. Stevens. Collegeville, MN: Hill Monastic Manuscript Library, Saint John's Abbey and University, pp. 27–63.

——— (1980), The figure of the earth in Isidore's *De natura rerum*. *ISIS*, 71:268–277.

——— (1981), Scientific instruction in early insular schools. In: *Insular Latin Studies*, ed M. Herren. Toronto: Pontifical Institute of Mediaeval Studies.

——— (1992), Sidereal time in Anglo-Saxon England. In: *Sutton Hoo: Voyage to the Other World*, ed. C. B. Kendall & P. S. Wells. Medieval Studies at Minnesota, no. 4. Minneapolis: University of Minnesota Press.

——— (in preparation), *A Catalogue of Computistical Tracts*, A.D. *200–1200*.

Strobel, A. (1977), *Ursprung und Geschichte des frühchristlichen Osterkalenders*. Texte und Untersuchungen, 121. Berlin: Akademie-Verlag.

——— (1984), *Texte zur Geschichte des frühchristlichen Osterkalenders*. Liturgiewissenschaftliche Quellen und Forschungen, 64. Münster in Westfalen: Aschendorffsche Verlagsbuchhandlung.

Suetonius Tranquillus, Gaius (ca. A.D. 69–125?), *Divus Julius*, ed. H. E. Butler & M. Cary, 1927. Introduction and notes G. B. Townend. Bristol: Classical Press, 1982.

Turner, C. H. (1895), The paschal canon of Anatolius of Laodicea. *Eng. Hist. Rev.*, 10:699–710.

Ussher, J. *Britannia ecclesiastium antiquitates*. Dublin: 1639. Rpr. London: 1687.

Van de Vijver, A. (1957), L'évolution du comput Alexandrin et Romain du IIIe au Ve siècle. *Rev. d'hist. ecclésiastique*, 52:5–25.

Van der Meer, F. (1961), *Augustin the Bishop*, trans. B. Battershaw & G. R. Lamb. London: Sheed & Ward.

Wilmart, A. (1933), Un nouveau texte de faux concile de Césarée sur le comput pascal. *Studi e testi*, 59:19–27.

Appendix
Luni–Solar Time Cycles

Within the concepts of Hellenistic science, attempts to coordinate observed cycles of the moon and of the sun were numerous and their tabular periods could extend for 8, 11, 16, 19, 28, 76, 84, 95, 100, 112, or 532 years. With the following selected list of cycles created in the Mediterranean basin and in Western Europe, the reader may be assisted in locating them.

Each cycle is identified by *Author or Cycle* as it is usually cited, by the *date* of first appearance and usage, as well as by place of *origin* in so far as these are known in modern scholarship. Greek forms are given for the names of Greek authors within limits of Roman typeface. Two or more dates and two or more place names are provided when the reappearance of a cycle is historically significant. The Roman province for a place name is sometimes indicated: thus Hippo Regis (Africa) is in the province of Roman Africa. Further information about most of these luni–solar time cycles may be found in the Introduction to C. W. Jones (1943), while more recent studies are cited in annotations of the present essay.

Author or Cycle		Date	Origin
Octaëteris	B.C.	ante 500	Babylonia
Euctemon, Meton		500–400	Athens
Kallippos of Cyzicus		370–300	Athens
Sosigenes		50	Alexandria, Rome
G. Plinius Secundus	A.D.	73–77	Italy
Laterculus Romanus		213–297	Rome
		412	Africa
Augustalis		213–297–312	Africa
Hippolytus		222–240	Ostia, Rome
Demetrios		232	Alexandria
Dionysios		247	Alexandria
Cyclus Hippolyti		241–256	Rome, Carthage?
		485	Gaul
		577	Gaul, Spain
Expositio Bissexti		241–304	Africa, Italy?
Anatolios		269–280	Laodicea

Appendix

Author or Cycle	Date	Origin
Romana Supputatio	312–342	Rome
Council of Arles	314	Arles, Provence
Eusebios	335	Caesaria
Athanasius	335	Alexandria
Romana Supputatio, revised	343–354	Rome
second revision	355–411	Rome
Epiphanios	367–403	Constantia, Cyprus
Ambrosius	374–397	Milan
Theophilos	395–412	Alexandria
Augustinus	395–430	Hippo Regis, Africa
Innocentius	401–417	Rome
Agriustia	412	Thimidia Regis, Africa
Ratio Paschae	439–449	Carthage
Acta Synodi Caesareae	400–450	Africa
Cyril	436–444	Alexandria
Dioscores	444–451	Alexandria
Leo	440–461	Rome
Pascasinus	442	Lilybaeum, Sicily
Zeitz Tafel	447	Italy
Proterios	451–457	Alexandria
Laterculus Romanus, revised	455	Africa
Victurius	455–457	Aquitaine
Dionysius Exiguus	520–525	Rome
Exemplum Bonifatii	523	Rome
Martin of Braga	580	Spain, Portugal
Cummian	633	Southern Ireland
Beda	701–725	Jarrow, Northumbria

3

Process as a Primitive Physical Category

C. J. S. Clarke

Abstract Sketches are provided of two theories, superstring theory and quantum cosmology, which illustrate the way in which attempts to bring together gravitation and quantum theory encounter inconsistencies at both the mathematical level (quantities become infinite) and the interpretational level (what is meant by space–time?). A description, based on Whitehead's earlier work, is then given of a possible method of deriving space and time from a network of overlapping processes, in a way that could provide a solution to these problems by making the concept of reality a relative property. Finally there are suggestions as to how the resulting view of the universe sheds light on the relation between the cyclic processes of nature and historical time.

Problems in Modern Physics

The impression is often given of a physics that is sweeping toward an ultimate explanation of the entire universe founded on the relativistic picture of geometrized space–time. Nothing could be further from the truth: the conventional view of space–time as the ground of physics is increasingly being called into question. In this section I shall describe the difficulties being encountered with the conventional view, which suggest that an alternative is required.

String Theory

The greatest challenge of theoretical high-energy physics today is the quest for the Theory of Everything, or TOE, uniting the four currently well-established forces of nature. The TOE is to be found beyond the now faltering GUT, or Grand Unified Theory, uniting only three of them. Many people believe that TOE must be different from all theories that preceded it, because the fourth force to be incorporated is gravity, and gravity is described not by particles or fields in a preexisting space–time geometry, but by the space–time geometry itself. According to conventional wisdom, physical theories are obtained by applying the principles of quantum mechanics to classical theories—a process called quantization. And so the Theory of Everything has to quantize everything, including geometry.

Unfortunately, there is little agreement as to what quantization is: but, speaking very roughly, it usually involves replacing the definite states of the classical theory by a diffuse wave function spread out over a fixed substratum. For example, to quantize a theory describing a particle moving in three-dimensional space, one will use a wave function spread out over three-dimensional space so that the substratum for the wave function is ordinary geometry; while, if the classical theory involves points in more abstract mathematical space, then the substratum for the wave function will be this abstract space. When it comes to quantizing geometry this idea encounters both conceptual and technical problems, due essentially to the difficulty of finding a suitable substratum against which to construct a quantized theory of geometry, given that geometry itself can no longer provide the substratum.

Many physicists would say that the best hope of constructing a Theory of Everything lies along the lines of the most famous of the recent attempts at unification, superstring theory. The epithet *super* is in this context not an upper-class British exclamation, but describes the fact that it is a theory in which multiplication is redefined so that *ab* is not necessarily the same as *ba*[1] So, let me point out very briefly how the difficulties that I have indicated are manifested in superstring theory.

The theory deals with a two-dimensional surface (the string) bending around in an eleven-dimensional space (the embedding space). It is based on an older theory in which strings are described as two-dimensional sheets in four-dimensional space–time. In the old theory the situation at any given instant of time was represented by the intersection of the two-dimensional sheet with the three-dimensional space defining that time–instant, this intersection being one-dimensional, that is, a curve or string. But whereas the old string theory regarded the space–time as a given arena and the string as the only dynamical object, in the new theory the geometry of space–time is also supposed to be dynamical.

Unfortunately, there is a pervading ambiguity as to what exactly the term *space–time* means, in superstring theory. And there is a related ambiguity as to what is a fixed substratum and what is a dynamical (and hence quantized) object.

Space–time is in fact described within the theory in two different ways. On the first description, the embedding space in which the string lives is thought of as being like a space–time. It is supposed to be curved, and it then transpires that requirements of consistency force its curvature to be governed by a higher-dimensional analogue of Einstein's theory of gravity, just like the space–time of classical physics. So it is hoped that the curvature of the embedding space might work out to be such that, when viewed from a distance, as it were, it would look like ordinary

[1]A supertheory is one in which the real numbers are replaced by a Grassman algebra: a Z^2-graded algebra comprising an odd part and an even part with the rule that elements in the odd part anticommute, while elements in the even part commute with each other and with the odd part.

four-dimensional space–time.[2] The snag with this is that the formalism as usually presented regards the embedding space as a fixed substratum which is not quantized. But if the curvature of the embedding space is determined by the string, and the string is a dynamical object, then the embedding space is a dynamical object as well and so its geometry should be quantized—it cannot be a fixed substratum.[3]

A second way in which space–time is described is by using the fact that the internal vibrations of the string have formal analogies with the vibrations of space–time geometry that characterize gravitational waves. Thus it is hoped that a collection of strings could build up a full gravitational field, and hence a full space–time. Although this is the relation to space–time that was envisaged when the theory was first developed, there have been few systematic attempts to work out the details of the idea.

Ambiguity arises because of uncertainty as to how these two ways of describing space–time are related, and as to how one is to obtain classical space–time from either of the ways. Furthermore, on the first approach there are deep formal difficulties involved in trying to use a theory in which the embedding space is both the background to a quantum theory and also is a dynamical entity within a quantum theory.

Because of these difficulties, superstring theory has not developed a consistent worldview, and has remained at a rather formal level. So in order to discuss how the concept of process might provide a way out of the difficulties, let me revert to an older picture which is representative of the sort of view which is likely to play a part in any TOE: the picture of quantum cosmology that was developed by Hartle and Hawking (1983). This provides a context for a theory in which both space–time and matter are dynamical objects.

Quantum Cosmology

In the Hartle-Hawking theory (see Figure 3.1), the entire four-dimensional space–time universe is first regarded as a time-sequence of three-dimensional spaces, each with its geometry (which describes gravitation) and matter contents.

The geometry and the matter change continuously through the sequence. This sequence is then regarded as a curve in a *hyperspace* in which each three-dimensional space is represented as a single point. Next one quantizes, passing to a quantum theory—the key step in the procedure. To do this, one maintains this hyperspace of all three-dimensional spaces as the fixed arena, the substratum for physics, but one abandons the idea that the history of the universe is represented

[2]More precisely, one assumes that *compactification* takes place, meaning that the embedding space can be written as a fiber bundle over a four-dimensional space with the induced metric on the fibers making them of a very small dimension, and with the projection onto the base inducing a well-defined space–time metric there.

[3]There are approaches due to Horowitz (1987) which try to remove the embedding space from the picture in the first place, allowing the embedding space to be defined at the end by the strings. But it is not yet clear how successful these are.

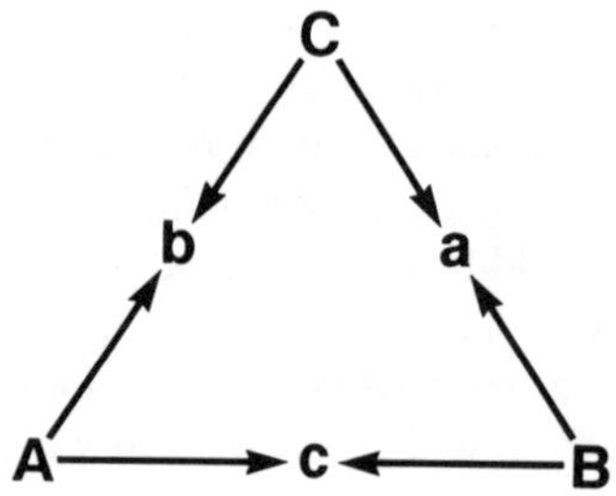

Figure 3.1. A sequence of spaces at varying times constitutes a curve in superspace.

by a precisely defined sequence of points in the hyperspace. (This idea is in any case unacceptable because it contradicts the principle, in general relativity, that there is no natural definition of any one particular time-coordinate.) Instead, one assumes that the state of the universe is given by a complex-valued function spread out over the hyperspace: the wave function of the universe.

The theory is supposed to provide a basis for calculating a priori what the wave function of the universe is. One can then use the wave function, not to describe an evolution of the universe, but to derive correlations between different aspects of the universe as we see it now, which can be compared with observation.

There are interesting aspects of this theory from the point of view of the study of time, which I can only touch on here, lest they take me too far from my goal of discussing process. For example, the chronological aspect of experimental prediction is expressed through the idea of records. Instead of asking, "if a particle is prepared in an excited state, what is the probability that two minutes later it will have decayed?" (which is a conditional probability of a future event) one asks: "given that there is now a record of a particle having been prepared two minutes ago in an excited state, what is the probability that the atom now is decayed?" (a correlation between present events).

We can see that this view does avoid some of the ambiguities that are so rampant in superstring theory. There is no problem about achieving three-dimensional space because it is explicitly fed in at the outset (whereas the more modern view is that this is not possible because the theories are only consistent in eleven dimensions). Moreover, there is a fixed arena, a substratum, in which the theory takes place, albeit a very strange one, namely the hyperspace of all three-dimensional spaces—the hyperspace of all instantaneous states of the universe. Despite these advantages, we shall see none the less that the problems of quantizing geometry cannot easily be avoided.

Before proceeding further, let me explain the sense in which a point in the hyperspace is an instantaneous state of the universe. This means that it is a set of events which are simultaneous with respect to some arbitrary time coordinate. For every time coordinate there will be different definitions of instantaneous states and all the different sorts of instantaneous states are represented in the hyperspace without favoring any one time coordinate over another. Students of time, particularly those of a Bergsonian persuasion, will immediately be put on their guard by the adoption of instantaneous states. Do we have a situation like that postulated by the Damascus school (Sorabji, 1983) and later by Descartes (1641) in which God creates the universe anew every instant, with only an illusion of continuity, or is there some underlying principle of continuity? On one extreme interpretation of the theory, the question is denied by holding that there exists only one instant (it is not clear which one) and all the previous instants, such as the instant when the excited particle was prepared two minutes ago, are conjectured from memories and records. Like solipsism, it is not possible logically to deny such a view, but neither is it possible to live on its basis.

I want to concentrate on two deficiencies of this approach to quantizing space–time, which will apply, I believe, to any approach which focuses on individual time instants and space–time events.

The first deficiency is a mathematical one. The geometry that is being quantized is founded on a three-dimensional topological manifold which is in turn modeled on the modern mathematical real number continuum. This continuum is supposed to contain an actual infinity of points[4] in contradiction to Aristotle's view that infinity can only be potential, never actual. As a result, the gravitational field can have structure at arbitrarily fine length-scales, or, equivalently, arbitrarily high energies. This is indeed true of every field: the difficulty is not confined to gravity but it takes a particularly acute form with gravity.

As far as the wave function is concerned (i.e., before observational considerations have entered) this fine substructure is actual, not potential. One cannot directly observe the large energy states in this substructure, because the energies available in laboratories are too low; but the fine structure enters into the definition of the wave function, which then determines the observations. States of arbitrarily high energies have to be taken into account in calculating the probabilities of those states that are observed. (Since they are not directly observable, they are often called virtual states, but this does not mean that they are merely potential and not actualized: they must be included in the description just as much as the state that is finally reached at the observation.)

The inclusion of states of arbitrarily fine structure leads to mathematical difficulties: because there is no limit to the energies that have to be taken into account, on most formulations the calculations that should in principle yield finite probabilities for different outcomes fail to yield any finite answer. In many cases procedures have been devised for extracting finite answers, but there is as yet no evidence that these methods can be applied successfully to any theory containing gravity. Moreover, in the case of gravity there are additional problems due to the fact that the topology of the manifold on which the space–time geometries are defined is not known in advance. Different topologies have to be included, and there is no agreed way of doing this.

[4]As Ian Thomson (unpublished) has pointed out, the term *actual infinity* is ambiguous. In mathematics *actual* tends to mean *definite,* whereas in physics, and in the philosophical uses that I have in mind here, it means *fully realized* or *not potential*. I shall return later to the relation between the mathematical and physical uses.

There is a second problem which is more concerned with the nature of scientific enquiry. One consequence of the quantum cosmology approach is the fact that anthropic arguments inevitably enter. Because the questions are being formulated in terms of correlations between presently existing states, part of the data that has to be taken as given is the fact that we are here as observers, with recording apparatuses and memories, and this fact is used to preselect one particular component of the wave function of the universe. Some would see this as an advantage, because it explicitly formalizes an input into the argument which is otherwise liable to be incorporated covertly. But others feel that the option should be open to discuss questions about the origin or nonorigin of life in the universe, without feeding it in as part of the data. This is difficult on a quantum cosmology approach because the wave function has no interpretation apart from the possibility of measurements being performed, and hence records laid down.

There is nothing new about this account of the problems inherent in the use of continuous manifold. For many years there has been a feeling that the quantization of gravity required a departure from the conventional model based ultimately on the idea of particles as points in a continuum-based space–time manifold. This was one of the motivations of string theory, which got away from the point particle but maintained the actually infinite continuum; as a result there was some improvement in the occurrence of infinities, but the early promise of a finite theory seems to be receding fast. Similarly, the twistor theory proposed by Penrose (Penrose and Rindler, 1986) offered the hope of a finite theory by replacing the points of space–time by extended objects (congruences of light rays); the continuum was still there, but was finitized by assumptions of analyticity. While this theory still exerts a strong mathematical fascination, and has proved a fruitful source of ideas in modern physics, it has not led to any change in world view.

A Process-Based Alternative?

We have seen that physics is faced with difficulties if it starts with a substratum based on the real number continuum, and in particular if it starts with the conventional model of space–time and builds up fields and extended bodies out of the infinity of points in the space–time continuum.

So the question arises, is there an alternative to starting with a manifold based on the real-number continuum? Could one instead start with a physics whose basic entities had some sort of extension in space and time from the start. This could be better expressed: to say that something is extended over space and time is to assume that there exists an independent space and time to start with, and this is precisely what we want to get away from. So let us, as a more suggestive form of words, look for theory in which the basic entities are processes, with the implication that processes are to define space and time, not vice versa, and that in so far as the space–time being defined turns out to be like conventional space–time, so it will turn out that processes are extended over the space and time which they define.

Process in the Early Whitehead

It is no accident that this way of speaking is reminiscent of the works of the early Whitehead (1919, 1926), whose aim was precisely what I have described (except that what I am here calling a process was termed by Whitehead at that time an event). Whitehead, however, seemed to be torn in two different directions. On the one hand, it appeared, at the time when he was writing, that the correct physical theory was special relativity (or some minor modification of it) and so Whitehead was drawn to get relativity as an outcome; but on the other hand, he was forming a deep metaphysical vision of a world based on process, which was incompatible with special relativity. Consequently, he first compromised his vision of process in order to get relativity, and then later repented, abandoned the argument of these early works and, in *Process and Reality,* developed his process vision independently of detailed physical theory.

Perhaps there is now enough new thinking in physics and philosophy to carry out the program that Whitehead was unable to achieve, building a physics of process. It will be instructive first to examine how far Whitehead himself got with the program before he compromised.

Whitehead called the basic entity of his approach an *event*. The term is unfortunate because in normal physicists' usage it means a point in space–time, and so I shall later replace the word by *process*; but for the time being let us stick to Whitehead's terminology, with the understanding that an event is not a point, but has both temporal and spatial extension. The key observation made by Whitehead, then, is that event a can *extend over* an event b; a relation that one thinks of, roughly, as meaning that b is a proper part of a. Although it is tempting to think of this process-theoretic extending over as being like the contained-in ($\subset$) relation in set theory, it is in fact different: it obeys different axioms.

Since Whitehead's axioms will be a kind of prototype for the sort of thing I have in mind for process physics, it will be useful to review them here. Instead of directly characterizing the relation extends over (which Whitehead denotes by K) on a set X, it is more convenient to work with a relation less-than-or-equal-to ($\leqslant$), where

$$a \leqslant b \Longleftrightarrow (b \text{ extends over } a \text{ or } a = b).$$

It is then easy to show that Whitehead's axioms are equivalent to the requirement that $\leqslant$ makes X a dense unbounded directed set with one additional axiom. Explicitly, we have

for any two events a, b there exists c with $a \leqslant c$, $b \leqslant c$ (1a)
if $a \leqslant b$ and $b \leqslant c$ then $a \leqslant c$ (directedness) (1b)
for any event a there exist distinct b, c with $b \leqslant a \leqslant c$ (unboundedness) (2)
if $a \leqslant b$ and $a \neq b$ then there exists c, distinct from
a and b

with $a \leqslant c \leqslant b$ (denseness) (3)

$\{ (\forall\ c\ (c \leqslant a \Longleftrightarrow c \leqslant \mathrm{b}) \} \Longrightarrow \{a = b\}$ (4)

The last axiom, an addition to the dense unbounded directed set axioms, is analogous to the Hausdorff condition in topology. It ensures that an event is uniquely specified by its subevents. While it will be needed for getting conventional physics, it is not clear that it is necessary: one could have a collection of subprocesses which, looked at in different ways, could be seen as belonging to two different larger processes. It is clear from these axioms that the resulting structure is quite different from set theoretic inclusion, which is a latticelike relation. That is why Whitehead, when possible, uses the phrase "*a* extends over *b*" rather than "*b* is a part of *a*."

Events are the basic objects in his approach, and space–time points and instants are derived from events by a limiting procedure. What Whitehead does is to consider nested sequences of events, each one extending over all the succeeding ones. A simplified, but I think essentially adequate, account of Whitehead's procedure is as follows. If X is the space of events, one supposes that there is defined a space X^* consisting of descending chains $(a_1, a_2 \ldots)$, $a_1 \geqslant a_2 \geqslant \ldots$, where X^* is required to include all descending chains that are eventually constant (i.e., are of the form $(a_1, a_2, \ldots, a_r, \mathrm{a}_r, a_r, \ldots))$. The relation $\leqslant$ can then be introduced into X^* by defining one sequence as greater-than-or-equal-to another if one can always find terms in the second sequence that are less-than-or-equal-to any given term in the first; in symbols:

$$(a_1, a_2, \ldots) \geq (b_1, b_2, \ldots) \iff (\forall i)(\exists j)(a_i \geq b_j).$$

Strictly speaking, this relation is a preorder on X^* rather than an order, in that some sequences a^*, b^* we might have both $a^* \geqslant b^*$ and $b^* \geqslant a^*$. This can be rectified by regarding such sequences as equivalent—that is, defining a relation $\sim$ (equivalent-to) by

$$a^* \sim b^* \iff a^* \geqslant b^* \ \&\ b^* \geqslant a^*$$

—and then grouping each class of equivalent sequences into a single abstract object. The set $\bar{X} = X^*/\sim$ of such equivalence classes is then an ordered set. Moreover, $\bar{X}$ naturally includes X, but it has in addition abstract objects not corresponding to elements of X under this inclusion. These added objects are ideal limits of sequences, and can correspond either to *points* or to *instants,* depending on how the original set X is chosen. Note that the order which we have defined on $\bar{X}$ can be described roughly by saying that one element of the enlarged set $\bar{X}$ is called *greater than* another if the processes in the sequence defining the first extend over some of the processes in the sequence defining the second.

It is at this stage that Whitehead seems seized with the desire to obtain special relativity. He focuses on those nested sequences of events that correspond to portions of the history of the entire universe (events that he calls *durations*). A nested sequence of durations will then define an instant of time, in some reference frame.

He then makes the further assumption that these instants of time satisfy the axioms for planes in a geometry, thus forcing the construction of special relativity.

From the point of view of modern physics, for which general relativity is the natural way of thinking, this sudden introduction of geometry seems quite foreign to an account which, up to that point, had been firmly based on epistemological considerations. Equally artificial, though perhaps less obviously so, is Whitehead's exclusive consideration of durations, by definition spanning the entire universe. This is a concept far removed from the original idea of an event, which was grounded in sense perceptions. A duration spanning the entire universe is just as much an idealization as is a point: it would be better to think of a duration as the ideal limit of an event as its spatial scope becomes larger and larger.

An Alternative View of Process

Let us then try to continue Whitehead's course from the point where he deviated toward special relativity. First, I propose a change of terminology, from *event* to *process*. This has the advantage of not being confused with the point events of normal physics, and also stresses the temporal extension of the things we are talking about (in contrast to *event* which has instantaneous implications). With Whitehead, we continue to assume that a process has an integrity of its own, while being able to extend over other processes. It includes subprocesses without being constituted by them. We shall for the time being assume that processes satisfy the same axioms as Whitehead's events. I shall use the term in a wider sense than Whitehead's *events,* to embrace both what he called *events* and what he called *objects* (chains of events linked by common features). For me the only "genuine" objects, which are not just fortuitous aggregates of similar instances, are those which have an internal unity which warrants the name process.[5]

Before continuing, I stress the main feature of the approach that sets it aside from conventional physics. In *process physics there is no substratum.* In conventional physics, everything happens in space–time, as modifications of that space–time; or in some generalized ether or field; there is a changing pattern of forms imposed on the substratum of this ether or space–time. In process physics, the processes are primary and independent of any other ontology. Space–time is derived from the processes, not vice versa. Since this is potentially the key to the problems of quantum theory, let me first enlarge on this.

Process and Space–Time. To be credible, any new approach must be able to make contact with the theories that have gone before. The way of doing this has, however, been established by Whitehead, though he did not carry the program through consistently. Namely, one enlarges the set X of processes to a bigger set $\bar{X}$, as described above, which includes (equivalence classes of) ideal limits of decreasing nested

[5]Emmet (1991) has persuasively argued that the central concept here is that of a thing-in-process, which would include this idea of an object.

sequences of processes; but, unlike Whitehead, one includes all such sequences in X^*, not just those that define durations. The enlarged set has an order on it, derived from the relation of extending-over for processes, in which (roughly speaking) one element of the enlarged set is called *greater than* another if the processes in the sequence defining the first extend over some of the processes in the sequence defining the second. Space–time points are then recovered as those elements of the enlarged set that have no elements smaller than themselves. The Hausdorff-like axiom 4 then ensures that processes can, if one wishes, be regarded as constituted from points.

If one can recover an underlying point-structure anyway, what is achieved by emphasizing process? The answer is that by thinking of the world in this reversed way, starting with processes and deducing points, one is led to try to formulate physical laws in a different way: as laws governing phenomena at the level of process, not at the level of space–time points. And at the level of process one is dealing with finite structures, free from the infinities that bedevil conventional quantum theory.

There are two aspects of this that I want to stress in particular. First, on this view it would be unnatural to impose the condition that the set of points should be a four-dimensional manifold. Instead, one looks for conditions on the laws governing processes which would result in the set of points having some sort of approximate relation to a four-dimensional manifold. For example, one could require that every process contained a network of subprocesses whose nerve (in the sense of the nerve of a covering) was a combinatorial four-manifold. As a result, physics inevitably becomes very different.

Second, we need to handle the question of actuality with care. The laws of quantitative physics have to be handled mathematically, but the implicit "quasi-Platonic" metaphysics of most mathematicians must not be allowed to infect the physical theory. Modern mathematics, with a few dissenting voices, is rooted in set theory, and its practitioners view the entire body of mathematical entities as eternally existent in some ideal realm. So, if we describe a process universe mathematically it is easy to slip over into regarding the points derived from processes as existent, and then as real, and then as actual. But a point, which is an ideal limit of an infinite sequence, can only be actual if every process defining it is actual; that is, if we have an actual infinity of processes. Our aim, however, is precisely to deny actual infinities, since that was the cause of the difficulties in quantum theory. Thus a point can never be actual, on the present picture.

The status of actuality is a major issue in quantum theory (or, more accurately, it should be a major issue but is often swept under the carpet), because the wave function cannot be regarded as fully concrete, but neither can it be regarded as totally unreal. I am still not clear how the issue will work itself out in a process approach; but it seems that a solution will be found in some sort of relative actuality. In some views of classical physics an object is actual relative to a particular space–time observation if that object is, at the time of the observation, fully formed and determinate and in communication with the observer. Future objects, by contrast, are only potential. About the past there can be debate. In process language, an observation involves a subprocess of the observer (namely, the act of observation) which extends over a subprocess of the object (namely, its participation in the observation). So the notion of one process being actual relative to another is already a familiar one.

But there are difficulties arising from the transitivity of extending-over. It would be tempting to say that *a* was actual relative to *b* precisely when *a* extended over *b;* but then all subprocesses of *b,* and their subprocesses *ad infinitum* would be actual relative to *a* and we would again have an actual infinity. An additional concept seems to be required, analogous to the "participation" of the later Whitehead, which restricts relative actuality in some way. The effect would be that not all the subprocesses of a given process need be actual relative to that process, but only those for which there was mutual participation.

For example, an atom of my body need not be actual to me unless I am examining my finger through a field-emission microscope. This may come as a surprise to those who are used to an atomistic view of the universe as being built up from elementary microscopic parts. But the process view is deliberately opposed to this sort of atomism. The evidence of the history of physics points to the conclusion that there are no ultimate microscopic parts, so that the attempt to realize matter as built up from smaller constituents leads to an infinite regress.

The Laws of Process. If there is to be a physics of process (and I stress that much of what I am describing consists of speculations whose details still have to be fleshed out) then there will be laws determining the ways in which different sorts of process can link through the extending-over relation. Quantum theory, and experiment, tells us that these laws must be probabilistic. So we need to try out various schemes in order to find a type of law that offers the possibility of agreeing with experience. Let me indicate two possible ideas that are being explored.

1. One can pursue a close analogy with quantum theory and associate a probability amplitude (a complex number whose squared modulus gave the probability of the given process occurring) with each network of subprocesses. This is reminiscent of the approach adopted by Penrose (1971) in his treatment of spin networks, except that in his case the aim was to justify the use of complex numbers by relating them to directions in three-dimensional space.
2. The most important formal structure on the space–time of general relativity is its topology. (We recall that a "topology" on a set X is a specification of those subsets of X that we are going to regard as open, in the sense that they are to be regarded as defining the concept of a *neighborhood* of any of their points.) Now the construction of space–time as a set $\bar{X}$ (see pp. 59–61) automatically gives space–time a topology, namely that generated by the processes constituting X. And this remains true even if we do not proceed to the limit of ideal points, but restrict ourselves to a given level of reality: the network of overlapping processes can be regarded as building up a topology on the smaller processes over which they extend.[6] Recently Isham (1990) has developed a theory of

[6]Formally, the processes generate a lattice that can be regarded as a topology over its atoms.

quantum topology where the topology alone can be regarded as defining a quantum-mechanical structure.

3. One can bypass quantum theory and look for laws that directly gave probabilities. One way of doing this has been suggested by Sheldrake (1988): the probability of a given pattern appearing is related to the number of times it has appeared in the past. This needs modifying in our case, where "the past" is in general not well defined, but one can define the relative probability of a given subnetwork of processes in terms of the number of occurrences of that subnetwork in the entire network of which it forms a part. This scheme seems quite hopeful, but there is no indication as to how it can be linked with conventional quantum theory.

Model calculations using the first approach have shown that such laws can be formulated so as to result, in some cases, in manifoldlike structures.

Causality. In formulating laws it will probably be necessary, however, to allow for two sorts of extending-over. The first is where the processes fortuitously happen to overlap (as when the city of Southampton extends over the valley of the river Itchen)—something which typically (but not necessarily) characterizes spatial overlap. The second is where one process "hands over" to another, or gives rise to another, including the case of a causal relation between the processes. This is the sort of extension that is peculiar to temporal order. It is not clear at this stage whether at least one other relation is required in order to distinguish these two sorts of extension. For the purposes, however, of describing time at the more primitive physical levels, where there is no intrinsic directionality, I shall provisionally take the view that a process A is *causally linked* with another process B if they both contain a subprocess c which can be designated as a *nexus process*. One can illustrate this graphically by a formula such as

$$A \longrightarrow \overset{n}{c} \longleftarrow B$$

in which $\longrightarrow$ denotes the relation of extending-over, and n designates a nexus process. If in addition one had directionality, then for A, c would be the subprocess whereby A generates another process; while for B, c would be the subprocess in which B came to birth.

Hierarchy. If we adopt the position that actuality is relative, in order to escape from an actual infinity, then in implementing this idea one is led to a hierarchical view of the universe. This is because, in computing the probability of any particular subnetwork of processes one should only include those other processes that are actual relative to that network. These other processes will in turn be related to the processes that are actual relative to them, and so on. This leads to various computational schemes, but there are no indications as yet as to which is more likely to be successful.

A hierarchical view such as this fits in well with the hierarchical structure that we in fact discern in nature, though we must accept a certain fuzziness as to what is admitted into the category of relative reality. Thus the process which is a human being extends over, and regards as real, physiological subprocesses such as the organs (which are also objects), the hormone system, digestion, and so on; and these in turn extend over, and regard as real, cellular processes—though it is not clear whether all cellular processes are real relative to the human process; and the cellular processes extend over the subcellular processes of protein synthesis, mitochondrial energy generation, and so on, which are probably not regarded as real by the human process.

It might seem to be overstating the case to say that such subsubprocesses are relatively "unreal," given that the whole of our structure and energy derives very directly from subcellular processes. The point is that relative to a given high level of the physical hierarchy the internal operations at a lower level that are sufficiently "far down" become totally irrelevant and unobservable, so that from the (empirical) point of view of the higher level they are unreal. This is not to say that they cannot be brought into reality by a change in the relationships of the processes, as when I put a microscope to my eye. It must always be remembered that "reality" here is a relative concept, relative not only to a (fluid) hierarchical structure, but also to the contingent and changing relations within that structure. It follows that reality is not only relative but also a matter of degree and not a hard and fast relation.

Organic and Physical Processes

This view is directly opposed to the traditional view of physicists, and of neurologically minded psychologists, who regard ultimate reality as residing in the smallest length scales, with higher levels of the hierarchy of the universe built up from this. If, instead, we accept the reality for us of processes at the organic, and particularly at the human level, as well as the reality of physical subprocesses that we can relate to, then we reach a radically new view of the nature of scientific explanation: it becomes just as legitimate to explain physical processes in terms of organic ones as the reverse.

Of course, while this may be legitimate, there is no reason to suppose that it will be effective, and in most areas the preferred approach will still be to argue from the physical to the organic. But we now have to rethink the whole relation between the two. In the matter of mental processes, for example, the question becomes, not "how can we explain those rare and barely repeatable telepathic events where the physical appears to be influenced by the mental," but rather, "how can we explain why they are in fact so rare."

Similarly, in the case of ecology it becomes entirely legitimate to introduce a Gaia hypothesis that the level of the entire biosphere is governed by laws peculiar to that level, provided that this approach leads to testable and useful conclusions. We are not forced to operate all the time only with the laws of the lowest levels that seem relevant.

Process Physics and Time

Given these underlying ideas, what would a process physics have to say about time? A priori no assumptions are made about the sort of time that will emerge: it depends on the level of the processes involved, as to whether they define any time at all, or whether they define the fully differentiated time of conscious experience.

By implication, a process has potentially a temporal element; but time as such only emerges through the relation between processes. A collection of processes, linked by the extending-over relation, can always define a certain formal structure, the mathematical structure defined by their connectivity, and the time-concept appropriate to that collection depends on this structure. We have already seen how one construction, based on a totality of processes, and ignoring questions of actuality, can lead to the space–time of conventional (general relativistic) physics and hence to a linear time for small enough regions.

This means that, if we restrict attention to the processes constituting one particular level of the hierarchy of the world, such as the processes of human society, or the processes within a particular human mind, then they will define their own particular structure of time, which may be only loosely related to the times that can be defined on other levels. This is in sharp contrast to the usual physicalist view, in which a single time permeates all levels of the universe; but would seem to fit well with the more pluralistic approach to time espoused by Fraser (1987).

Consider the time of human experience. A physicalist would take it as axiomatic that human time must be subordinated to physical time, whatever the appearances to the contrary. Take, for example, the psychologist Julian Jaynes, who writes of the apparent continuity of consciousness: "But what could this continuity mean? If we think of a minute as being sixty thousand milliseconds, are we conscious for every one of those milliseconds? If you still think so, go on dividing the time units, remembering that the firing of neurons is of a finite order . . ." (Jaynes, 1982, p. 24).

On the view I am proposing, however, human time is defined by human processes, and it is quite illegitimate to bring processes at a much lower level into the understanding of human time. Where the human processes are linked to the inorganic world—in the case, for example, of my perception of the movement of some physical object—then the formal structure of the processes will inherit the formal structure of time at the lower levels, in addition to the richer structure characteristic of human time. Thus, if I am looking at a scene containing many interacting objects, it will be possible to order the events of their interactions on a linear time continuum. But if we take my purely internal mental processes, then there is no reason whatever to suppose that they should be so ordered: an internal world could have a very strange temporality, once it is freed from the external, as indeed we see with dreams.

This conclusion is, it must be stressed, actually independent of the view taken of "reductionism." It could be that mental processes will at some stage be reduced, in some useful and constructive sense, to structures of microscopic interactions

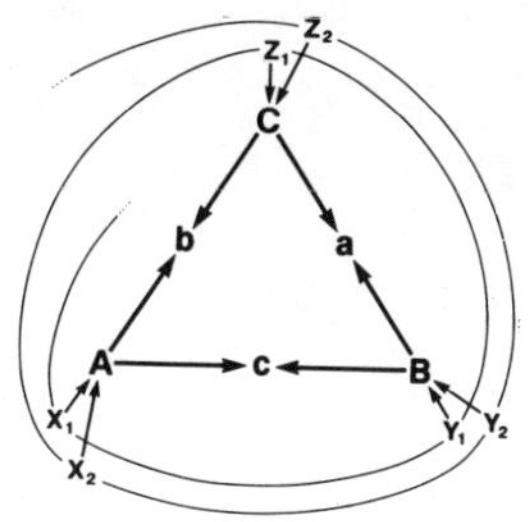

Figure 3.2. A circular process comprising three subprocesses *A, B, C* and three nexus subprocesses *a, b, c*. Arrows denote "extending over."

(though I personally doubt this), in which case mental processes could be located in the physical time of the microscopic world. But that time would still be quite irrelevant to human time, which is defined directly from the mental processes themselves. The conventional physicalist view would be that the time of the physical level is the "real" time and the human time is "merely subjective," with the presupposition that the microscopic world and everything that pertains to it is ontologically prior to the higher levels. But on the view I am advocating this is not the case: since there is no underlying continuum on whose time everything is built; since, in other words, the idea of the time continuum only arises by a process of abstraction from higher levels of processes, then no sort of time has any privileged claim on reality. I am entitled to say that the personal time of my own experiences, internal and external, and, more importantly, the transpersonal time of those less conscious parts of my being that link me with other humans, is the most real time that we can know.

An interesting aspect of this view, on which each system of processes had its own time, is that there can be a plurality of time-structures *even on a given level*. It may be that human time has cyclic aspects as well as linear aspects, for example, depending on the dispositions of the internal processes out of which it arises.

To demonstrate the possibility of cyclic aspects to time consider the simple example of processes exhibiting this structure shown in Figure 3.2, where as before the letters denote processes and an arrow from one letter to another denotes the relation of extending-over. Nexus processes are indicated by lower case.

What is depicted here, where we have a finite number of processes returning to the start, must be distinguished from the usual concept of periodicity, where there is an infinite sequence of events in which similar events periodically return. To make the distinction clear I shall refer to what is shown in the diagram as a circular process. In a dream time, as in the sort of dynamic eternity described by Plotinus (Clarke, 1973), events could dynamically circle round in this sort of way.

Once we admit the possibility that in the internal world there may be both circular and linear processes, then we have a very natural way of thinking about periodic processes, as being an interlinked combination of the two. That is, the circular processes have subprocesses that participate in (that is, are actual relative to) another collection of processes which form a linear sequence. This is illustrated

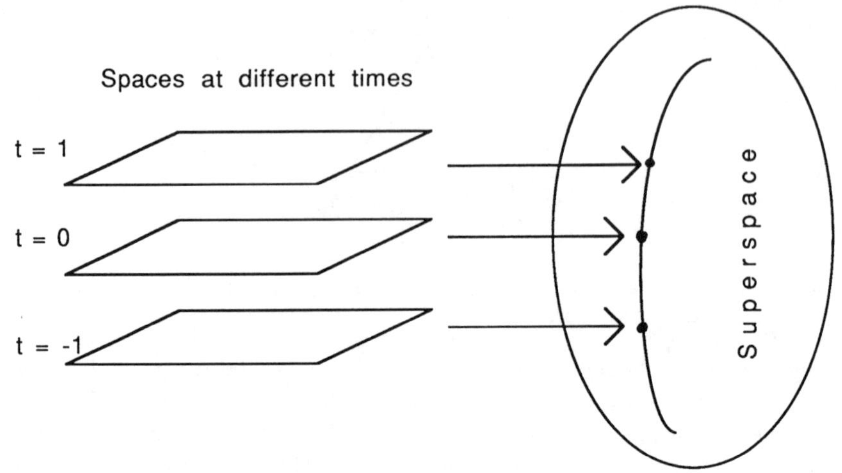

Figure 3.3. Part of a periodic process $\{. . . , X_1, Y_1\ Z_1, X_2, Y_2, Z_2, . . .\}$ where periodicity arises from extension over a circular process such as that in figure 2. Lines indicate connections through extension over nexus subprocesses (not shown for the outer processes).

in Figure 3.3, where a linear sequence has been added and related to the circular process in the center.

What we then have is a truly circular process being observed within a linear time; that is, a sequence of subprocesses *A, B, C* which recurs not just in the sense that similar features recur in an infinite sequence, but in the sense that they are actually the same subprocess on each recurrence.

There is thus a place within a process-physics view of the universe for a mingling of different temporal structures within a single world view.

References

Clarke, C. J. S. (1973), Eternal life. *Theoria to Theory,* 8:317–332.

Descartes, R. (1641), *Oeuvres,* ed. C. Adam & P. Tannery. Paris: Cerf, pp. vii, 48–49, 369–370, 1904.

Emmet, D. (1991), *The Passage of Nature*. London: Macmillan.

Fraser, J. T. (1987), *Time the Familiar Stranger*. Amherst: University of Massachusetts Press.

Hartle, J. B., & Hawking, S. W. (1983), *Phys. Rev. D.,* 28:2960–2975.

Horowitz, G. (1987), *Phys. Lett. B.,* 199:176–182.

Isham, C. J. (1990), *Quantum Topology*. Preprint, University of London.

Jaynes, J. (1982), *The Origin of Consciousness in the Breakdown of the Bicameral Mind*. Boston: Houghton Mifflin.

Penrose, R. (1971), Angular momentum: An approach to combinatorial space-time. In: *Quantum Theory and Beyond,* ed. E. W. Bastin. Cambridge, U.K.: Cambridge University Press.

——— Rindler, W. (1986), *Spinors and Space-Time,* Vol. 2. Cambridge, U.K.: Cambridge University Press.

Sheldrake, R. (1988), *The Presence of the Past*. London: Collins.

Sorabji, R. (1983), *Time, Creation and the Continuum*. London: Duckworth.

Whitehead, A. N. (1919), *An Enquiry Concerning the Principles of Natural Knowledge*. Cambridge, U.K.: Cambridge University Press.

——— (1926), *The Concept of Nature*. Cambridge, U.K.: Cambridge University Press.

Perspectives

Social process—the fabric of human beliefs, desires, choices, actions, and transactions—is the theme of the two following papers. Their fields are radically different, comparative mythology (Akhundov) and sociology (Elchardus), and yet a number of common threads run through each of their arguments, among them (1) process regarded as "an interweaving of the invariant and the variant"; (2) the rejection of historicism and the authority of religion; and (3) the suggestion that rituals provide an appropriate forum in which to express and integrate our common understandings of time.

Each of the two authors addresses the question of how individual human beings live, choose, and act—specifically, what authority is invoked for their way of life, their choices, actions, and interactions. In one way or another, human conceptions of time are deeply involved in this quest for authority, whether time is seen as invariant (as in myth), a variant from which to escape (religion, in Akhundov's interpretation), a rational model of quantitative measure that is at once a resource and a measure of all other resources (as in economics and politics), or the aesthetic product of social action. Elchardus proposes that we examine important points of transition in order to see how people attempt to solve their problems and reorient themselves to time when their world has been turned upside down. Akhundov's suggestion is that our culturally given frame of mind orients us in a particular direction, accepting or rejecting the apparent direction of time and seeking justification for our actions from either the past or the future.

With respect to ancient man, Akhundov (in the gradual transition from a disordered to an ordered world, that is, from Chaos to Cosmos) identifies a process in which the original oscillatory, dualistic model of time is gradually transformed into a cyclic, a spiral, and finally a linear model. In his analysis of modern man's attempts to define reality and understand the manifold processes within which he lives, Elchardus examines in turn (1) a mechanical, cause-and-effect paradigm that assigns priority and authority to the past (in effect, a myth); (2) a game model in which communication and deliberation within the forum or marketplace serves as the means by which consensus and group preferences are forged; and (3) a new aesthetic model based upon the affective bonding within a group, creating thereby a basis for relations sufficiently stable for agreement and understanding to be achieved. In a shifting, unstable, and increasingly dangerous world (his argument runs), our common, binding emotions may be the only form of invariance available to us!

L.R.

4

Transformations of Kronos

M. D. Akhundov

Abstract An examination of the distinctive temporal conceptions of mythology, ancient natural philosophy, and monotheistic religion reveals a gradual transition from Chaos to Cosmos, as well as a progression from an oscillatory model to cyclic, spiral, and linear models of time. In mature mythology we encounter a multilayer, multilevel model of the world, similar to the perceptions of a child at the preconceptual level of intelligence. The mythological world is situated in a time before the beginning of time, and events in the mythical past are embedded in an invariant structure that is simultaneous with the past, present, and future. Mythology subsequently developed the notion of cyclic time, in which the present and future are regarded as little more than different hypostases of the mythical, archetypal past. The final section contrasts the temporal attitudes that are manifest in myth and religion, viewing them as two essentially different ways of thinking that are organically connected with two different models of time: the mythological worldview is oriented toward the past, but the religious worldview is oriented toward the future and seeks to escape from time—in effect regarding time either as an attribute of a deity or a substance whose existence is independent of the objects, events, and processes with which it is filled. The Christian river of time, as Augustine conceived it, flows from the future, through the present, and into the past, thus forming the model of an hourglass that superficially resembles the contours of the light cone in relativity theory. In modern physics, however, the flow of time has once again changed direction, returning to the ancient mythological model: from the past toward the future.

When we seek to assess the mythological picture of the world and, in particular, mythological notions of time, we must from the outset address ourselves to one basic question: Does primitive thought differ from modern Western thought, and, if so, is the distinction a qualitative or a quantitative one, with reference to thought processes and thought content? One finds very different answers to these questions. Thus, on the one hand, Franz Boas (1911) contends that "the functions of the human mind are common to the whole of mankind" (p. 122), and Levi-Strauss (1963) firmly declares that the principles by which reason functions are the same in all cultures and throughout all ages. On the other hand, a majority of observers, including Levy-Bruhl (1935) and Vygotsky, conclude that mental processes exhibit significant intercultural differences.

The development of human consciousness is vitally dependent upon man's social being. But because sociohistorical progress proceeds by qualitative leaps rather than as a steady quantitative evolution, the development of the human psyche may at certain stages similarly take the form of a qualitative change. A majority of scholars whose work has been guided by the methodology of dialectical materialism (including V. S. Vygotsky, A. P. Luria, and A. N. Leontyev) or by ideas close to Marxism (H. Wallon and E. Meyerson) have taken just such an approach, which does not, of course, exclude other paths leading to the notion that thought differs quantitatively at different stages of cultural and social development.

Here the following factors are to be taken into account. First, a child in his or her individual development (ontogeny) passes through an entire series of qualitatively different levels of intelligence (sensorimotor, preconceptual, and the like). This may be seen as an indirect confirmation of our views on the existence of qualitatively different types of thought on the different levels of sociocultural development (within the frame of the well-known correspondence of structural aspects of human phylogeny and ontogeny).

And second, our notion is not in any way connected with claims that mental abilities differ between races or ethnic groups. If children from as widely divergent peoples as Russian, French, Chinese, Bushmen, and Australian aborigines are taught in a single group, the results will support the idea put forward by Boas that "the functions of the human mind are common to the whole of mankind." But if our population is taken from different societies at essentially different levels of development (some, for example, representing the tribal level of organization and others, the modern industrial world), one is sure to notice considerable differences in their thinking. Accordingly, one can speak about mythological thinking, the mythological picture of the world, and mythological conceptions of time. The latter will be the problem under discussion in this article.

Let me first turn to the mythological world image and attempt to determine its characteristic spatiotemporal concepts; while the present paper will focus on mythical concepts of time, it is quite impossible to consider them without reference to spatial concepts. I will not be concerned with the essence of myth, in which we have discovered an entire spectrum of ingredients corresponding to the classifications prevalent in our culture: myth is a mixture of allegory, poetry, symbol, science, philosophy, archetype, structure, and worldview. Myth is a peculiarly primitive syncretism of these components, and yet it is something more, for "to its creators it was objective reality" (Steblin-Kamensky, 1976, p. 30).

In this connection a person whose experience is informed by the mythological consciousness receives a much wider selection of syncretic information than does a modern observer of mythology. But, from the opposite perspective, a modern observer is able to reconstruct conceptions, structures, and models from ancient mythology, about whose existence the very creators of that mythology knew nothing and could not possibly know. Accordingly, when we speak of the significance of ancient mythology or Eastern wisdom for the development of modern science (as, for example, the applicability of Buddhist ideas to the development of "bootstrap" models of elemental particles in modern microphysics), we refer to this reconstructive type of modeling, one that turns out to be a very important form of innovation. In this way a new and original idea appears, one that has no independent existence either in ancient mythology or in modern science. The idea is the result of their mutual efforts. In this productive process we see a consolidation of the intellectual efforts of people of all countries and ages, no matter how far separated from each other in space and time. The process defines the actual character that arises from our study of the mythological picture of the world, and makes it possible to assign high significance to any elements in the conceptual fund of mankind.

The dynamics of the mythological world can be described in the most general terms as a transition from chaos to cosmos. Chaos in the initial category of mythology and the initial state of the mythological universe: "Verily at the first Chaos came to be" (Hesiod, *Theog*. 116). Moreover, notions of chaos have not only a spatial component but also a temporal one.

But, significantly, already in Pherecydes (ca. 550 B.C.) the role of Hesiod's chaos is played by Kronos, who symbolizes both space and time. Also, if we leave the ancient Greek mythology of Homer and Hesiod, in which we find a relatively generalized notion of chaos (expressed in such images as the abyss, the void, ocean, infinite space, and darkness), and turn instead to archaic Australian, African, Siberian, and other world mythologies, we find that the image of the primary unordered world is more concrete and generally connected with water. Water epitomizes the indefinite beginning which, with the aid of *dike*, *nous*, the idea, god, and the like, is structured and transformed into the spatially and temporally ordered cosmos.

Cosmos is the final category and the culmination of mythology; in the oldest mythological conceptions there is no notion of the world as a definite, unified whole. On the contrary, relative order is connected only with the tribal territory, which is surrounded by and permeated with a mysterious and pernicious force that can in Levy-Bruhl's terms be called the supernatural world. The powers at work in this world are unordered and nonhierarchical, a "fluid multiplicity of vaguely related elements" (Levy-Bruhl, 1935, pp. 19–36).

Relative order within this narrow little world of primitive mankind (analogous to the feeling for near space that develops initially in the infant) was unstable; it was not so much natural as it was moral and social in character, because only the observance of definite rules and taboos could ensure its preservation. Here, of course, it must be remembered that in ancient times the world was not neatly divided into nature and society. The two were united, so that the developing picture of the world was sociomorphic.

As mythology (and intertribal relations) developed farther, we find that the oases of ordered existence multiply and participate mystically with one another, often so much so that they appear to be different hypostases of a single oasis of being. In its spatiotemporal aspect this picture of the world is similar to the perceptions of a child on the level of preconceptual intelligence: the space of the local world is continuous, whereas that of the "big" world beyond is discontinuous and multiply connected. Thus the mythological world gradually comes to acquire a multilayer, multilevel character.

The evolution of notions of space was accompanied by a corresponding development of temporal ideology. Thus the following conception was characteristic of the early stages of human culture: "We know that the primitive mind does not 'sense' the successive moments of time as homogeneous. Certain periods of the day or night, of the moon's phases, of the year, and so on, are able to exert a favorable or a malignant influence" (Levy-Bruhl, 1935, pp. 48–49). Here, mystically colored, we encounter the temporal conception of preconceptual and representative intelligence as explained by J. Piaget, in which each change is assigned its own time and temporal interrelationships are understood vaguely.

Intuitive time, as analyzed by Piaget (1971), displays similar features (pp. 136–137). In intuitive (or representative) thinking, time is connected with individual objects and movements; it is not homogeneous, nor does it flow uniformly. Thus, if two bodies begin simultaneously to move in different directions and at different speeds away from point A and then stop simultaneously at points B_1 and B_2, respectively, a four- or five-year-old child will be unable to grasp the simultaneity of their arrival. The child refuses to understand that their movement ended "at the same time," because he or she has no concept of a universal time for different speeds (and in this respect the child is closer to Einstein than to Newton). "Before" and "after" are distinguished in terms of a spatial rather than a temporal sequence. To the child, "faster" means "farther," and this involves "more time." Even when temporal measurement becomes possible at the operational stage, a seven- or eight-year-old child will still be unable to perceive time as a relation independent of concrete objects and changes.

In this connection it is instructive to note one particular curiosity involving the measurement of temporal durations in ancient Mesopotamia. Babylonian scientists had reached the conclusion that different periods of time should be measured by different units: the hours of the day and of the night differed in magnitude and underwent changes according to the season. In this concept we encounter the idea of heterogeneous time with an uneven flow. The Babylonian solution to the question of time measurement demonstrates how practical problems can arise as a result of the mythological framework of ideas, problems that at times required peculiar solutions and led to constructive results. In this case the idea of flexible time units helped the Babylonians accomplish an important task, that is, regulating the duration of the unified year and eliminating the need for an additional day for the "leap" year.

As has been noted, in the multilayer model of the world that is characteristic of mature mythology, each level has its own time distinguished by such parameters as rhythm and duration. Thus the *kamlaniye* seance of the shaman (his travel to other worlds where he would fight with evil spirits) may take several hours of earthly time, but in this short period he manages to visit several levels of the world, pursue the evil spirit, do battle with it, and return victorious to earth, that is, to the territory of his tribe. Reckoned in the internal hours of magical action, a long period may have passed—more than a human lifetime; but after the seance the shaman stands before his fellows tired, but not years older. It was not the shaman himself who traveled through other worlds but his imperishable supernatural double (his soul or some similar concept). If the body of the shaman, or of any human, lives in ordinary earthly time, his soul exists in a distinct, sacred time and in a different, supernatural world. We find clear parallels to this idea in many idealistic philosophical doctrines and religious systems of antiquity, and in later ages.

One consequence of this line of reasoning is that the soul does not disappear at death but is reincarnated in a newborn baby of the same tribe. It is this notion that underlies the oscillatory or cyclic model of time characteristic of mythology. I shall return presently to this model, but first let us consider farther the question of multiple times in the multilayer worlds of mythology.

During the shaman's *kamlaniye,* the timeless component of mankind sets off for other worlds; but what happens, with respect to time, to the shaman on such a visit? Is he still subject to the time of his world, or does he break free and live according to another system of time-reckoning? A typical Chukot-Eskimo myth provides an answer to this question: a traveler departs for distant lands on different levels of the world, wanders about them for two or three years, and then returns home to find striking changes. His home has crumbled away with age, and his children have become old and gray, because many decades of local time have passed. Seeing his aged son, the young wanderer falls dead and disintegrates into dust—earthly time has claimed him.

The main stress in this myth would seem to be not so much on the difference between the temporal measures of different worlds (although this aspect is also important), as on the concept of human mortality and our inability to exist in different times. It may even be correct to say that once a human has left his own time, any return will destroy him. Thus the exit from earthly time is in itself tantamount to death. Significantly, the ancient Hindus believed that a man returning from a long journey was born anew.

In a number of more archaic cultures, such a wanderer must undergo a symbolic ritual of new birth, in which his mother (his own mother or one chosen for the occasion) gives birth to him in great pain. We should not exaggerate the level of symbolism of such a birth for the wanderer himself; for him it is simply next in sequence after his real birth.

Underlying these mythological notions is the germ of a thesis that was not clearly articulated until the rise of natural philosophy in ancient Greece: one cannot enter the same river (of time) twice. Reincarnation may seem to violate this thesis, but in fact it is quite another mechanism that involves not a single river of time, but rather different temporal cycles.

It must be taken into consideration that, in different ages and different literary genres, the relationship between the times of "one's own world" and the world of "others" has undergone significant changes. Thus, in the ancient Greek adventure romance (for example, *The Adventures of Leucippe and Cleitophon* by Achilles Tatius of Alexandria), the hero and heroine fall in love instantaneously and passionately (moment t_1), but their happy marriage is delayed by tribulations: the abduction of the bride, wanderings about the world, shipwrecks, war, being sold into slavery, taken prisoner, and the like. An enormous amount of time is expended in overcoming these difficulties. Finally the pair is united and married (moment t_2). Characteristically, however, moments t_1 and t_2 are interpreted as contiguous; the tribulations occurred in the "alien" world and are measured on its temporal scale. As M. M. Bakhtin (1975) observes, "This is a temporal hiatus between two moments of biographical time" (p. 240). An exception occurs in Voltaire's *Candide*: at the end of his numerous adventures Candide is reunited with his beloved Cunegonde, who has turned into a peevish, ugly hag with swollen eyes and a dry neck, wrinkled cheeks, and weather-beaten hands. In this case Candide's adventures occurred during a period of unified, homogeneous time with a steady flow.

The only aspect of mythological time that I have considered here is the aspect that is intrinsically related to the distinctive features of preconceptual, representative intelligence. This concept of time was not the exclusive achievement of the mythological world image—mythology merely arrays it in vivid, mystical garb. Time in mythology, however, has a number of other aspects. Myth, for example, is characteristically oriented toward the past. By this I mean not the simple past of yesterday or "last year in Marienbad," but the distant mythological past, the time in which mythical ancestors accomplished their deeds. This mythical past precedes the ordinary time of primitive mankind and its surrounding world. The mythical world is thus situated in a time before the beginning of time.

On closer examination, however, this mythical universe proves to be not so distant after all, separated from the present by a mere four or five generations. From this it can be concluded that so-called primitive peoples lack the distant temporal perspective, and that in many respects time and the past are dependent upon memory, because the dead were remembered by children and grandchildren to whom they could appear in dreams. Only vague stories about prominent persons (who by that time had become reincarnate in the world of the living) were handed down to subsequent generations. Stretching even farther back is the world of the mystical ancestors, a world that is qualitatively different and not located in ordinary time.

But mythological time has still more features: despite its bond with the past, mythical pretime was also the present and even the future. Such a statement strikes us as extremely illogical, but the contradiction is only an apparent one; modern science proceeds from a model of linear time, whereas primitive notions are based upon a cyclic model. Once this is realized, it is easy to understand A. P. Elkin (1932) when he states that the mythical *dzugur* epoch of Australian aborigines—an archaic epoch when mythological personages were alive—". . . must not merely be thought of as past time, but as present and future, and a state as well as a period" (pp. 135–136). Events in the mythical past are embedded in an invariant structure that is simultaneous with past, present, and future. Levi-Strauss is therefore justified in defining mythical time in terms of such dualistic features as reversibility and irreversibility, synchroneity and diachroneity.

As many observers have noted, mythical time and the general orientation of primitive thought toward the past have no analog in Piaget's developmental psychology. Thus Meletinsky (1970) remarks that "a description of mythical events involving supernatural beings acting in distant times answers not so much questions [typical of childhood] as certain spiritual need, providing a kind of 'poetic' key to them. 'Before' is the sphere of primary causes, the source of everything that came 'after' " (p. 173). It would be erroneous, however, to ignore totally the psychological component and attempt to explain mythological time solely on the basis of such purely sociological phenomena as ancestor worship. It is quite true that developmental psychology does not posit a level of intelligence oriented toward the past, but psychopathology has shown that the separate hemispheres of the brain are responsible for different modes of cognition and are oriented toward different times. Thus the right hemisphere controls representative, imaginative cognition and is directed toward the past. Significantly, primitive thought is in fact preconceptual, representative, imaginative, and transductive, and thus dominated by right-hemispheric activity. Another reason for taking psychological features into account is that many of primitive mankind's channels of communication with the mythical past—rituals, rites, the shaman's *kamlaniye,* the fits of the sorcerer, hallucinations, divination by dreams, and the like—were clearly psychopathological in nature.[1]

Each time a ritual was performed, a situation from the mythical past was acted out, and thus the present and future were perceived through the structure of the present. These actions took on a periodic character in calendric rituals, thereby reinforcing the cyclic model of time that is inherent in the idea of reincarnation, even though the mechanism corresponds to a more archaic temporal model. In the matrilineal kinship structure of the Australian aborigines, for example, the same terms are used to denote both grandfathers and grandchildren. As E. Leach (1961) observes, this stage reveals an archaic sense of time as a kind of oscillation between life and death, day and night (pp. 124–136). Mythology subsequently developed the notion of cyclic time, in which the present and future are merely different temporal hypostases of the mythical, archetypal past. This is the tribe's guarantee of continued existence, on the condition that the mystical rites and rituals and the taboos are strictly observed.

Mythological space and time can be defined in the most general terms as a multilayer spatial isomorphism coupled with a cyclic temporal structure, a conception that has proved heuristic. As Y. M. Lotman (1973) has noted, "The exceptionally strong tendency of mythological thought to establish homeo- and isomorphism has made it scientifically fruitful and accounts for the fact that it is periodically revitalized in different historical periods" (p. 87). These features have both stimulated the development of myth itself and made it a productive source of different forms of subsequently differentiated social consciousness. Although myth tends to canonize the mythical past, its temporal cyclicity serves as an intertemporal structure that of necessity brings the past, present, and future into organization.

[1]In his *Man, God and Immortality* (1927) James Frazer notes that a certain tendency to worship insanity and insane people is characteristic of primitive societies, a tendency that has not been sufficiently recognized by historians but has exerted an enormous influence on the development or decay of social institutions of various peoples. More recent examples (Stalin and Hitler) demonstrate that the cult of the insane is not limited to primitive societies.

To the extent that it conforms strictly to the requirements of the above model, archaic society is doomed to stagnation; and indeed such societies can exist even today. Any innovations are considered as breaches of the archetypical past and disruptions of the temporal cycle; that is, they stipulate the emergence of history. But the archaic mythological world is antihistorical. These features have been well-described by M. Eliade (1954) as a refusal of the archaic man to recognize himself as a historical being, and similarly a refusal to assign any value to memory and unusual events (i.e., those events that lack an archetypical model) from which duration is composed. In the long run, in all these rites and conceptions we recognize the striving to devaluate time. Accordingly, if a distortion of the archetypical past penetrates into the mythological world (as, for example, when man's "sins" cause him to deviate from the archetype and he falls into duration), this distortion is not corrected, but leads instead to the next act of world creation in which the world begins its undistorted existence afresh, thanks to the rites and traditions that provide such an opportunity.

The life of archaic man exists, as it were, in time but it does not carry the weight of time, nor does it notice its irreversibility. In other words, it does not pay attention to the features that are characteristic of and important to the realization of time (Eliade, 1954). Thus the desire of archaic man was to "devaluate" Eliade's model of linear irreversible time, or, to be more exact, to annihilate it along with other perceived distortions of his world. It is precisely this distorted archetypical world that gives birth to the concept of linear time. We imagine that archaic man was conscious of this kind of time and tried deliberately to annihilate it, replacing it with an archetypical world of renewable, cyclical time.

However, K. Jaspers and other scholars have pointed out that this very procedure (repeated frequently) in which the rudiments of irreversible linear time are annihilated and cyclical time reborn, resulted in the fact that man has managed to break through the closed cycle of time and emerge into axial time. Such a breakthrough became possible in a number of ancient civilizations, including ancient Greece and India, in which man worked out sly and intricate mechanisms with which to neutralize the conservative cyclic paradigms.[2]

This in turn made it possible to reexamine the canons of the past in accordance with the altered sociocultural metrics of the present. The wheel of time rolled forth from the canonized past, gathered up the advances of the present, and—passing through the future—bore them off into the past once again. Although the past remained canonized, it underwent changes as the achievements of primitive thought and cognition continued to accumulate. In other words, by rejecting the archaic oscillatory model, mythological time effects the transition to a cyclic model; but it does not stop there. After all, the cyclic model operates in cultures that are either

[2]The expression "sly and intricate" is used intentionally, because mythology features not only courageous heroes but also cunning tricksters. Their rebellion against the gods takes on a joking tone, but they nevertheless manage to introduce new reforms into the world in their own tricky way. Many scholars, including G. Dumezil and M. Eliade, regard tricksters as spiritual forerunners of the first philosophers and scientists.

stagnant or have an unchangeable past, that is, an immutable present and future. The transition continues to a spiral model (Akhundov, 1986, p. 45).

This process can be traced in the cosmological and mythological conceptions reflected in the ornaments of ancient cultures, in which the spiral gradually replaces the circle. One point merits attention here: scholars have frequently linked the image of the spiral to the idea of time. Thus B. A. Rybakov (1965) interprets the typical ornamental "running spiral" of third- and fourth-millennium B.C. Tripolitanian culture as a representation of the course of the sun, emphasizing that "the main idea of the Aeneolithic spiral-solar design, whose rhythmical repetition of the course of several suns consummately illustrates the continuity of the movement, is the idea of Time" (p. 43). Within the cyclic model, time did not always flow in a single closed circle, and at a certain stage in the evolution of temporal conceptions we encounter concentric circles of time. This model contains, albeit in inchoate form, the idea of development in terms of a succession of cycles, clearly expressed in a "building of bridges" between the different concentric circles. Here, already, we see the spiral model. The spiral nature of time was also expressed in a similar way in the tenth-century Buddhist doctrine of *kālachakra,* the highest level of the four tantric systems, in which the unity of cyclic and linear temporality is emphasized. This conception found expression in the spiral model of the Wheel of Time.

In later forms of social consciousness the spiral of time was transformed into a linear construction (the appearance of history), but mature mythology continued to operate with the spiral model. The idea of the cosmos corresponds to the development of sociopolitical organization into expanding, unified structures such as the polis, state, or empire. This development is manifest in late mythologies, in which primary time is occupied not by totemic animals or spirits but by the cosmogonic process. The rise of productive forces changes mankind's view of nature. Mankind begins to see itself in opposition to animals, and this trend strongly stimulates the conception in which the gods are visualized in anthropomorphic form.

The anthropomorphosis of mythological figures gave rise to two ideas that are diametrically opposed in their relation to the supernatural world. One of these is the notion that humans can actively influence that world. Actually, in its later stages, mythology itself arrives at this idea when it assumes the genre of the heroic epic, in which the hero fights with the gods (Diomedes at the siege of Troy) or defies their orders (Prometheus). This heroic atmosphere furthered the development of the idea that supernatural forces could be predicted, put to use, or even controlled. Such activities become part of the domain of magic.

It should be noted that such an independence of the hero is not characteristic of the epic. More typically, the hero does not create history but only participates in that history, under the control of the iron laws of necessity and situated in linear time. The two above-mentioned types of heroic behavior are evidently reflections of two different conceptions of time that are embodied in the ancient heroic epics. A view of time from the perspective of the temporal world yields the model of dynamic time that flows from the past, through the present, and on into the future, thus determining the sequence of comings-into-being and destructions and revealing the feats of the hero in gradual sequence. In contrast, a view of time from the perspective of the extratemporal world results in a "spacelike" model in which past, present, and future can be grasped in a single glance, as it were. In fact, such a division of time becomes meaningless in the static model, because the world is constantly present in its totality and, accordingly, the deeds of the hero are accurately dispersed in time from his birth to the moment of his heroic death.

In a world operating with such temporal conceptions, necessity was understandably a significant force, serving as the absolute law of events. The concept of fate did indeed play an enormous role in the cultures of antiquity, and although it underwent important changes throughout the history of Greek mythology, the heroic epic, and natural philosophy, its first manifestation was as necessity (*ananke*). Fate (or necessity) and time were the two sides of the same coin; together they determined the organization of the changing world.

Already in the writings of Thales of Miletus we find the following two interrelated theses: "Most powerful of all is necessity, for it prevails over everything" and "Wisest of all is time, for it reveals everything" (Diogenes Laertius, *Thales* Book i, Sec. 35). These tenets, in combination with the unity of the dynamic and static aspects of time, give us the following picture of the early pre-Socratic world. Although the future was no less real than the present, it lay ahead and was revealed to mankind only in the course of time. Situated in time, humans could not see beyond the temporal horizon of the present, but from their vantage point in eternity—"above" and outside of time—the immortal gods could view all of time at once. In a state of trance or ecstasy, the soothsayers, oracles, Pythians, and mantics that were a permanent feature throughout the history of ancient Greek culture could gain a glimpse into the future. The world of the future, in other words, was precisely located in time and accessible to madness. Gods or oracles could reveal to the hero what tomorrow had in store, but this changed little in the order of things, because everything was determined by the stern laws of necessity. The temporal order remained immutable, and, because it enjoyed the protection of fate, anyone violating it—even the sun itself—was subject to punishment by the Erinyes. Achilles, for example, learns that death awaits him in the impending battle, but he makes no effort to alter anything. He strides boldly into combat, filled with an unquenchable thirst for his enemies' blood, yet understanding that nothing can be changed, that his death will not only occur but that it in fact already is—not today but tomorrow.

These conceptions of time and fate characteristic of the epic and tragedy were adopted by ancient Greek natural philosophy, which through a rational process of abstraction separated the idea of time from the events with which it was filled. At first there was no such division: Chaos appeared in the unity of spatial and material principles, and Kronos in the unity of temporal and material principles. Subsequently, these components were differentiated so that in the mature natural philosophy of antiquity the ideas of space and time are separate, as are the objects and processes that fill them (atoms and the void, for example), and are set in motion and ordered by means of organizing principles such as *nous* and *logos*. Instead of its original role as fateful destiny, necessity becomes a law—*ananke* gives way to *nomos*—and attitudes toward the future change accordingly. Now the future could not only be probed through the unintelligible mutterings of the oracle, but also measured with precision. The future proves to be accessible not to madness, but to rational calculation and speculation.

In this way the initial development of mythology led to magic and natural philosophy. With these means one could influence the supernatural world, learn from it, and understand its connection with the appearance of history and linear time, as distinct from the objects and processes that were its contents. Another stream of thought similarly proceeded from a clear division of the world into natural and supernatural levels, but its attitude toward the supernatural was based on worship and passive faith—in a word, religion. N. Zederblom (1932) has brilliantly formulated the difference between these two modes of thought: "Magic is a matter of calculation, whereas religion is by contrast [a matter of] humility and submission to the mysterious and inscrutable" (p. 172). This has led certain scholars to regard magic as prescience (Ducasse, 1953, p. 63).

As for religion, its basic premises have always been and will always remain antiscientific. At the same time, however, the atmosphere in which science was born and evolved was determined for many centuries by religion, particularly Christianity, and this fact obliges me to give careful attention to the spatial and temporal notions that have developed in the course of religious history. In addition, it must be remembered that the evolution of these conceptions did not follow the linear pattern that led from mythology to natural philosophy and science, but included many sterile currents that nevertheless furthered ideas that influenced the development of more fruitful notions. Many contemporary theologians and philosophers of religion have attempted to prove mankind's innate religious impulses by equating mythology and the preanimistic concepts of religion. But despite their many points of intersection, however, mythology and religion are essentially two different ways of thinking, and the principal criterion for this distinction lies in their attitude toward time.

Furthermore, mythology and religion are organically connected with different models of time. In the monotheistic religions we see for the first time that the mythological concept of cyclic time is abandoned in favor of a sacralized history in linear time. Already in Judaism God appears not simply as a creator of archetypical actions but as a distinct personality, constantly interfering in history and displaying his will in the form of invasions, sieges, battles and the like. In this way historic facts become transformed into "situations" where man confronts God, and as such these situations acquire a religious meaning that they had never had before. History is thus understood as a theophany, or a series of epiphanies. Further development of these progressive ideas occurred in the Christian tradition, although we have also pointed out above that the development of notions of linear time took place in natural philosophy. Many scholars, including V. Toporov, give credit to Christianity for being the first religion to place God completely in historical time, insisting on the historical character of Jesus Christ, his human birth, and his life and suffering at the time of Pontius Pilate.

Here, for the first time, the time of liturgy coincides with historical time. In this way historical time acquires a new value, a value that is incommensurable with the value of the old cosmological time, and becomes, like history itself, the object of sacralization. The difference in their temporal orientation becomes the main criterion of distinction between mythology and religion. Hultkrantz (1963) has emphasized that the bond which mythology maintains with the distant past distinguishes it from religion, which is instead oriented toward a continuously experienced present (p. 33). But although one can agree with Hultkrantz that the temporal orientation of religion is different from that of mythology, it is difficult to accept the latter part of his statement. The strict sacred ordering of the present, which religion does indeed demand (and here we must take into account Kierkegaard's two forms of religiousness), is not an end in itself but merely a means for ensuring that certain future goals or a certain future will be realized. What is the religious person's dream of the future? He hopes to free himself from time! Religion acquires its power over the souls of humans by promising them the ability to conquer time: the ancient Egyptian's blissful fields of Ialy, the Christian's Paradise, the Islamic Djjanna, and the Buddhist's Nirvana are all located not in time but in eternity. Arising simultaneously with the world, time is the form of existence of the world between creation and the day of judgment. For example, the Heavenly City, as described in *Revelation,* needs neither the sun nor the moon for light. Night will not exist there, for sacred time lacks any temporal reckoning. After all, it is the sun and the moon, the alternation of day and night, on which relative time is measured in the ordinary world, whereas sacred time (the prototype of Newton's absolute time) participates in the Godhead.[3]

It is interesting in this connection that a wide variety of religions refer to divine acts as "instantaneous." Already the hymns dedicated to Amon-Ra characterize him as "arriving immediately from afar to those who call upon him," and in Homer the "immortal" beings are described as moving with the speed of thought. This concept of sacred action-at-a-distance is a necessary attribute of an omnipresent and omniscient deity whose sensorium is absolute space and time. This notion of sacred action-at-a-distance is picturesquely embodied in the literature of Islam: in one legend the prophet Mohammed is instantaneously transported from al-Buraq to Jerusalem to the altar of the Allah, but when he returned home he discovered that not a single drop of water had run out from an overturned jar.

A clear-cut distinction between time and eternity, which in scientific terms appears as a division into relative and absolute time (which are in turn based on the attributive and substantial notions), represents a significant advance in the development of generalizing and abstract thought. This is amply illustrated by Feuerbach, who notes in his famous "Lectures on the Essence of Religion" (1967) that "Eternity itself is nothing other than the concept of time, abstract time, time disjoined from temporal differences. Thus it is not to be wondered that religion should have made time into an attribute of God or into an independent God" (p. 118). This subtle observation explains why time (in the person of Zurvan, Kala,

[3]However, in certain noncanonical texts describing medieval visions of the other world, we read that temporal differentiation exists even in Hell, where, for example, certain types of sinners are given a rest from tortures on Sundays; and similarly, in Paradise, Sundays and church holidays are suitably celebrated. In this connection see Patch (1950).

and the like) was regarded as a supreme deity in the ancient religions of Persia, Babylonia, Phoenicia, and in India. Summarizing Feuerbach's study, Lenin wrote that "time outside temporal things = God." Thus evolved the notion of time as a substance existing independently of the objects, events, and processes with which it is filled.

In conclusion, let us touch upon some of the variations in the concept of linear time, variations that were worked out in the transitions from mythology to ancient natural philosophy and to monotheistic religion, particularly Christianity. In this connection, let us consider one of St. Augustine's characteristic arguments: "As for the present, should it always be present and never pass into times past, verily it should not be time but eternity. If then time present, to be time, only comes into existence because it passeth into time past, how can we say that also to be, whose cause of being is, that it shall not be; that we cannot, forsooth, affirm that time is, but only because it is tending not to be?" (*Confessions,* book 2, p. 239). With respect to the reality of the present, Augustine repeats the reasons given by Aristotle, but with a new twist: time has reversed its course. Notions of time in mythology and early Greek natural philosophy were connected with the image of a river flowing out of the past, through the present, and into the future, taking us in its flow from childhood to old age. This metaphor is obviously related to the orientation of myth toward the past, which is the point of reference in reckoning the flow of time. Religion, on the other hand, is oriented toward the hope of salvation and escape from time, so that the Christian river flows out of the future. It is in this sense that we must understand Augustine's thesis that the present is exhausted as it moves continuously from the future to the past.

The future leaks through the present into the past; it becomes exhausted. Actually, the image of the river is no longer applicable, and we should use instead the image of the hourglass. God created time (He is the "creator and constructor" of time, as St. Augustine puts it), and, after measuring a suitable quantity of it, he "poured" it into the upper chamber of the glass. This is the future, which trickles through the opening of the present into the lower chamber of the past. The contours of this model resemble the light cone in the theory of relativity, but in modern physics the flow or "arrow" of time has once again changed direction and returned to the ideal of antiquity—from the past toward the future.

References

Akhundov, M. D. (1986), *Conceptions of Space and Time: Sources, Evolution, Directions,* trans. C. Rougle. Cambridge, MA: MIT Press.

Augustine, Saint (1631), *Confessions,* trans. W. Watts. Cambridge, MA: Harvard University Press, 1951.

Bakhtin, M. M. (1975), *Questions of Literature and Aesthetics* [in Russian]. Moscow: Khudozhestvenaya Literatura.

Boas, F. (1911), *The Mind of Primitive Man.* New York: Macmillan, 1938.

Ducasse, G. J. (1953), *A Philosophical Scrutiny of Religion.* New York: Ronald Press.

Eliade, M. (1954), *The Myth of the Eternal Return,* trans. W. Trask. New York: Pantheon Books.

Elkin, A. P. (1932), The secret life of the Australian aborigines. *Oceania,* 3/2:119–138.

Feuerbach, L. (1967), *Lectures on the Essence of Religion,* trans. R. Manheim. New York: Harper & Row.

Frazer, J. (1927), *Man, God and Immortality: Thoughts on Human Progress.* London: Macmillan.

Hultkrantz, A. (1963), *Les religions des Indiens primitifs de l'Amerique.* Uppsala: Almqvist & Wiksell.

Leach, E. (1961), *Rethinking Anthropology.* London: Athlone Press.

Levi-Strauss, C. (1963), *Structural Anthropology,* trans. C. Jacobson & B. G. Schoepf. New York: Basic Books.

Levy-Bruhl, L. (1935), *Primitives and the Supernatural.* New York: E. P. Dutton.

Lotman, Y. M. (1973), *On the Mythological Code of Subjective Texts* [in Russian]. Tarty: Tarty University Press.

Meletinsky, Y. M. (1970), *The Poetics of Myth* [in Russian]. Moscow: Nauka.

Patch, H. R. (1950), *The Other World According to Descriptions in Medieval Literature.* Cambridge, MA: Harvard University Press.

Piaget, J. (1971), *The Psychology of Intelligence,* trans. M. Piercy & D. E. Berlyne. London: Routledge & Paul.

Rybakov, B. A. (1965), *Cosmology and mythology of Eneolithic farmers. Soviet Archaeology,* 1:24–47 [in Russian].

Steblin-Kamensky, M. I. (1976), *Myth* [in Russian]. Leningrad: Nauka.

Zederblom, N. (1932), *The Origin of Religion in the Understanding of Bourgeois Scholars* [in Russian]. Moscow: Moscovskiy Ralochy.

5

The Revolutionary Transition and the Modern Problem of Process

Mark Elchardus

Abstract This article uses the experience of the French calendar reformers (1793) in order to study the way in which the distinction between *aeternitas* and *tempus* (or, we may say, the relatively invariant and the relatively variant) was redefined at the end of the eighteenth century. It is argued that the calendar reformers faced the problem of process in its modern form: how can we understand history as an interweaving of change and permanence, variance and invariance, when the factors that can be called upon to explain the variance are themselves considered to be subject to change? The contemporary sequels to that legacy of the French Revolution are studied in two contemporary social science paradigms (or root metaphors): instrumental rationality (utilitarianism, rational choice theory) and achieved rationality (Habermas). The conclusion is reached that both paradigms solve the modern problem of process by appealing to the residue of essentially religious attitudes. In the final section of the paper the implications of this finding are discussed with regard to the likelihood and possible consequences of increased reflexivity. The article concludes with the expectation that the future will see a greater sensitivity to the ritual aspects of everyday temporal practices and social rhythms, and a concomitant shift in our conceptions of time.

''Our Legislators! The Revolution blooms in the glow of your genius.'' This was how Isnard, an obscure but enthusiastic citoyen from Vence, near Nice, addressed the members of the Etats Géneraux who were then gathered at Versailles (Castellan, 1978, pp. 173–174). His letter dates from the middle of July 1789. It must have been written just before or just after the ''prise de la Bastille.'' What makes it interesting is its author's conviction that the events to which he was a witness were a revolution. That same certainty was succinctly expressed by the many patriots who on the eve of what was to become the French Revolution started using new chronologies. From the fall of the Bastille on, many revolutionaries dated their pamphlets in new ''years of Freedom,'' ''of Reason,'' or ''of the world reborn.''

Later, after the King was beheaded, the Paris Commune would date its papers "in the fourth year of Freedom and the first year of Equality." Tha rather cumbersome formulation would soon afterwards yield to the briefer "first year of the Republic" which was to become part of the official calendar reform of 1793.

When presenting the new calendar to the members of the Convention, Gilbert Romme, its main architect, was emphatically clear about the meaning of the new way of counting the years. "The old chronology," he told the gathering, "is a count of cruelties and lies, it counts the time of falsehood and slavery; it had to go together with the King. . . ." With the new chronology, Romme continues, "time opens a new book to history."[1]

More than the turning of a page, the revolution is to be the opening of a new book. The thesis that the revolution constituted a clear break with the past was later adopted by many historians, especially by those of Jacobin and Marxist persuasion. They too tended to view the revolution as the "triumphant beginning of a new history" (Soboul, 1965, p. 124). Alexis de Tocqueville's *L'Ancien Régime et la Révolution,* which first appeared in 1856, offers the classic critique of that view). For Tocqueville the Revolution completed the work of the Kings and should be considered the result of a slow, quasi universal change of ideas and ideals; the product of a tectonic drift of political and economic relations. Subsequent historical research seems to have confirmed Tocqueville's, rather than the Jacobin's thesis (Reichardt and Schmitt, 1980).

I do not intend to pursue the question as to what degree and how suddenly the revolution changed French history. What interests me here is the behavior of people who are convinced that they are participating in an absolute turning point: ending one era and opening another. I want to listen to the revolutionaries while they are reforming the calendar, and are caught in one of those rare moments in which men and women, in the midst of social upheaval, have to directly address the philosophical and political issues of time. The ideas the French patriots put forward are of course not entirely new. The revolutionaries are heir to the ideas of their century. Yet, in the course of the calendar reform they formulate and discuss those ideas with a clarity and concreteness that is still instructive and anticipates the modern philosophical debates of the nineteenth and twentieth centuries.

I shall first use the experience of the calendar reformers to illustrate what I have elsewhere called the dimensions of temporality (Elchardus, 1988a). These dimensions will then be used as an analytic tool to look at the way in which the revolutionary "timesmiths" formulated and faced the specifically modern conception of time that made process problematical. In the second part of the paper I describe how some strands of contemporary social theory deal with the problem of

[1]From Romme's declaration of motivating reasons for the calendar reform, first presented to the Convention on September 20, 1793, and published, in a revised version, in the procés-verbaux of the meeting of the Convention of the 19 Brumaire II. All quotes used in this paper are based either on the *Procès-Verbaux du Comité d'Instruction Publique de la Convention Nationale* edited by M.-J. Guillaume (1891–1907), or on the *Réimpression de l'Ancien Moniteur,* edited by L. Gallois (1840–1845). They will henceforth only be identified with the date of the session.

process. That discussion ends with a hypothesis concerning the way in which contemporary society is likely to solve that problem.

Republican Time

The reformers wanted their new, republican calendar to be rational and express the progress of Reason. That goal, so they thought, could be achieved through strict conformity with the findings of natural science.[2] Such conformity is given as one of the main reasons for placing the beginning of the republican year on the day of the exact equinox of autumn, as determined by the Paris observatory. In the name of the same rationality, Romme also proposed keeping the astronomical year and the calendar year as closely synchronized as possible, even if that implied the irregular intercalation of days. Of course, such provisions are not rational in the pedestrian sense of making life easier. A conventional new year, which always falls on the same calendar day, and the regular recurrence of leap days are simply more convenient. Several people were quick to point out these disadvantages of the new calendar to Romme. After consulting with some astronomers, the latter amended his calendar. His new project never got implemented. According to some historians Romme carried it with him when he was arrested and jailed for his sympathies with the popular uprising of June 17, 1795. According to others the project was introduced, and placed on the agenda of the Convention on the very same day that Romme, condemned to the guillotine, took his own life.[3]

For Romme, the calendar that was going to be used to record the new French history had to derive its form from the eternal regularities of Nature as these were revealed through scientific Reason. The Republic inaugurated an era in which Man (l'Homme) would be able to live in accord with the laws of Nature. The new chronology had to express that transition, and the new calendar had to provide a frame in which the new time could unfold. The new calendar symbolically embedded history within the eternal laws of Nature that the new Republic would henceforth respect. The place of intermediary between Nature and history, which the republican calendar reformers assigned to natural science, is reminiscent of the way in which the sacred kings of Mesopotamia and ancient Egypt linked the cosmos and society: "Civilization found its sustenance in the authority of the king, through whom, like the constricted center of an hourglass, the divine and constant measure of the cosmos permeated society" (Gunnell, 1987, p. 34).

The Dimensions of Temporality

The core issue the revolutionaries address in their calendar reform involves the division of reality into two distinct realms: a realm of permanence, eternity, and

[2]Romme before the Convention, September 20, 1793.

[3]The main secondary sources concerning the calendar reform I rely on for this paper are: Andries (1988), Baczko (1984), De Rey-Pailhade (1908), Elchardus (1989), Ozouf (1988).

law, and one of contingency, history, and society. These realms are hierarchically ordered. If presented graphically or spatially one would place them on a vertical axis. The realm of the eternal is thought of as Nature and treated with awe. It is, for the revolutionaries, the realm of things sacred, that can and may not be touched. For them the defects of the old Gregorian calendar were due to insufficient knowledge of the measures of nature, as well as to the intrusion of irrelevant particularities such as the tyrants' personal pride (e.g., the 31 days of the months called after Julius Caesar and Augustus) or the era of the particular Christian faith. The link between the two realms, as they try to establish it, runs both through scientific reason and the individual, intuitive sentiments about what is just and reasonable. The Republic is the form of government in which consistency with the laws of Nature and expression of the true sentiments of justice, truth, and beauty become possible. For this reason the new calendar is presented as part of the regeneration, purification, and edification of the soul of the people, that the establishment of the Republic will make possible.[4]

Reduced to its most simple, mathematical expression, this activity of the revolutionaries can be described as the making of a distinction or the drawing of a line. For Spencer Brown that is the first operation we have to perform in order to be able to talk about reality, and, one might add, that reality has to perform in order to be talked about: "a universe comes into being when a space is severed or taken apart. . . . The skin of a living organism cuts off an outside from an inside . . ." (1971, p. v). One of the most fundamental differences concerns the division between the relatively variant and the relatively invariant (Elchardus, 1988a, pp. 47–48). This is, like the skin Spencer Brown uses as an example, the mark of the boundary between things that are placed in the flux of time and things that are, in an absolute or relative sense, placed outside of time. This formulation is not very rigorous, since it uses the term *time* in a way that makes it almost synonymous with sequence, change, or history. Using the language of system theory, one could say that the distinction between the (relatively) invariant and the (relatively) variant marks the boundary of any system that can maintain itself as distinct from its environment (compare Luhmann [1987]).

I use the terms *relatively variant* and *relatively invariant* because they seem neutral and general, as opposed to, for instance, the Augustinian *aeternitas* and *tempus,* which denote one specific way of making that distinction and defining the relation between the realms that have been separated. The difference between the (relatively) invariant and the (relatively) variant is considered as an essential and fundamental distinction by means of which the world can be interpreted. As such it is part of the patterns of culture, and can be expressed in many different, culture-specific ways. Whenever people are, like the French revolutionaries, in a situation that deviates strongly from "business as usual," and are challenged to (re)think

[4]These are recurrent themes of the texts concerning the calendar. They are quite succinctly expressed in the declaration of motivations written by Romme, and introduced before the Comité d'Instruction Publique on December 21, 1792, as well as in the report concerning the nomenclature, presented by Fabre d'Eglantine, on October 6, 1793.

their relation to time, they will have to consciously draw that distinction, specify what can be considered invariant, and develop accounts about the way in which the relatively invariant and the relatively variant are related.

That distinction between the relatively variant and the relatively invariant should be considered a temporal dimension, along with the more familiar distinction between before and after. The distinction between before and after, when replicated, creates the familiar image of time as a sequence of moments. Using the logic of distinctions once more, one can try to clarify this second dimension by saying that if only the first dimension (invariant vs. variant) were given, reality could only be described as being in one of the states that have been distinguished. Statements about changes from one state to another would obviously be impossible. A hypothetical change from state A to state B and then back to A, using merely the first distinction, can only be called "a being in state A." The introduction of the before/after distinction, and recurrences thereof, allows one to describe the latter state as "a return to A after having been B." It is when systems are able to make both distinctions—invariant/variant, and before/after—that they become fully temporalized, and gain a past, a present, and a future.

The two temporal distinctions (invariant/variant and before/after) and corresponding dimensions (relative invariance and sequence) are not mere matters of logic. They were suggested to me by Gerald Clemens's definition of a time keeper as "any repetitive phenomenon, whatever, the recurrence of which can be counted. . ." (1981, p. 406). Recurrence does imply identity or relative invariance, and to count implies a sequence. Any statement about time therefore presupposes the possibility of making those two essential temporal distinctions. Only a system in which we can make those distinctions (and count recurrences) can be used as a measure of time and compared to other such systems in the act of measuring time (compare Fraser, 1982, p. 14). The two temporal distinctions constitute the possibility of conceiving of sequence and recurrence, or, more correctly, of conceiving of sequence through recurrence, and of recurrence through sequence (Elchardus, 1988a, pp. 45–46).

I submit we talk about a temporal view of reality from the moment people use the two temporal distinctions or dimensions to cope with and apprehend the world. It seems to me that in that sense the human species is a temporal species. Around that universal core of temporality, cultures differ along three axes of variation. The first of these concerns the concepts of time. A concept of time is a category that fuses the dimensions, for example, in the sense that clock time is thought of as an invariant sequence of identical (invariant) elements. Between cultures such time concepts can vary in their degrees of discursive articulation, abstraction, formalization, and connotations. Only in a very specific sense can one say, as for instance Gunnell (1987, p. 26) does, that the prephilosophical Greek had no concept of time. This can only mean that the old Grecian way of bringing the temporal distinctions to bear on daily life and thought was characterized by a low degree of abstraction. If one intends to mean more by such a statement, then one has to face the problem of denying a concept of time to a culture that practices time reckoning, which is exactly the problem Gunnell gets himself into (1987, pp. 25–26).

The second axis of variation concerns the way in which different cultures, ideologies, and theories specify the temporal distinctions and conceive of their relationships. This variation concerns for instance the specific way in which a culture divides the world between the invariant and the variant or defines the past, the present, and the future. The way in which the relationship between the relatively invariant and the relatively variant is conceived of, as well as the conceptions about the relationships between past, present, and future, or the theories about the way in which the invariant reveals itself in a sequential order, also belong to this type of social and cultural variation.[5]

The third axis of variation emerges around the transgression of those distinctions. As soon as the distinctions are clearly elaborated they can be transgressed in attempts to experience mythical union with the origins, transcend the difference between past and present, experience the eternal in the fleeting moment, and so on. Such practices and the systems of beliefs built around them seem to derive part of their rapturous character and emotional significance from the transgression of the most fundamental temporal distinctions.

With the conceptual tools just gained, I now turn again to the calendar reform which belongs to the second of the three types of variation just sketched.

The Problem of Process

In the case of a dramatic transition, such as the French Revolution, the ways in which the temporal dimensions are represented and related are likely to change. The first temporal dimension, which is usually latent or taken for granted, will come to the fore. To some extent it could well be that we are only willing to apply the label *revolution* to a series of events when these involve a problematization of the taken for granted, and a new definition of the (relatively) invariant. Therefore revolutionaries are also faced with the problem of process, that is, with the problem of how the relatively invariant will work itself out in the historical and contingent sequence of events, or how to go about "the interweaving of change and permanence . . ." as Whitehead put it (1968, p. 53).

When the new calendar project was debated for the first time in the Convention, the discussion primarily concerned the meaning of subsequent French history. The commission charged with the development of a new calendar proposed to name some months after important revolutionary events ("prise de la Bastille," founding of the Republic), and dedicate the others to the republican virtues (Equality, Justice, Freedom). In that way, as the representative of the commission pointed out, French history would each year repeat the history of the revolution and recall the main

[5]To the extent that the relatively invariant is taken as the ultimate foundation of order, the relationship between the relatively invariant and the relatively variant—between the foundations of order on the one hand, and history and society on the other—emerges as a preeminent political problem. Pocock (1975) and Gunnell (1987) laid the, by now almost classical, foundations for such a temporalist approach to political philosophy.

republican virtues (e.g., the discussion concerning the nomenclature of the new calendar in the Convention, October 5, 1793). That proposal was opposed by some of the members, not because they did not like the particular choice of events or virtues, but because they had quarrels with the chronosophy (i.e., the particular conception of sequential order or theory of history they sensed behind that proposal). We do not know, one of the critics said, "what history will bring us." "We do not know whether later and better generations will still hold our virtues in high esteem." The proposal is rejected, primarily because it is felt to suggest a closure of history. That closure was symbolically realized in two ways: first, by the suggestion of a cyclical pattern (French history passing each year again through the main events of its founding revolution); second, by a definitive listing of the republican virtues. The other argument against the proposal was based on the view that names would always introduce particularism while the new calendar "had to be as universal as reason itself" and could, consequently, only speak the universal language of numbers. In order to avoid any suggestion of closure and particularism the proponents of an open future propose to designate the months and the days by their ordinal position only. This proposal carries the day, and the debate itself is dated on the "14th of the first month of the second year." This rather austere way of recording the time will only be practiced for a couple of weeks. Several members of the Convention wanted to use the calendar for educational purposes and were in favor of names that speak more directly to the popular imagination than "those cold numbers." The months will finally receive names inspired by nature and the course of the seasons. Nature does indeed appear as the only possible compromise between the two positions: it speaks to the imagination of an agrarian society, and it is the eternal essence that will manifest itself all the more clearly under the new republican order.

When we consider the calendar reform as a discourse about the relationship between time and history, we see a number of rather clear positions emerging. The relatively invariant is in fact considered absolutely invariant, and represented as Nature. Its relationship with society and history runs through Reason. The revolution is a turning point, a new beginning or genesis. The revolutionaries themselves frequently call it a "regeneration." It sets in motion a new history that will be in conformity with Nature and Reason, and will allow these to fully unfold. Within that discourse the problem of process manifests itself as a conflict between a "closed" and an "open" view of the republican future. It is important to note that this conflict does not concern the question of whether the future can or cannot be made through human agency. It is rather a conflict between those who think that the factors that guide human action and history—the virtues, according to the revolutionaries—are fixed, and those who think that those factors are themselves a product of history. The first tend to see history as determined by the revolutionary moment and the project it sets in motion. The latter do embrace temporality in a more radical way, and announce the form in which the problem of process poses itself in its modern form. For them even virtues (or values) are only relatively invariant. Their position announces Chateaubriand, and Proust, and the all-encompassing feeling of passing: "Man has not one life, but different lives, placed one after the other, that's his misery'' (Chateaubriand, 1849, Vol 1, 3, p. 17). It should not surprise us that in a culture of outspoken individualism and subjectivism, the feeling of passing should most forcefully be expressed as a detachment and alienation from a former self. The core of the position I want to highlight here does not, however, concern that individualism, but the conviction that those factors which in a given discourse about society and history are held to be fundamental, and ultimate—henceforth called ''action-guiding factors''—are themselves contingent and historical. The invariant is no longer placed outside of history and human action.

It is noteworthy that in the debates concerning the calendar reform, both the opponents and proponents of a closed future locate the action-guiding factors in the ''virtues.'' Even though the republican reformers thought that these virtues would naturally reveal themselves, their language retains a medieval flavor and seems to hark back to the opposition of *virtus* to *fortuna,* rather than to be solidly grounded in eighteenth-century social theory. In that century the pursuit of self-interest increasingly came to be looked upon as the very Nature of Man. Mandeville and other precursors of liberal theory had ceased to focus on virtues as the hard core of a self-directed moral being. Under their influence, passion and self-interest had begun to emerge as the new fundamental action-guiding factors. It is possible that the proponents of a closed future, and a definite listing of the main republican events and virtues, were still holding on to the classical, and already somewhat old-fashioned civic notions, while the proponents of an open future had more completely assimilated the dynamics of capitalism and merely spoke of virtue where, in fact, they meant passion or interest. If that was the case, then the latter would naturally look upon future history as potentially as volatile and unpredictable as the passions and interests from which it sprang. That reading of the calendar discussion would be consistent with the view, advanced by Pocock, that eighteenth-century political thought abandoned the search for timeless universals in favor of an acceptance of history (1975, pp. 466–486). Pocock's thesis, however, is based on a one-dimensional concept of time that conflates the dimensions of relative invariance and sequence. In analyzing the ways in which men and women think about the relationship between time, world, and history, such a conflated conception of temporality can but record shifts between a pole of invariance on the one hand, and a historicist pole of mere sequence on the other.

Looked at through a two-dimensional conception of time, the calendar reform appears in an entirely different light. It does not so much announce a shift from timelessness to historicism, as a redefinition of the invariant. The reformers do not conflate the dimension of relative invariance, reducing the invariant to a chronologically ordered sequence of events. On the contrary, they redefined the relatively invariant, and achieved a different view of its relation to the variant.

For the proponents of an open future, the invariant, be it Nature, Reason, or ''the universal and timeless propensity of man to pursue his self-interest'' no longer showed itself directly in history, but revealed itself in the guise of changing action-guiding factors, be they changing virtues, passions, or interests. When such a position is taken, process—the interweaving of invariance and variance—becomes problematic: it then becomes difficult to make sense of action and history because the action-guiding factors that can be called upon to explain are themselves subject to change. How can a self comprehend its previous selves, if they have nothing left in common? How can we explain actions if we are used to explain them in terms of interest, and interests have changed? How can we understand change when the stable reference points that allowed us to master it are swept away? It is in that specific sense that one can say that their vision of an open future defined the specifically modern problematic of time and process. It created a new modern present that had to make sense of itself, had to provide its own invariance (Habermas, 1987, pp. 7–16). It is to some contemporary sequels to this legacy of the revolution that I turn in the next section.

Social Theory and Temporal Dimensions

Without trying to be exhaustive, I would like to distinguish three contemporary ways of dealing with the problem of process, and shall approach these by looking at the way in which they specify the temporal dimensions. These solutions can be called paradigms or, maybe more appropriately, world models or root metaphors (Michon, 1986). They consist of general presuppositions, models, concepts, and ways of defining reality. They are also normative and prescriptive in the sense that they tell people how they should explain the world and how institutions should be designed. In that sense they constitute theories of sufficient scope to influence both social action and the way in which it is observed. They are, in other words, self-verifying. When they take on the form of scientific theories, and are confronted with disturbing observations, they are likely to try to accommodate these through peripheral adaptation, keeping their core elements intact and/or to correct the defective world, bringing it more in line with the theory. That does not mean, however, that their propositions cannot be subject to sustained and reasonable discussion about validity claims. These world models have to meet the challenge of pertinence. People should be able to relate them to some matter under consideration. When an insufficient number of people is able or willing to do so, when too few of them are willing to shield the core by peripheral corrections and/or world improvement, then the theory is likely to die of lack of pertinence.

The Paradigm of Instrumental Rationality

When it is used to describe and explain, rather than prescribe, the first of those paradigms gives priority to the past. Concepts such as cause and effect, and stimulus and response occupy a prominent place. Within that model the search for meaning is looked upon as a somewhat childish or primitive obsession. Since only the past can explain, any search for meaning becomes suspect of what is, within that model, rather demeaningly called "consequence explanations." That form of explanation is held to be problematical because, as one critic put it, it is "wholly foreign to the idea that there could be such a thing as sound and fury in social life, unintended and accidental events that have no meaning whatsoever" (Elster, 1983, p. 101). This quote clearly states the general presupposition of the paradigm of instrumental rationality, I want to highlight: events and actions can only have meaning if that meaning coincides with their causes and was present before the actions or events were. If such causal meaning cannot be discovered, then one has to acknowledge a break in time, in the sense that a sequence of events has to remain meaningless: sequence without sequential order. Closely related to that view of sequential order is a strong preference for the root metaphor of mechanism. An explanation is acceptable when it acknowledges the priority of the past and specifies some mechanism in which a cause can produce effects. In the economically inspired approach to social science in which this model is prevalent, the cause is often an individual's preference and the mechanism consists of mental wheels imparting movement to the interested subjects and wheeler-dealers.

Yet, while it apparently carries the burden of explanation, the past does not really enter the world of this paradigm. The action-guiding factors (in casu the interests) that are supposed to causally explain human behavior and history are either considered as invariantly given or else left unexamined. The first position, which is still premodern in character, is taken by classical utilitarianism which assumes that there are natural, ultimate, and invariant ends (i.e., maximize pleasure, minimize pain). All other definitions of ends are considered to be instruments, rationally designed in order to achieve those invariant, ultimate ends.

In the modern variant of the paradigm—applied in neoclassical economic theory and sociological rational choice theory—the relative invariant is constituted by consciously neglecting the formation of action-guiding factors (interests, wants, or preferences). The past, in this case the formation of the action-guiding factors, is left unconsidered. Only its product, the formed wants or interests, are brought into the picture.[6] This is illustrated very clearly in the so-called theory of revealed preferences. According to this theory, first developed by economist Paul Samuelson, it is the act of choosing that reveals the preference, and it is that revealed preference that explains the choice (Sen, 1986). In this case it is the observed behavior that creates the definition of the past that in turn explains the observed behavior. This shows clearly how the formation of preferences is in fact excluded from the world of this theory. The processes and mechanisms leading to the formation of preferences are supposedly based on the invariant human nature, consisting of the individual's propensity to rationally pursue his or her self-interest.

That relative invariance is not explicated, however. Only its results are said to be "revealed." In that way the invariant human nature is linked to action-guiding

[6]That neglect of past history is usually linked to methodological individualism: to the idea that action can only be explained in terms of individual motive, interest, or preference. Essential, for our present purpose, however, is not that emphasis on the individual subject, but the built-in blindness for the way the action-guiding factors are formed. It is conceivable to develop theories that are both nonindividualistic and blind for the formation of action-guiding factors. Yet, in most concrete theories, this blindness is formulated with regard to individual want formation.

factors and to action and history, even though it remains clouded in mystery. The relative invariant is in fact gained by a willed blindness for the processes by which the action-guiding factors are formed. In that way one salvages the relative invariance of the mechanisms of want formation, by simply shielding them from view, and locating them in the blind spot of the theory.

The low value this paradigm accords the past is clearly brought out when it is used prescriptively. It then radically devalues the past as it is condensed in tradition, habit, knowledge, and so on, and resolutely turns toward the future. Instrumentally rational decisions only take expected future cost and benefits into account, and disregard the past as a "sunk cost" (Elchardus, 1988b).

The willed blindness for the formation of action-guiding factors leads to ethically relevant consequences as soon as this paradigm becomes pertinent and is used prescriptively (Elchardus, 1988b). It does so under the form of generalized utilitarianism (Caillé, 1989, pp. 39–52) which holds that individual preferences constitute the ultimate touchstone of value, be it in the realms of truth, beauty, or justice. In relation to that conception of justice, beauty, and truth, the theory of revealed preference acts in fact as a sacred doctrine. By declaring the formation of wants inscrutable, it shelters that process from explicit consideration. Within the scientific discourse which neoclassical economics affects, this doctrine often takes the form of the argument that change in preferences cannot be studied scientifically. The processes through which individual preferences are formed, and hence the contingent character of those processes, remain hidden and mysterious. The religious connotations of the term *revealed* reveal themselves to be quite appropriate indeed. Religion is understood here in a Durkheimian way, as comprising both beliefs and rites. The beliefs and myths are essentially statements about the sacred. The rites are rules regulating attitudes and behavior with regard to the sacred. The sacred, as opposed to the profane, is attached to objects, ideas, places, or time slots that are felt to have a special power, hence the necessity of the rites (Durkheim, 1912). One of the most commonly prescribed attitudes toward the sacred, and a visible sign of sacredness, is its untouchability: it is that which cannot be touched, known, or controlled.

The Paradigm of Achieved Rationality

When the mechanisms of want formation are profaned one can raise sacrilegious questions such as "why should the choice between feasible options only take account of individual preferences if people tend to adjust their aspirations to their possibilities?" (Elster, 1983, p. 109). In answering that question Jon Elster gives a long list of mechanisms through which action and actual choice can influence preferences, wants, and more generally, action-guiding factors. That list of how history touches upon its own foundations and of how conclusions mold their premises does not bring him to a critique of utilitarian morals, but to an attempt at peripheral improvement. The individual wants that are taken as a touchstone should be autonomously arrived at. This moralistic position cannot be derived from the theoretical premises Elster accepts, and must be imported from other sensibilities, and as such stands as a proof of theoretical failure.

It does, however, signal an important procedural turn. It is indeed quite unlikely that autonomous wants can be defined substantively. Already the French revolutionaries saw that an "open future" excluded the definitive listing of virtues. The autonomy of wants can therefore only be grounded in the correctness of the procedure that was followed in developing them. A want or preference—or more generally an action-guiding factor—will be autonomously arrived at when certain procedural qualities have been respected. When applied to law, this principle can lead to the conclusion that legality can no longer be distinguished from legitimacy, since respect for procedures is the ultimate form of legitimacy. Generalized into a universal basis for normative grounding, we find this procedural turn not only in normative legal theory, but also in the definition of ideal forms of organization and communication.[7] It is, obviously, an attempt to find a point of relative invariance in a world in which a substantive definition of value or virtue has become impossible. In that sense it is another solution for the modern problems of time and process. The formation of action-guiding factors is no longer sheltered from inspection, but defined as the outcome of a procedure. The invariant is now defined as a set of procedural rules.

The way in which the two rational models specify the temporal dimensions can now be stated more succinctly. Knowledge of sequential order and process emerges from a causal analysis of individual want satisfaction. The basis of order—both prescriptively and descriptively—or the relatively invariant, is constituted by a nature that reveals itself in those wants or preferences. In one variant of the theory those preferences are formed in an inscrutable way, and then revealed through the market. Within that version of the rational model one cannot distinguish between the good, the popular, and the efficient. This has led some people to inquire into the origins of wants. Because good wants can no longer be distinguished from bad ones, this has led to a sanctification of procedure. In this second version of rational theory the link between the relatively invariant and the relatively variant is no longer occupied by the "blind" market forces, but by the forum: taste and truth have to emerge from communication free of constraint (Habermas, 1987, pp. 294–326). The first version can be called "blind" or "instrumental rationality," the second "communicative" or "achieved rationality."

The paradigm of the blind rationality has been widely applied in the creation of institutions and the evaluation of performance. Its extensive use has even given rise to the concept of a "greedy economy" in which market success effectively acts as the final touchstone of worth. It is important to note that the relevance of this model is not restricted to things economic. Liberal democracy is based on much the same set of principles. The people express themselves through the vote and the

[7]Recently David Sciulli (1986) analyzed Lon Fuller's concept of procedural legality, Talcott Parsons's collegial organization, and Jürgen Habermas's communication theory as respectively the specification, organizational framing, and theoretical grounding of this procedural turn.

interpretation of the election concerns the preferences and wants that were expressed, not the way in which these preferences were formed.[8]

According to the logic of the institution, the preferences are expressed, and that expression should serve as the basis for policy formation. The preferences should be freely and autonomously expressed, hence secret voting. Yet, the secret ballot, together with the simultaneity of the vote, also excludes deliberation. Nobody is held to offer a justification for his or her vote, and the decision not to communicate partial results before the voting booths are closed is inspired by the same desire not to have the vote of some influence that of others.

This willed blindness for the political preference formation does lead to problems when deviating opinions are expressed, for instance in support of racist parties. European politicians giving in to racism and the anti-immigrant sentiment, in the hope of drawing votes away from the right-wing parties that profit from that sentiment, can be charged with dangerous opportunism, but are also acting within the logic of liberal or utilitarian democracy. The vote is a signal they should abide by since individually expressed preferences are the final touchstone of legitimacy. On this issue too, the critical version of rational theory—the paradigm of achieved rationality—is forced to place its hope in communication and deliberation as a means of preference formation, and to call for absolute respect for procedural rules concerning the generality and universality of rights and duties.

These two versions of rational theory stand in a very ambivalent relation to the religious. They are, on the one hand, the obvious expression of the disenchantment with the world and with the modern conceptions of time and process. The relatively invariant is no longer projected outside of society and away from man: on the absolutely other. The relatively invariant action-guiding factors are, on the contrary, seen as man's products. While this shift could be taken to signal the end of the religious (e.g., Gauchet, 1985), we see that in order to legitimate private wants or collective rules of procedure as the ultimate touchstones of value and order, the two versions of rational theory nevertheless appeal to essentially religious attitudes: acceptance and absolute respect. In the paradigm of instrumental rationality the market (either economic or political) has taken the place of the ancient divine kings. It mediates between the want formation, clouded in mystery, and the revelation of wants in social exchanges. Within the logic of that theory one cannot but say of the market, "thou art the living image on earth of thy father, Atum of Heliopolos. Taste (authoritative utterance) is in thy mouth, intelligence in thy heart; the seat of thy tongue is the shrine of truth (justice)" (Breasted, Ancient Records of Egypt, cited in Gunnell [1987, p. 35]).

The paradigm of communicative or achieved rationality less blatantly appeals to religious authority, and is certainly more critical. It does not expect revelation from the market, and doggedly attempts to find a theoretical foothold from which

[8]Our democratic institutions are also based on the principles of methodological individualism. It is the individual that votes. There is no institutional room for groups or for multiple selves. I will, however, bracket that aspect because I think that the link between this individualism and the willed blindness for preference formation is contingent.

to reasonably discuss and evaluate the verdicts about truth, justice, and beauty rendered by the market. In doing so, however, it does not escape appealing to an essentially religious attitude. Even the paradigm of achieved rationality has a sanctum sanctorum into which questioning and reflective men and women should not or cannot penetrate. The procedural rules which make possible a communication free of constraint have to be accepted, and as far as I can see, the reason for doing so is not fundamentally different from the respect accorded the sacred. Habermas argues differently: he claims that the authority of the sacred has been replaced by the authority of achieved consensus (1987a, p. 77). That what was rendered invariant by placing it out of the reach of human control, has in other words been reappropriated by Man, and consensus now replaces invariance. When arguing that position on the basis of linguistic theory, Habermas does reintroduce the sacred under the form of "that which cannot be questioned." The achievement of consensus rests on a "pretheoretical grasp of rules" (Habermas, 1987b, pp. 297–298). The possibility of achieving agreement using language presupposes "idealizing suppositions" which the participants to the conversation know to be idealizing (1987b, pp. 198–199). The meaning of utterances, in order to be really understood, must be complemented by "the background of an implicit knowledge that participants normally regard as trivial and obvious" (1984, p. 336). Those background assumptions are part of what Habermas calls the lifeworld, that is, the repository of knowledge that is "intuitive, unshakable certain, and holistic" (Habermas, 1987b, p. 326). In this view the possibility of invariance or consensus no longer directly rests on respect for the sacred but on a kind of inability to question the ultimate action-guiding factors: on pretheoretical understanding, implicit and intuitive knowledge. Yet, the rationalization process Habermas describes will inevitably lead to a questioning of this understanding and knowledge. The very possibility of doing that in a reasonable way depends on the invariance of procedural rules. The appeal to respect those rules is still essentially religious: it demands the kind of respect usually accorded the sacred, and warns that dire consequences (e.g., the impossibility of raising validity claims and discussing in a clear and reasonable way) will follow upon profanation.

We see that the two rational theories have a very ambivalent way of solving the modern problem of process. The invariant that is relevant for human action is seen as a human product, yet its very possibility is still seen as dependent upon the attitude that was developed toward the sacred. For rational, disenchanted theories to appeal to the residue of religious attitudes, is an admission of failure. Why should players respect the rules of a game they have themselves designed? Rousseau already asked that question, thought that answering it was like squaring a circle, and remained a Deist. Different positions and corresponding theoretical strategies are possible. One can hope for or believe in the end of the sacred, as Habermas does, or one can consider the sacred as an irrepressible category. The failure of rationalist theories to avoid an ultimate appeal to the sacred and its rites suggests that the latter road should be explored.

Elements for an Aesthetic Paradigm

In a disenchanted world, respect for the sacred often has the character of an ironic and knowing suspension of disbelief. The relation to the (relatively) invariant, that once had the trappings of religious ritual, then appear as a game played by players who know that they change the rules while playing, yet are serious and careful to maintain a difference between the rules of the game and the game being played. That attitude of serious playfulness is not found equally in all spheres of activity. Some ways of making a living offer a much more fertile ground for its emergence than others. Two conditions seem to be important. First, people have to be involved as creators. They have to believe that they realize themselves in the creation of works and deeds, and they have to dispose of the necessary freedom to be able to live with that aspiration. Second, they have to be involved with the definition or creation of the relatively invariant, be it in the form of truth, beauty, or justice. In contemporary society science, or more precisely the activity of sciencing, seems to meet these conditions. Scientific theories have been expected to yield truth in the sense of a relative invariant constellation of elements that can account for the observable variations in the world. The pursuit of science has moreover been regarded as a quite respectable and fulfilling occupation. A relatively great number of scientists have been able to work in relative freedom, some of them to the extent of being able to reflect upon their own activities. It is therefore interesting to look at science, and at the way science is looked at, to see what concrete form the attitude of serious playfulness can take.

The empirical study of the way science is made has revealed that an important part of the "knowledge" that allows scientists to come to an agreement about the validity and/or relevance of their products turns out to be practical. It is not discursive; cannot be rendered explicit without a great effort of interpretation, and at the risk of rendering the practical rules inoperative (Pinch, 1981). In that sense science does not differ fundamentally from everyday life. The act of reflecting, of searching, and investigating presupposes a relatively invariant background of temporarily unquestioned concepts, judgments, and principles (Gadamar, 1976, pp. 18–43; Habermas, 1987b). Many scientists no longer believe that these enabling background assumptions are fixed or noncontingent. They are instead increasingly viewed as variable sets or language games.[9] One cannot play or communicate in such a game without accepting the basic rules. On this score the sociology of science is consistent with the theory of achieved rationality: procedural rules have to be respected, otherwise communication becomes noise. Even if a lifeworld and practical and discursive rules are shared, the interpretation of data, the choice of methods, and the understanding of theories remain subject to uncertainty. They are not determined by reality, but subject to serious discussion and real choice, ultimately grounded in conventionally accepted background knowledge (Bloor, 1976). These observations can of course be used as an argument against the neopositivist's appeal to

[9]The expression *language game* is Wittgenstein's. Both Kuhn (1970) and Philips (1975) present different paradigms as different language games.

''reality'' as the final arbiter of truth. Yet, even if one stops short of such a strong conclusion, these various observations and interpretations point toward the central role of the communication between scientists. If science can approach ''truth,'' then it can only do so by virtue of certain characteristics of the organization of sciencing and the nature of the scientific communication, rather than on the basis of a method that can be individually applied. Science is no longer a monologue with the world, but a dialogue between men and women or, to borrow Rorty's metaphor, no longer a mirror image of reality, but an ongoing conversation (1980).

In such a view of science and truth, the definition of the relatively invariant is quintessentially modern. It is no longer a reality that exists independently of the historical and social processes that lead to a certain ''(re)presentation'' of it. This new definition of the relative invariant in the realm of science is consistent with, and a continuation of, the disenchantment of the world. The invariant which was projected outward, and conceived of as ''Reality,'' has been placed within the community of scientists, and now appears as the result and condition of their pertinent communication. The very possibility of science is now recognized to depend on the relative invariance of the ''rules of the game''—procedural rules of communication, shared lifeworlds, practical rules, horizons of expectation, unproblematical background assumptions—that make sciencing possible. That recognition is, however, becoming very explicit. With science becoming more reflexive, scientists are becoming increasingly aware of the conventional and provisional character of the background assumptions on which the possibility of their practice depends. They know that they are in a sense playing a game that is performed within a set of relatively invariant background assumptions that are not entirely discursive, not entirely clear, and certainly without any absolute validity. The game metaphor is in fact not really appropriate here. A better metaphor would be the work of art. In the absence of rules, no art can be produced. Yet, an academic submission to rules produces something less than art. Few today would claim that valid art can be produced by following the correct procedures; even less would they try to formulate (or reconstruct) such procedural rules.

Social life has been considered as a mechanism, an organism, and a game. If science is progressive, and indicative of attitudes and modes of thought that are likely to become more broadly diffused, then the social is likely to become increasingly viewed from an aesthetic point of view.

While the root metaphor of art easily comes to mind, it is much more difficult to specify its consequences for social theory. Sociology has little to offer by way of an aesthetic paradigm, certainly nothing that in scope and refinement can be compared to the products of its rationalist traditions.[10] An aesthetic paradigm should still be viewed as an emerging world model, stimulated by a number of societal

[10]The shift in metaphor from machine over game to art could bring sociologists to model their work less after the methods of natural science and cybernetics, and to look with more respect and interest toward art and the humanistic disciplines. The reapproach to the humanities and the social sciences already seems underway. For elements of an aesthetic paradigm, see, for instance, Maffesoli (1985, chapter 7).

developments about which and for which people try to develop a set of pertinent theories. The use of the expression "aesthetic paradigm" is moreover very likely to call forth contradictory associations. Modern usage, from Schiller to Nietzsche, has associated the aesthetic with the idea of absolute (individual) freedom. In that sense the aesthetic has a meaning that is analogous to what I have earlier called the third dimension of variation of temporal concepts and experiences. It is related to transgression and negation, of the temporal dimensions in the case of the aesthetics of time, of the everyday and the normal in general. Such a meaning seems quite useless for an aesthetic paradigm because the latter should not explore the possibilities of transgression of the existing temporal dimensions but yield insight into the way a rationalized, reflexive culture maintains the enabling temporal distinction between the relatively invariant and the relatively variant.

It seems to me that the changes I have sketched point toward a new locus of social integration. Under a locus of social integration I understand the place in which the relatively invariant is realized and expresses itself. In ancient Egypt and Mesopotamia that locus was the body of the divine king. Medieval Europe retained something of that notion in the idea that the king was "human by nature and divine by grace" (Kantorowicz, 1957, p. 87). That idea was later elaborated in the doctrine of the king's two bodies, one human and transient, one eternal or sempiternal. In that way the king participated in the two realms and united the invariant to the variant. For the French revolutionaries thinking in the republican tradition, the Republic was the setting or cosmos in which men could, through science and individual reason, reach the eternal laws of nature. For the Utilitarians the market fulfills that function. It is based on the invariant principles of human nature and reveals the ultimate touchstones of order. For the critical rationalists the forum, and communication geared toward unforced rational agreement, should replace the blind forces of the market. The invariant has then to be sought in some procedural properties of the communication. The developments I have sketched seem to point toward a new locus of integration: the affective bonding of the group, based on emotions and renewed by the experience of the bonds' creative powers. It is that feeling that seems more solid than willed blindness, more profound than rules, constitutions, and procedures. The affective bonding creates a level of stability, and therefore allows for changes in the rules, conceptions, convictions, procedures, and so on, without destroying the possibility of the group's continued existence. This does not mean that groups are long lasting or sentiments more stable than rules. Yet, in a world that becomes increasingly reflexive and ironic, sentiment and affect—ways of feeling together, and feeling in common—might well provide the most solid forms of relative invariance left.

A Model of Temporality for an Aesthetic Paradigm

Let me briefly recall some of the arguments and observations that have been presented. The way in which the relationship between time, history, and social action is conceived was drastically altered in the course of the eighteenth century. The new ideas played an important role in the French Revolution and were clearly expressed in its calendar reform. However, they are not clearly understood when one uses a conflated, one-dimensional concept of time, for then the change can only be understood as a shift from a belief in a world governed by invariant and eternal laws to historicism. Time is two-dimensional: it is a synthetic expression of two fundamental distinctions. The modern conception of time as it was formulated in the calendar discussion signals a change in both the way the relatively invariant is conceived of and in the conception of process, that is, in the way the relatively invariant is supposed to express itself in the variant or in history. Since the invariant was disenchanted, placed in the hands of men and women, and in history, the very possibility of maintaining a distinction between the relatively invariant and the relatively variant seemed threatened. The collapse of that distinction has been the theme of most antimodernist thinking. Yet, without that distinction recurrence, meaning, and identity become impossible. Short of losing the possibility of understanding the world and each other, the temporal distinctions have to be maintained. Their disappearance means chaos. They can only be transgressed in an act that seems extraordinary and rapturous, when experienced against the gray background of the ordinary, everyday practices that maintain the distinctions. Even in contemporary, rational theories the distinction between the relatively invariant and the relatively variant has been maintained by appealing to the residues of an essentially religious attitude: respect for the sacred. But that solution has become unmentionable. When pointed out it looks like a critique or indictment of the paradigms in which it is uncovered. This suggests that the invariant has to be based on those elements that escape instrumental human control. That is probably the reason why Habermas situates it in the pretheoretical, intuitive, and holistic. That solution does not seem very solid, however, for as reflexiveness increases, the pretheoretical, intuitive, and holistic is bound to erode. Thus I expect the new locus of integration to shift toward the emotional and affective bonding that creates a basis of relations stable enough for agreement and understanding to be achieved. In that sense the modern problem of process, the disenchantment of the world, and the increased reflexivity lead us toward an aesthetic paradigm: an understanding ultimately based on common feelings.

This hypothesis has both a descriptive and normative significance. Descriptively it offers one possible explanation for the late twentieth century or postmodern disfavor of theory and triumphant return of the rhetoric. It could explain the growing scientific attention to emotions and for the conditions of their instrumental manipulation. It seems consistent with a world in which the distinction between the everyday world and the world of soaps is often blurred and in which the role models of television heroes circulate as in mythical times. By way of conclusion I would like to explore the normative, practical potential of the hypothesis.

The temporal dimensions are fused and synthetically represented in a concept of time. It is under the form of such a concept that time is used as an everyday category: that it can be used to time, to coordinate, plan, date, measure, analyze.

The rationalist and aesthetic models I have described correspond to specific conceptions of time. It is, in fact, the specific way in which time is conceived and operationalized that allows for such models to be practically implemented and realized.

The instrumental–rational conception of time is adequately realized in absolute, Newtonian time and in the more familiar clock time. On the basis of such a conception people see an (almost) indefinite stretch of empty time before them. The time of the rationalist model is also "of its own nature." It is a quantity that can be used as a resource and hence as a measure of other resources. In the course of Western history, and especially since the development and spread of capitalism, time came to be viewed as the most general of media, that is, as a measure of not yet committed possibilities. Once spent it can be converted into money, or power or knowledge or love or into any other definition of a valuable. Saved, it has the potential to become each and all of those things. On the basis of such a conception the "use of time" will in the first place concern the quantity of time spent or invested. Pursuing the logic of such a conception, one inevitably reaches the conclusion that human potential is expanded by learning to save time. The bulk of the practical, technological applications of the rationalist model are based on that thesis.

The emergence of a model that uses art rather than the game or the machine as a root metaphor challenges that conception of time. This new model emphasizes the created character of time. Time is the product of acting together, of keeping some things relatively invariant of creating sequential order, of allowing for a degree of predictability, of managing beginnings and endings, closing the past and opening the future. It is not an objectively given resource, but the product of social action. The mediatization of time, however important and fundamental to contemporary life, is not sufficient to allow that activity to become reflexive. Mediatized clock time does not allow people to become aware of and master the production of time. Another, complementary conception of time is called for. An expression like "the use of time" should take on a more complex and realistic meaning. It should no longer be understood as referring merely to the expenditure of a naturally given resource, but should also cover the way in which the creation of relative invariance and sequential order are dealt with. One of the ways in which that goal can be pursued involves a greater attentiveness to the creation of a present. Time should be used to mark a present and to create beginnings and endings. The way in which those boundaries in time are marked should moreover strengthen the feeling of social integration, because it is precisely that feeling that will allow a group to master the development of action-guiding factors.

While it is relatively easy to list the logical prerequisites of a new conception, it is somewhat more difficult and hazardous to point toward practices in which the potential for the called for innovations is present. One practice that seems able to meet this requirement is ritual. Ritual can infuse the everyday with a sense of the aesthetic, the extraordinary, and the symbolic. It also unites people in a simultaneous present. It has clearly marked boundaries and can create a collective rhythm that makes the pulse of social life audible. While the chronological, mediatized time tends toward the flattening of the temporal landscape, ritual and rhythm create relief. This is not, of course, a call for the creation of new rituals, but for a greater awareness of and attention to the ritual character of so many of our activities. In that sense I join Young (1988) in calling for a greater sensitivity to habit, the cyclical and the rhythmic, and for the particular experience of time that comes with that sensitivity. Ritual is not a separate and exceptional category of behavior (Wuthnow, 1987, pp. 100–102) but rather a suppressed dimension of everyday behavior, and a forgotten practice of time.

Ritual has the property of awakening awareness for the temporal distinctions. The ritual present marks the distinction between past and future and between the variant forms and the variant behavior. In that sense it can highlight the produced, poietic character of time. At the same time the ritual can strengthen the common feelings and the social bonds. Durkheim suggested that ritual's special force is due to the fact that it gives expression to and permits one to experience the capacities or "force" people derive from participating in collective life (Durkheim, 1912, pp. 300–312). With regard to the problem of time, one could say that ritual has the potential of fusing emotion and cognition in the insight that the acceptance of the temporal distinctions and the bonding, binding power of social affect, enable us to deal with the world and each other in a somewhat reasonable way. In that sense the exploration of the modern problem of time and process brings me to the seemingly paradoxical conclusion that enlightenment now depends on ritual.

References

Andries, M. (1988), L'illustration dans les almanach révolutionnaires ou le triomphe de l'allégorie. In: *Les images de la révolution française*, ed. M. Vovelle. Paris: Sorbonne, Histoire Moderne, Vol. 21, pp. 299–306.

Baczko, B. (1984), Le calandrier républicain. Décréter l'éternité. In: *Lieux de mémoire*, Vol. 1. Paris: Gallimard, pp. 37–83.

Bloor, D. (1976), *Knowledge and Social Imagery*. London: Routledge & Kegan Paul.

Brown, S. G. (1971), *Laws of Form*. London: George Allen & Unwin, 1969.

Caillé, A. (1989), *Critique de la raison utilitaire*. Paris: Editions La Découverte.

Caillois, R. (1950), *L'homme et le sacré*. Paris: Gallimard.

Castellan, G. (1978), *Une cité provençale dans la revolution. Chronique de la ville de Vence en 1790*. Paris: Flammarion.

Chateaubriand, A. De (1849), *Mémoires d'outre-tombe*, Vol. 1. Paris: Flammarion, Edition du Centenaire, 1948.

Clemens, G. (1981), Time measurement for scientific use. In: *The Voices of Time*, ed. J. T. Fraser. Amherst: University of Massachusetts Press.

De Rey-Pailhade, M-J. (1908), Etude historique sur l'emploi du Calandrier Républicain et sur le temps décimal à Toulouse pendant la révolution. *Bull. Soc. de Géographie de Toulouse*, 27:429–457.

Durkheim, E. (1912), *Les formes élémentaires de la vie religieuse*. Paris: Presses Universitaire de France, 1968.

Elchardus, M. (1988a), The rediscovery of chronos: The new role of time in sociological theory. *Internat. Sociol.*, 3:35–59.

——— (1988b), Austauschtemporalitäten. Selbstorganisation zum Zweck gesellschaftlicher Steuerung. *J. für Sozialforschung*, 28/4:391–416.

——— (1989), De Republikeinse Kalender . . . niets minder dan een verandering van religie. In: TOR, *De opstand van de intellectuelen. De Franse revolutie als avant-première van de moderne cultuur.* Kapellen/Amsterdam: DNB/Uitgeverij Pelckmans/Wereldbibliotheek.

Eliade, M. (1987), *Histoire des croyances et des idées religieuses,* Vol. 1. Paris: Payot, 1976.

Elster, J. (1983), *Sour Grapes. Studies in the Subversion of Rationality.* Cambridge, U.K.: Cambridge University Press.

Fraser, J. T. (1982), *The Genesis and Evolution of Time.* Brighton: Harvester Press.

Gadamar, H.-G. (1976), *Philosophical Heremeneutics.* Berkeley: University of California Press.

Gauchet, M. (1985), *Le désenchantement du monde. Une histoire politique de la religion.* Paris: NRF–Gallimard.

Gunnell, J. G. (1987), *Political Philosophy and Time, Plato and the Origins of Political Vision.* Chicago: University of Chicago Press, 1968.

Habermas, J. (1984), *The Theory of Communicative Action,* Vol. 1. London: Heineman.

——— (1987a), *The Theory of Communicative Action,* Vol. 2. Cambridge, U.K.: Polity Press.

——— (1987b), *The Philosophical Discourse of Modernity.* Cambridge, U.K.: Polity Press.

Kantorowicz, (1957), *The King's Two Bodies. A Study in Medieval Political Theology.* Princeton, NJ: Princeton University Press.

Kuhn, T. (1970), *The Structure of Scientific Revolutions.* Chicago: University of Chicago Press.

Luhmann, N. (1987), *Soziale Systeme, Grundriss einer allgemeinen Theorie.* Frankfurt: Suhrkamp.

Maffesoli, M. (1985), *La connaissance ordinaire. Précis de sociologie comprehensive.* Paris: Librairie des Méridiens.

Michon, J. A. (1986), J. T. Fraser's "Levels of Temporality" as cognitive representations. In: *Time, Science and Society in China and the West,* The Study of Time, Vol. 5, ed. J. T. Fraser, N. Lawrence, & F. C. Haber. Amherst: University of Massachusetts Press, pp. 51–66.

Ozouf, M. (1988), Calendrier. In: *Dictionnaire critique de la Révolution Française,* ed. F. Furet & M. Ozouf. Paris: Flammarion, pp. 482–490.

Philips, D. (1975), Paradigms and incommensurability, *Theory & Soc.,* 2/1:37–61.

Pinch, T. (1981), The sun-set; the presentation of certainty in scientific life. *Soc. Studies of Sci.,* 11:131–158.

Pocock, J. G. A. (1975), *The Machiavellian Moment, Florentine Political Thought and the Atlantic Republican Tradition.* Princeton, N.J: Princeton University Press.

Procès-Verbaux du Comité d'Instruction Publique de la Convention Nationale, ed. M.-J. Guillaume, 7 vols. Paris. Imprimerie Nationale, 1891–1907

Reichardt, R., & Schmitt, E. (1980), Die Französische Revolution—Umbruch oder Kontinuität. *Zeitschr. für Hist. Forschung,* 17/3:257–320.

Réimpression de l'Ancien Moniteur. Depuis le réunion des Etats-Généraux jusqu'au Consulat, ed. L. Gallois, 31 vols. Paris: Bureau Central, 1840–1845.

Rorty, R. (1980), *Philosophy and the Mirror of Nature.* Oxford: Basil Blackwell.

Sciulli, D. (1986), Voluntaristic action as a distinct concept: Theoretical foundations of societal constitutionalism. *Amer. Sociolog. Rev.,* 51/6:743–766.

Sen, A. (1986), Behavior and the concept of preference. In: *Rational Choice,* ed. J. Elster. Oxford: Basil Blackwell, pp. 60–81.

Soboul, A. (1965), *La Révolution française.* Paris: Presses Universitaires de France.

Tocqueville, A. de (1956), *L'ancien régime et la Révolution,* ed. J.-P. Mayer. Paris: Gallimard, 1967.

Whitehead, A. N. (1968), *Modes of Thought.* New York: Free Press.

Wuthnow, R. (1987), *Meaning and Moral Order. Explorations in Cultural Analysis.* Berkeley: University of California Press.

Young, M. (1988), *The Metronomic Society. Natural Rhythms and Human Timetables.* London: Thames & Hudson.

Perspectives

"Cognitive science," wrote Simon and Kaplan in their introduction to Posner's massive Foundations of Cognitive Science *(1989), "is the study of intelligence and intelligent systems, with particular reference to intelligent behavior as computation," (p. 1). Today it is quite common, they remark, to attribute intelligence to both human and nonhuman systems and, in particular, to programmed computers. With such a broad base to work on, the task of constructing a model of human time becomes an attractive challenge.*

There are many approaches to an understanding of time in psychology, as R. A. Block carefully points out in his Cognitive Models of Psychological Time *(1990). Indeed, "models of psychological time have assumed a number of orientations: biopsychological, behavioral, cognitive, developmental, psychoanalytic, and social-psychological" (p. xiii). One may add that some of these approaches, and yet others dedicated to an understanding of the psychology of time, might not see their work as calling for a description that could be called a model.*

Attempts to define the nature of time as seen through cognitive psychology and permit, thereby, a comparison between the time of human experience and the nature of time as identified in physics, biology, and social process, are very rare. But a definition of time proposed by John A. Michon (1990) would probably be broadly accepted.

"Time," he wrote "is the conscious experiential product of the processes that allow the (human) organism to adaptively organize itself so that its behavior remains tuned to the sequential (i.e., order) relations in its environment" (p. 40). He added that the merit of this definition was that it described time as a process or ensemble of processes capable of generating the entire conceptual structure that we conventionally call time*. The task of the cognitive psychologist, then, is to "look for a process or ensemble of processes that can explain (or reproduce) the entire spectrum of conscious temporal experience" (p. 41).*

The "entire spectrum" envisaged by Dr. Michon is a very broad one. The almost 900 pages of the Simon and Kaplan (1989) work suggest but do not define that breadth. Dr. Macar's paper exemplifies the classical scientific protocol. She inquires into a narrow problem area within the field of cognitive psychology, and does so in depth. Specifically, she asks: What biological mechanisms control laboratory timing tasks, and how do the processes of timing depend on external conditions?

J.T.F.

References

Block, R. A., ed. (1990), *Cognitive Models of Psychological Time*. Hillsdale, NJ: Lawrence Erlbaum.

Michon, J. A., (1990), Implicit and explicit representations of time. In: *Cognitive Models of Psychological Time*. Hillsdale, NJ: Lawrence Erlbaum.

Simon, H. A., & Kaplan, C. A. (1989), Foundations of cognitive science. In: *Foundations of Cognitive Science*, ed. M. I. Posner Cambridge, MA: MIT Press.

6

Central and Peripheral Mechanisms in Timing

Françoise Macar

Abstract Time is not only an abstract object of philosophical inquiry, it is also a source of information whose processing is necessary for organizing motor acts. Human sport activities as well as animals' hunting behaviors demonstrate that success often depends on careful timing of movements. Motor timing tasks in which responses of precise durations are required provide experimental copies of such natural situations. They offer a good basis for studying time estimation processes.

The mechanisms responsible for timing are still mysterious. It has been postulated that multiple time bases are available in an organism and that some are selected for particular situations. This hypothesis attributes a major role to central mechanisms and implies that, in motor timing tasks, the accuracy of a response should be relatively independent of the motor conditions of response execution.

Thus, the first question raised is whether movement accuracy is affected by motor parameters such as force or velocity. A positive answer would argue for a peripheral conception of motor control, according to which sensory feedback resulting from response execution plays a major role in the mechanisms of timing. Experimental evidence both against and for this view is discussed.

The second question is that of the specificity of time processing. It can be discussed in relation to the central programming hypothesis, which states that rapid movements are precisely planned before being executed. That is, various motor parameters are specified in a motor program which is constructed as soon as sufficient information is available. Recent data indicate that temporal parameters can be programmed, among others such as force and amplitude. These data underscore the role of central mechanisms in timing.

In conclusion, it is suggested that both peripheral and central mechanisms are involved in control of timing, their respective contributions being condition-dependent.

For anyone who is interested in the mechanisms of behavior, one of the most exciting questions about time is how temporal information is processed to regulate motor acts. Even when it is so automatic that we are unaware of it, accurate timing of movements is necessary for success in a variety of tasks. In tennis, without a proper evaluation of the moment when a ball will hit your racket, you will no doubt be unable to return it adequately. If you feel like taking off quickly when a traffic light turns green, you become more and more ready to step on the gas as time passes. In your kitchen, you need not look at your watch to be sure that spaghetti is cooked. The durations of events in the external world and of one's own movements are continuously processed; this is an essential adjustment of living organisms to their environments. It can even be a question of life or death, as it is in Central America jungles for those hummingbirds that feed on nectar and must reach the proper flowers, scattered at various places, just in time: if flowers have not yet renewed their nectar stock, they are too early; if other feathered visitors have already emptied the calyxes, they are too late. Repeated timing errors would lead these birds to starvation (Gill [1988] cited in Friedman [1990]).

Preparing to move at a particular moment is a behavior which reveals the adaptive function of temporal information processing, reminding us that, before being an abstract object of philosophical inquiry, time is both an essential dimension of the world where we live and an inherent property of all our acts. As such, time must be processed by our nervous system, and much evidence indicates that it is indeed processed efficiently. This is not to say that a sensory organ devoted to time perception exists. Such an organ has never been identified, and its existence has never been seriously considered plausible. Researching the mechanisms of time perception consists, rather, in trying to assess whether some of the multiple cues that vary as a function of time could be taken as direct sources of information for elaborating time judgments. This means that our impressions of duration would not be inferred from external cues, such as the number of auditory or visual events that occurred, but would be based on internal variations, such as modulations in muscle tension or specific patternings of neuronal discharges. Such internal processes could admittedly be influenced by external changes. Accordingly, an important question in the search for mechanisms concerns the conditions of matching between internal and external temporal sequences. The study of biological rhythms, which, albeit endogeneous, may present modifications in frequency under the influence of external synchronizers, such as cyclic variations in luminosity, may lead to an understanding of this topic.

This view conforms to the parallelism of formal structures that Gallistel (1989) postulates between the brain and the world as a result of natural selection on brain mechanisms: time, like space, is a basic aspect of our environment, and as such its formal characteristics should be mirrored in those of our brain processes. Placing time in the perspective of the adaptation of living organisms to the world does not face the thorny question of its origin, however. Whether time existed before and independently of life is a question that goes beyond the limits of psychophysiological research. How a living organism, endowed with internal rhythms, can respond adequately to external temporal constraints is a problem subject to experimental analysis. As is true of any other parameter, time seems to constitute a source of information; as such, it should be coded so as to regulate behavior.

In the course of evolution, selection must have functioned on the basis of temporal as well as other criteria. An animal which, in its hunting activities, does not correctly anticipate the movements of its prey, has a good chance of being rapidly eliminated. This example not only gives an idea of the interactions that exist between time and behavior, it further suggests that the numerous manifestations of animal and human capacities to cope with time parameters could draw their origin from motor anticipation. Along this line of reasoning, research on temporal parameters of motor acts should contribute to the discovery of the mechanisms of time estimation.

Two models of time perception are well known: that of the timer and that of storage size. They differ on the question of whether an internal chronometric device, requiring attentional resources to function, does or does not exist.

For Ornstein (1969), the timer model has no utility. Rather, in a small book entitled *On the Experience of Time,* he proposed a cognitive model which still informs much of contemporary research. Subjective duration is supposed to increase with storage size, a metaphor referring to the memory of the information processed while objective duration elapses. Storage size depends on the amount of information delivered and on the duration of the processing operations. Thus, following two tasks of identical duration, subjective time is hypothesized to be longer for the task conveying the larger amount of information or requiring more complex processing.

Some of Ornstein's predictions have been confirmed, but his theory has two important limitations. First, the positive relation postulated between information processed and subjective duration remains vague, since Ornstein has not designed any parametric study. The complexity of information processing, in particular, is hardly accessible to quantification. Besides, Ornstein did not take accuracy indices into account, because, he said, accuracy can only be measured according to an objective duration which is conventional and arbitrary. Most of the researchers inspired by Ornstein's conception persist in making no mention of absolute errors relative to objective time; their analysis is focused on the lengthening or shortening of subjective duration as a function of experimental conditions. It is as if, in studies of motor control wherein subjects are required to point to a target, only the tendency to point to the left or the right of it, but not the absolute distances from target, were measured. Centimeters do not seem less arbitrary than seconds. A second limitation is that Ornstein only considered retrospective judgment of duration obtained from subjects who were not forewarned that they would be requested to estimate time. He asserted that ordinarily one is not conscious of time passing and suspected that subjects would try to be as precise as possible (which, in his opinion, was meaningless) if they knew the aim of the experiment. Once again, the analogy with spatial pointing tasks makes it easy to imagine distance errors would probably increase and become more variable if instructions directed subjects' attention toward other parameters. Nowadays, it clearly appears that Ornstein's theory, contrary to his expectation, cannot be generalized to all situations in which one estimates time. Prospective and retrospective methods, differing by whether temporal requirements are or are not specified prior to the task performance, offer quite distinct data (Hicks, Miller, and Kinsbourne, 1976; McClain, 1983), stressing thereby the important relation between attention and time.

Focused attention is considered as a limited-capacity system. If certain aspects of a situation benefit by attention, it is at the expense of the neglected aspects, which are processed less satisfactorily. This should be equally true of temporal parameters. As emphasized by Michon (1972), there is no reason to suppose that temporal and nontemporal information involve processing mechanisms of a different nature. Thomas and Weaver (1975) maintain that attention is shared between an "information process," devoted to the processing of nontemporal information, and a "time processor," for temporal judgments. Subjective duration is supposed to shorten whenever attention is focused on other than temporal parameters of the situation. This conception originally applied to a duration of around .1 second, but it has since been extended to longer intervals. Indeed, with the prospective method, several studies examining periods of 8 to 120 seconds, during which categorization judgments were done on objects or on linguistic material, revealed a shortening of subjective duration, with greater shortening as the categorization task increased in difficulty and required enhanced attention (Hicks, Miller and Kinsbourne, 1976; Hicks, Miller, Gaes, and Bierman, 1977; McClain, 1983; Zakay, Nitzan, and Glicksohn, 1983; Zakay and Fallach, 1984; Arlin, 1986; Zakay, 1989; Macar, submitted). These results contradict Ornstein's generalizations. A possible interpretation is that some of the indices necessary for processing time are lost when attention is turned away from the time processor.

The nature of such hypothesized indices is unknown. Multiple sources of impulses, rhythmic and otherwise, exist in the central nervous system. It has been proposed that, during time estimation, impulses are summed in a counter such that their total number at the end of the interval determines subjective duration (Creelman, 1962; Treisman, 1963). Biological rhythms are possible candidates for yielding cues of various frequencies. Numerous physiological variables, such as the chemical composition of blood, body temperature, or neuronal excitability, fluctuate with very different periods, from microseconds to years. There would be no discontinuity between "biological" and "psychological" time if biological rhythms provided time bases for subjective duration (Richelle, 1968).

Whatever the exact nature of possible temporal cues, the relation that has been found between attention and time judgments under prospective conditions favors the old "internal clock" theory, albeit with serious reshaping. Rather than a single, permanent, and unchanging clock that has long been postulated, multiple, transient, and variable time bases would seem more adapted to the wide range of durations that one has continuously to process. The proper time bases for any particular situation could be selected from the complete arsenal and combined into an efficient chronometric system that would progressively strengthen with learning (Figure 6.1). The convenient metaphor then is not a clock, but a family of chronometers that can be started, stopped, and reset at will. This revised internal clock conception attributes a major role in timing to central mechanisms.

Several types of data support the central view of timing processes. This paper will present some of the available evidence. The central view, however, seems to attribute a minor role to peripheral factors in motor execution. It suggests, for instance, that if one has to press a button for exactly three seconds, the force needed for pressing it will not influence timing accuracy. This question should be carefully examined, since an alternative hypothesis is that timing proccsses may draw their

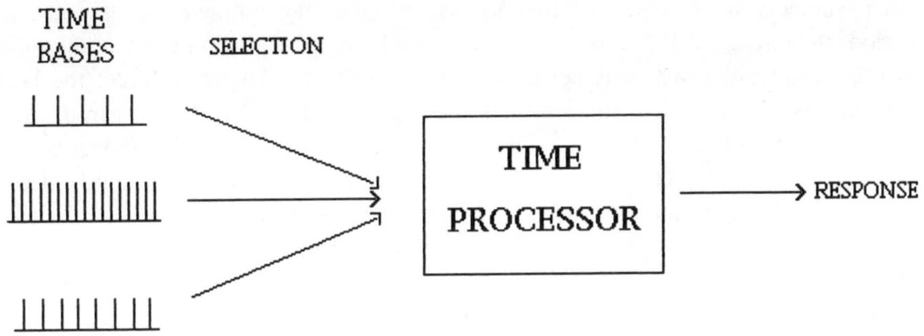

Figure 6.1. Does the internal chronometric system function by combining selected time bases proper to a particular situation?

origin from motor activity. When a response is performed, sensory cues deriving from muscles, tendons, and joints provide information about movement, muscle tension, and position in space. This proprioceptive feedback may be essential for subsequent responses. It has been reported that, in motor timing tasks, increasing the proprioceptive feedback derived from response execution increases spatial and temporal accuracy (Schmidt, 1971).

Feedback manipulations address parameters such as force and the velocity of movement; thus, if required to lift a weight for exactly 10 seconds, one would be more accurate with weights of 3 than of 2 kilos. This may be understood in the light of either the closed-loop or the chaining hypothesis, each coming from research on the motor control of skilled movements (Schmidt, 1980). The closed-loop view stipulates that sensory feedback resulting from movement execution is compared with an internal model of the intended movement; if there is no match, the next response is corrected. The chaining concept holds that, in a sequence of movements, sensory feedback derived from one movement triggers the next one, and so on until the end of the chain. While you are pressing a button for a specified duration, a chain composed of subtle changes in sensation of muscle tension might unfold. At the end of the chain, a specific pattern of proprioceptive cues would serve to signal response termination. This pattern would become more and more associated with the correct duration as training proceeds. Adams (1977) has suggested that the slow adapting receptors of the joints are crucial elements in the chaining mechanism. The variations of receptor discharges, as a function of time, would regulate the chronometry of successive movements in rapid motor sequences: the receptors fire in relation to the first movement; the discharge decreases; the second movement would occur as soon as a certain level of decay is reached.

The influence of force manipulations on timing accuracy, however, is not entirely convincing. Certain data suggest that other factors may be responsible for the observed trends. An increase in force is supposed to improve timing accuracy because it increases the rate of change in the proprioceptive discharges derived from response execution, thereby enhancing the quantity or quality of the cues used for timing (Schmidt, 1971). This hypothesis was not sustained by Weybrew's data (1963), which showed an inverse relation between force and subjective duration when subjects were requested first to grip a hand dynamometer for a constant period, then give a verbal estimate of that period length. With an objective duration of 43 seconds, the task was performed at five different levels of force; the best accuracy was obtained at the lowest force level. A study by Warm, Smith, and Caldwell (1967) under similar conditions compared four different durations: 6, 12, 24, and 48 seconds. There were no differences between 6 and 12 seconds at the five levels of force that were used. Equivocal results were obtained with the two longest delays, where a ''zero-load'' condition yielded a longer objective duration than the ''loaded'' ones. Thus, it may be that the specific level of force required is of less importance than whether or not there are force constraints.

Three other studies support this proposition. The timing task of Ellis, Schmidt, and Wade, (1968) consisted in subjects pushing a sliding carriage along a track in exactly 2 seconds. Weights were added to the carriage under certain conditions, so that the level of resistance varied. There was no effect of resistance on timing accuracy. In Parker's (1973) work, subjects moved a handle in 2 seconds with the right hand, while the left one gripped a dynamometer with various levels of tension. This procedure was meant for involving one hand for timing and the other hand for feedback manipulations. Tensions close to minimum or maximum strength on the dynamometer yielded less accurately timed movements with the right hand than did the tension fixed at 50 percent of maximal strength. Difficult as it was, the latter condition could have mobilized more attentional resources than the other ones; this view seems to account for the results better than an interpretation which focuses on force parameters. Finally, Macar and Vitton's (1989) data indicated that subjective duration did not differ according to whether one handled a dynamometer with a weak or a stronger level of force throughout a period of 5 seconds precisely determined by the subjects. The level of force required changed from trial to trial, but remained constant within trials. Such conditions, on the other hand, gave better timing accuracy than two other conditions during which force varied within trials, thereby increasing task complexity. This again suggests, as in Parker's task, that factors other than the force level were responsible for the results obtained. Variations in task complexity or in the level of attention plausibly influenced the timing performances.

More consistent is the effect of movement velocity upon timing accuracy. Within the set of experiments by Newell et al. (Newell, Hoshizaki, Carlton, and Halbert, 1979; Newell, Carlton, Carlton, and Halbert, 1980), movement times ranging from 1.0 to 1 second can be compared for distances ranging from .75 to 67.5 centimeters. A ballistic timing task was used: subjects moved a slide along a track. Timing error was found to decrease as the movement velocity increased. This result appeared even when identical movement times were compared, that is, when greater velocities were elicited at greater distances. We can assume that the variability in the durations of the muscular actions that compose the entire movement decreases as velocity increases, and that greater consistency implies more efficient proprioceptive cues (Schmidt, 1987). Thus, it appears that movement velocity can be manipulated better than force so as to attest the influence of feedback on timing performances.

Some problems remain, however: why should the cues be proprioceptive? First, there is no certainty that other sources of peripheral feedback do not play a role as well. In particular, visual feedback may intervene, except when special precautions are taken to prevent the subject from watching his or her hands (as in Ellis et al. [1968] and in Macar and Vitton's works [1989]). Second, the manipulation of force or velocity parameters designed to reveal feedback control may produce mechanical modifications such that timing is facilitated in one condition relative to the others. It was in awareness of this objection that Parker chose a paradigm that involved both hands, one for the timing task performed in unchanging conditions, and the other for feedback manipulations. Third, efferent processes are as plausibly involved as proprioceptive ones in all timing experiments. Passive movements can be used (Quesada and Schmidt, 1970) as an attempt to rule out this possibility; but still, subjects may resist against passive displacements.

These difficulties in the interpretation of the data have received serious consideration. After a period of intense debate between opposing views, a restricted conception of feedback control was proposed. Both central and peripheral mechanisms for the control of motor timing are accepted as possible, their relative recruitment depending upon the type of task involved. When movement time is long and accuracy constraints are strong, sensory feedback of proprioceptive origin does appear to be a major factor influencing response accuracy (Schmidt, 1980).

Perhaps the relative weight of central and peripheral mechanisms may also vary as a function of an individual's age and of the amount of training. Only a modicum of data yield further information on this point. The problem has seldom been investigated in children. However, two studies suggest that increasing proprioceptive and/or motor cues facilitates young children's timing accuracy.

Video games offer a good means of obtaining temporal regulations from children. For instance, if the displacement of a train on the screen is made dependent on a sustained button press, the task may consist in pressing the button long enough to ensure that the train reaches a certain place within a tunnel that eliminates visual cues. Children three to five years old, when required to regulate the duration of their responses in such a game, either by pressing a supple rubber bulb or by grasping a similar but rigid bulb (filled with plaster) for 5 to 10 seconds, tended to exhibit a higher level of performance in the first than in the second condition (Macar, 1988). Moreover, a larger proportion of subjects in the three-year-old group, for which the task was undoubtedly difficult, succeeded with the supple but not with the rigid bulb. It was hypothesized that the allometric press induced by the supple bulb was particularly favorable to timing because it involved movement-related cues. In a second study, five-year-old children had to produce durations of 5 to 6 seconds by pressing one of two compressible bulbs that differed in resistance (Macar and Grondin, 1988) so as to direct the activity of a boy on the screen. Greater accuracy was obtained with the low compared to the high resistance bulb (Figure 6.2). The more supple bulb, which could be easily compressed, induced finger movements of weak force but of relatively large amplitude, and several children pressed it rhythmically on certain trials. This was impossible with the higher resistance bulb, which could not be compressed more than one third, and

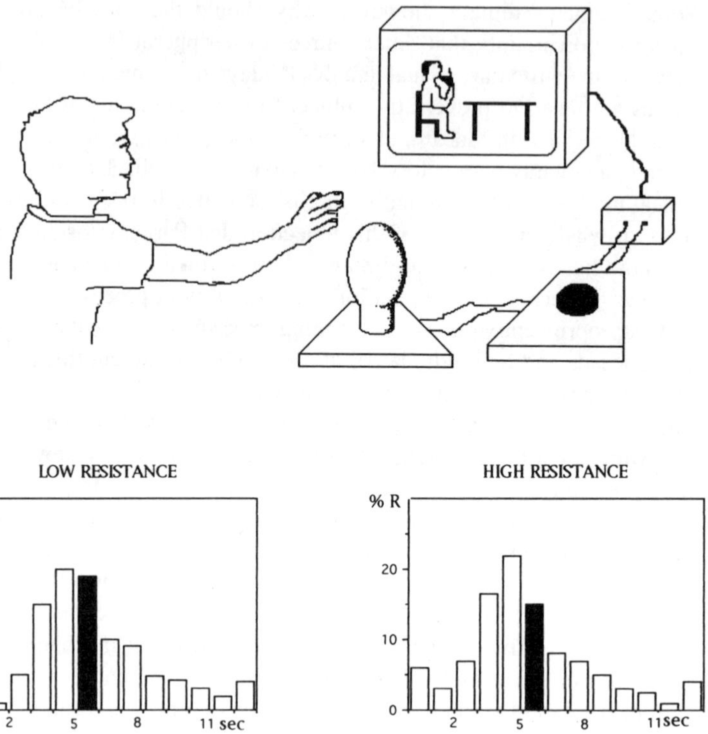

Figure 6.2. Distribution of response durations in a visual-motor game in which the child pressed a bulb of either low or high resistance (10 children in each section). Response percentages plotted along the ordinate, durations in 1-second increments from the start of the test along the abscissa. The solid columns correspond to criterion durations (5 to 6 seconds). The level of performance was significantly higher with the low-resistance bulb: note the difference in classes 4–5 and 5–6 seconds between conditions (data from Macar and Grondin [1988]).

thus was associated with less ample and more stereotyped movements. Once again, it appears that the manipulandum which enhanced action-derived cues facilitated timing.

In both visual–motor games proposed to the young subjects of those two studies, the bulb had to be gripped for at least 5 seconds, and then released without the help of an external chronometric signal. It is possible to speculate, along the lines of the chaining hypothesis for motor control, that the moment of release was triggered by a specific pattern of proprioceptive cues, which became more and more consistent with training, and was particularly detectable and efficacious as a result of numerous and well-differentiated movements. Another argument in favor of this conception is that the level of performance observed in those studies was higher than the one obtained from comparable age groups in globally analog timing tasks (Pouthas, 1981, 1985; Stein and Landis, 1978; Weisberg, 1970). This result seems to depend, among other factors, on the specific nature of the response involved. A sustained motor response concomitant with the critical delay that has to be estimated appears to facilitate timing, relative to a discrete motor response given at the end of the delay, as required in the other works. Apart from the fact that a sustained response provides a memory cue which can remind one that the delay has begun, and reduces the probability of distracting activities unrelated to the task, it may also provide indices of time elapsed as previously argued. Accordingly, the advantages of this type of response have been evidenced at several ages (see the longitudinal investigation of Matsuda and Matsuda [1983]).

Possibly such studies indicate that the amount of feedback control on timing accuracy is larger in children than in adults. This might be due to maturation as well as learning. During childhood, motor control mechanisms progressively develop and cognitive processes grow in complexity. Motor inhibition capacities, which appear essential to temporal regulations (Richelle and Lejeune, 1980), also mature gradually. Besides, typically, the amount of training on timing parameters is greater in adults than in young subjects. The following data stress the relevance of this factor. As mentioned previously, large differences were obtained between conditions of constant and variable force in adult subjects whose task was to grip the handle of a dynamometer for precisely 5 seconds (Macar and Vitton, 1989). However, such differences reduced dramatically with training, despite the fact that force, amplitude, and speed of movement changed from trial to trial (all conditions were mixed), the only unchanging parameter being the required duration. In addition, sequential dependencies were found from one trial to another: subjects tended to shorten or lengthen the actual duration when it was preceded, respectively, by a too long or a too short response, and the probability of a precise response increased when the preceding one was accurate. These sequential effects, thus, were a function of duration but were independent of the muscular changes imposed across trials. Such data suggest that the control of proprioceptive cues upon performance may be important when one begins to learn a new timing task, but diminishes in importance with training. Also, the eventuality that central mechanisms intervene with greater and greater efficacy is underlined. It is as if an internal temporal reference common to all experimental conditions were progressively elaborated, until it were finally to overcome the effects of feedback. Subjects seemed to rely on this internal model of the goal: to produce responses of 5 second duration whatever the actual conditions of motor execution.

At this point, the question is whether "duration" may really be abstracted from the other aspects of the situation so as to benefit from independent processing. Specifically, is it possible to argue that the duration of a response constitutes a dimension in motor processing that can be separately defined, to the same extent as force or amplitude?

This problem can be best investigated in the light of motor programming conceptions (Keele, 1981). The notion of the motor program was first applied to movements performed under strong time constraints, such as sequences of rapid acts that are elicited by visual–motor tracking tasks: trying to keep the index finger close to a fast-moving target on a screen with many direction changes, for instance. The program, which is completed before the index starts, is viewed as a set of

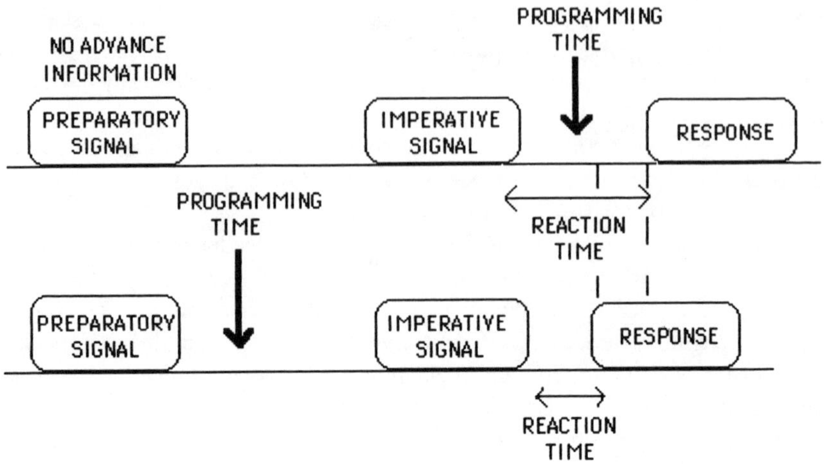

Figure 6.3. The relation between advance information, programming time, and reaction time, according to motor programming assumptions. Reaction time shortens when advance information allows programming to take place before the imperative signal.

abstract instructions regarding the various dimensions of the movement to be performed. That is, when information concerning one or another dimension is available before the moment of execution, this dimension is selectively specified in the program so as to be ready previous to action. The advantage of prior information is evidenced in those reaction time experiments which present subjects with a choice between different types of responses on each trial. An "imperative signal" indicates which response should be executed, as rapidly as possible. In a paradigm known as the "precueing technique," a "preparatory signal" delivers or not information on one or several dimensions of the response, depending on the trial. A dimension which is, in this way, unveiled in advance, is said to be "precued": one can specify it in the motor program, and thus gain time at the moment of movement execution. This gain in time is indexed by the shortened reaction times, while the movement itself remains unchanged. In summary, the movement is planned even before execution, on the basis of a central program in which its dimensions are specified, and the programming time is part of the reaction time that procedes movement onset. Reaction time is reduced as a result of precueing, because advance information transfers some or all programming operations to the period preceding the imperative signal (Figure 6.3). Interesting deductions regarding programming processes are drawn from a comparison of the reaction times obtained when none, one or several dimensions of the response are precued (Rosenbaum, 1983).

Studies on precueing have suggested that force, direction, and amplitude, as well as the muscular effector involved in a response, can be programmed (Rosenbaum, 1980; Bonnet and MacKay, 1989; Lepine, Glencross, and Requin, 1989). Duration, however, raises particular problems. Motor acts unfold in time; at first sight, it seems difficult to imagine that their duration may be processed prior to their occurrence. This eventuality appears contrary to the peripheral hypothesis, wherein movements are timed with the help of proprioceptive cues collected in the course of their execution. Another current argument against the possibility of programming time is that the duration of a movement is not processed per se, but rather derived from other parameters, such as the ratio of speed and amplitude.

The first indications that programming time is not a fanciful speculation came from an experiment involving two short response durations, .2 and .6 seconds, respectively (Klapp, Wyatt, MacLingo, 1974). Those durations corresponded to the "dit" and "dah" of Morse codes, produced by key presses. When only one of those response durations was associated with the imperative signal within a block of trials, subjects who were requested to execute it as fast as possible yielded similar average reaction times before "dit" and before "dah." On the other hand, when the task implied a choice between the two durations, since either one or the other could be cued by the signal, reaction times were longer for "dah" than "dit." Such reaction time differences, linked to the fact that the duration parameter was or was not known in advance, indicate the involvement of programming mechanisms along the lines reasoned above.

However, it has been proposed that such mechanisms might apply only to responses of very short duration, perhaps because the cessation of the movement, when it follows the onset quite rapidly, must be prepared prior to its execution. Also, one may wonder whether the timing of responses other than those consisting of pressing a key for a defined period also would require programming. In other words, the generality of timing mechanisms is questionable if they could not be evidenced with a larger range of durations and movements.

Recent data obtained in a precueing choice reaction time paradigm underline the relevance of time programming conceptions for durations of a few seconds. In a first study, subjects were provided with two lateralized small buttons, and emitted one of four possible responses on a trial: a button press with either the right or left hand, for either a short or a long duration. The effector (right or left hand) and response duration (.2 and .7 seconds within a block of trials, .7 and 2.5 seconds within another block) were precued or not by the preparatory signal, depending on the trial (Vidal, Bonnet, and Macar, 1991). Two seconds later, the imperative signal gave total information for the required response, which was executed as fast as possible. Programming effects on reaction time were seen in relation to duration as well as to effector parameters. The preliminary data from work in progress extend these conclusions to a range of three durations, of which 5.5 seconds was the longest one. In addition, they reveal identical trends with another type of response, consisting of the production of two successive brief presses separated by a delay; the duration of the delay had to be accurately estimated. Thus, duration can be programmed even in the case of an "empty" interval in motor sequence, not solely when it constitutes a dimension of the required movement. Programming time, definitely, does not appear to be a phenomenon limited to particular categories of responses.

Finally, a quantitative analysis of brain event-related potentials in the precueing paradigm indicated that local changes in cerebral activity are associated with time programming (Macar, Vidal, and Bonnet, 1990). Brain activity was recorded from various areas while subjects processed the information conveyed by a preparatory

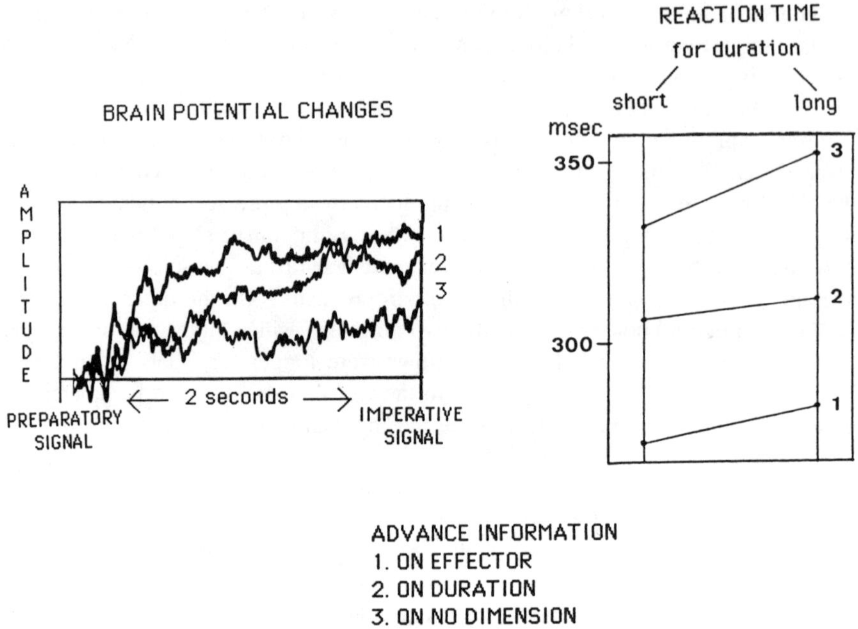

Figure 6.4. Brain potential changes (on the left) and reaction times (on the right) as a function of the type of information given by the preparatory signal. Larger potential changes and shorter reaction times were obtained when advance information was available; in addition, differences were seen according as information concerned effector or duration.

signal. In agreement with the reaction time data, greater potential changes were induced by conditions which yielded information on one or the other dimension of the response than by the condition which did not, and the observed amplitudes differed according to whether the advance information concerned effector or duration (Figure 6.4). In addition, distinct cortical sites were more or less activated as a function of the type of information processed. The primary motor cortex was most involved when the effector was precued, whereas the supplementary motor area, together with more parietal sites, were predominantly activated in relation to the precueing of duration. Still more specifically, local modifications also appeared depending on which of the two durations was cued. Although the interpretation of such results is not wholly clear, tentatively it can be suggested that either distinct mechanisms or distinct regions are involved in the programming of different durations.

The finding of possible localized brain activities in relation to the processing of time is undoubtedly surprising, and further research is needed before this eventuality can be accepted. Little evidence in the literature would point to such specific localizations. Lesions of temporal and occipital areas are associated with timing deficits, but the latter generally drown in a sea of diverse symptoms including memory-related and attentional troubles. Selective deficits regarding time estimation are thus difficult to assess (see Macar [1980], for review). The question of localization is perhaps of lesser interest, however, than the possible existence of specific timing mechanisms as suggested by the programming data. Other arguments in favor of such a specificity could be drawn previously from works with brain potentials. It has been shown that a particular slow potential change, the contingent negative variation, is larger when timing requirements constitute the major aspect of the task, and that its amplitude varies in relation to temporal accuracy (Macar, 1977; Ruchkin, McGalley, and Glaser, 1977; Macar and Vitton, 1980, 1982; Macar and Besson, 1985). Such data have stressed the existence of bioelectric indices associated with processes of time estimation "on-line." The programming data, in addition, indicate that even preparation for time estimation, prior to response execution, corresponds to specific mechanisms. The data also suggest that such mechanisms are not fundamentally different from the ones involved in the programming of other dimensions of the movement: similar types of processing should be at work, up to a certain point.

Obviously, the exact nature of the programming of time is still an open question. A possibility in line with the hypothetical chronometric system described above is that, as soon as the information regarding duration is delivered, the proper time bases are selected from the potential stock, or even constructed piece by piece, such that they are ready for use at response onset. The chronometric system may consist of neural networks which would group various sources of impulses and combine rhythmic events of multiple distinct frequencies. Klapp (1977) has imagined that longer neural networks might be elaborated for measuring long compared to short durations. He deduced this idea from the short–long effect that he has found repeatedly: reaction time is longer before a response of long than of short duration in conditions where reaction time includes programming time. Thus, it would take more time to program a long than a shorter duration. Such an observation has been confirmed by others (Vidal, Bonnett, and Macar, 1991; Zelaznik, Shapiro, and Carter, 1982). However, a particular duration may be considered short or long depending on the position it occupies within the range of durations concerned; to be precise, our data indicate that the short–long effect is based upon relative rather than absolute durations. When a range of only two durations is used for eliciting programming processes, it is as if one of those durations, the shorter one, was taken as a reference for processing, and thus necessitated a minimum programming time, in comparison with the longer duration which would require additional programming operations. Such operations would consist, for instance, in adding a new component to the reference network, or in multiplying it by a constant. Whatever they might be, they would take an additional amount of time, and, consequently, would lengthen the reaction time.

Such an interpretation is admittedly speculative. No neural networks implicated in time processing have yet been discovered. It is promising that models of neural networks capable of recognizing, producing, and memorizing temporal sequences elicit much interest nowadays. Certain models have been derived from research on the acquisition of song by birds. Simulated networks that spontaneously produce "prerepresentations" of temporal sequences have been proposed (Dehaene,

Changeux, and Nadal, 1987). Despite their originally random character, those prerepresentations are sensitive to finer and finer selection as a result of their possible matching with external constraints. Learning occurs when produced and perceived sequences present a high level of resonance. The synaptic connections which stabilize the system are selected, whereas those which perturb its equilibrium are excluded; in this way, the system becomes progressively strengthened. Such propositions can help in understanding how chronometric mechanisms could, step by step, acquire consistency.

In conclusion, the role of central mechanisms in timing processes is worth emphasizing. It is attested to by data collected in various fields: motor programming, consequences of learning upon the production of durations in changing muscular conditions, and effects on subjective duration depending on whether attention is directed to temporal as compared to nontemporal information. However, its importance does not preclude a function for proprioceptive cues in temporal regulation. The central and the peripheral hypotheses should be considered as complementary rather than exclusive: it is likely that both mechanisms exist, and that their relative contributions to timing depend on duration, type of response, accuracy constraints, and other such factors. Further research is needed to determine the exact conditions under which each type of mechanism is recruited.

References

Adams, J. A. (1977), Feedback theory of how joint receptors regulate the timing and position of a limb. *Psychol. Rev.*, 84:504–523.

Arlin, M. (1986), Effects of quantity and depth of processing on children's time perception. *Experiment. Child Psychol.*, 42:84–98.

Bonnet, M., & MacKay, W. A. (1989), Changes in contingent negative variation and reaction time related to precueing of direction and force of a forearm movement. *Brain, Behav. & Evol.*, 33:147–152.

Creelman, C. D. (1962), Human discrimination of auditory duration. *J. Acoustic Soc. Amer.*, 34:582–593.

Dehaene, S., Changeux, J.-P., & Nadal, J.-P. (1987), Neural networks that learn temporal sequences by selection. *Proc. Nat. Acad. Sciences, U.S.A.*, 84:2727–2731.

Ellis, M. J., Schmidt, R. A., & Wade, M. G. (1968), Proprioception variables as determinants of lapsed time estimation. *Ergonom.*, 11:577–586.

Friedman, W. J. (1990), *About Time: The Construction of Temporal Experience*. Cambridge, MA: MIT Press.

Gallistel, C. R. (1989), Animal cognition: The representation of space, time and number. *Ann. Rev. Psychol.*, 40:155–189.

Gill, F. B. (1988), Trapline foraging by hermit hummingbirds: Competition for an undefended renewable resource. *Ecology*, 69:1933–1942.

Hicks, R. E., Miller, G. W., Gaes, G., & Bierman, K. (1977), Concurrent processing demands and the experience of time-in-passing. *Amer. J. Psychol.*, 90:431–446.

——— ——— Kinsbourne, M. (1976), Prospective and retrospective judgements of time as a function of amount of information processed. *Amer. J. Psychol.*, 89:719–730.

Keele, S. W. (1981), Behavioral analysis of movement. In: *Handbook of Physiology, Motor Control*, ed. V. Brooks. Bethesda, MD: American Physiological Society, pp. 1391–1414.

Klapp, S. T. (1977), Response programming, as assessed by reaction time, does not establish commands for particular muscles. *J. Motor Behav.*, 9:301–312.

——— Wyatt, E. P., & MacLingo, W. (1974), Response programming in simple and choice reactions. *J. Motor Behav.*, 6:263–271.

Lépine, D., Glencross, D., & Requin, J. (1989), Some experimental evidence for and against a parametric conception of movement programming. *J. Experiment. Psychol.: Hum. Percept. Perform.*, 15:347–362.

Macar, F. (1977), Signification des variations contingentes négatives dans la dimension temporelle du comportement. *Année Psycholog.*, 77:439–474.

——— (1980), *Le Temps—Perspectives Psychophysiologiques.* Bruxelles: Mardaga.

——— (1988), Temporal regulations in children of 3 to 5 years old. *Cahiers de Psychologie Cog.*, 8:39–51.

——— (submitted), Subjective duration in dual task depending on whether attention is focused to temporal or nontemporal parameters.

——— Besson, M. (1985), CNV in processes of expectancy, preparation and time estimation. *Biolog. Psychol.*, 21:293–307.

——— Grondin, S. (1988), Temporal regulation as a function of muscular parameters in 5-year-old children. *J. Experiment. Child Psychol.*, 45:159–174.

——— Vidal, F., & Bonnet, M. (1990), Laplacian derivations of CNV in time programming. In: *Psychophys. Brain Res.*, 1:69–76. ed. C. H. Brunia, A. W. Gaillard, & A. Kok. Tilburg, Netherlands: Tilburg University Press.

——— Vitton, N. (1980), CNV and reaction time task in man: Effects of interstimulus interval contingencies. *Neuropsychol.*, 18:585–590.

——— ——— (1982), An early resolution of CNV in time discrimination. *EEG & Clin. Neurophysiol.*, 54:426–435.

——— ——— (1989), Effect of learning on production of duration in variable motor conditions. *Acta Psycholog.*, 72:247–261.

Matsuda, F., & Matsuda, M. (1983), A longitudinal study of learning process of duration estimation in young children. *Japan. Psycholog. Res.*, 25:119–129.

McClain, L. (1983), Interval estimation: Effect of processing demands on prospective and retrospective reports. *Percept. & Psychophysics*, 34:185–189.

Michon, J. A. (1972), Processing of temporal information and the cognitive theory of time experience. In: *The Study of Time*, ed. J. T. Fraser, F. C. Haber, & G. H. Müller, Vol. 1, Berlin: Springer Verlag, pp. 242–258.

Newell, K. M., Carlton, L. G., Carlton, M. J., & Halbert, J. A. (1980), Velocity as a factor in movement timing accuracy. *J. Motor Behav.*, 12:47–56.

——— Hoshizaki, L. E., Carlton, M. J., & Halbert, J. A. (1979), Movement time and velocity as determinants of movement timing accuracy. *J. Motor Behav.* 11:49–58.

Ornstein, R. E. (1969), *On the Experience of Time*. Harmondsworth, U.K.: Penguin Books.

Parker, N. K. (1973), Influence of induced-muscle tension on a time estimation motor task. *J. Motor Behav.*, 5:111–120.

Pouthas, V. (1981), Adaptation à la durée chez l'enfant de 2 à 5 ans. *L'Année Psychologique*, 81:33–50.

——— (1985), Timing behaviour in young children: A developmental approach to conditioned spaced responding. In: *Time, Mind and Behavior*, ed. J. Michon & J. Jackson. New York: Springer, pp. 100–109.

Quesada, D. C., & Schmidt, R. A. (1970), A test of the Adams-Creamer decay hypothesis for the timing of motor responses. *J. Motor Behav.*, 2:273–283.

Richelle, M. (1968), Notions modernes de rhythmes biologiques et régulations temporelles acquises. In: *Cycles biologiques et psychiatrie*, ed. J. de Ajuriaguerra. Paris: Masson, pp. 233–255.

——— Lejeune, H. eds. (1980), *Time in Animal Behaviour*. Oxford: Pergamon Press.

Rosenbaum, D. A. (1980), Human movement initiation: Specification of arm, direction and extent. *J. Experiment. Psychol.: General*, 109:444–474.

——— (1983), The movement precuing technique: Assumptions, applications and extensions. In: *Memory and Control in Motor Behavior*, ed. R. A. Magill. Amsterdam: North Holland.

Ruchkin, D. S., McGalley, M. G., & Glaser, E. M. (1977), Event-related potentials and time estimation. *Psychophysiol.*, 14:451–455.

Schmidt, R. A. (1971), Proprioception and the timing of motor responses. *Psycholog. Bull.*, 76:383–393.

——— (1980), Past and future issues in motor programming. *Res. Quart. for Exercise & Sport*, 51:122–140.

——— (1987), *Motor Control and Learning. A Behavior Emphasis*, 2nd ed. Champaign, IL: Human Kinetics.

Stein, N., & Landis, R. (1978), Effects of age and collateral behavior on temporally discriminated performance of children. *Percept. & Motor Skills*, 47:87–94.

Thomas, E. A., & Weaver, W. B. (1975), Cognitive processing and time perception. *Percept. & Psychophysics*, 17:363–367.

Triesman, M. (1963), Temporal Discrimination and the Indifference Interval: Implications for a Model of the Internal Clock. *Psychol. Monogr.*, No. 576. Washington, DC: American Psychological Association.

Vidal, F., Bonnet, M., & Macar, F. (1991), Programming response duration in a precueing reaction time paradigm. *J. Motor Behav.*, 23:15–26.

Warm, J. S., Smith, R. P., & Caldwell, L. S. (1967), Effects of induced muscle tension on judgment of time. *Percept. & Motor Skills*, 25:153–160.

Weisberg, P. (1970), Effects of reinforcement history on timing (DRL) performance in young children. *J. Experiment. Child Psychol.*, 9:348–362.

Weybrew, B. B. (1963), Accuracy of time estimation and muscular tension. *Percept. & Motor Skills*, 17:118.

Zakay, D. (1989), Subjective time and attentional resource allocation: An integrated model of time estimation. In: *Psychological Time: A Lifespan Perspective*, ed. I. Levin & D. Zakay. Amsterdam: North-Holland, pp. 365–395.

——— Fallach, E. (1984), Immediate and remote time estimation—A comparison. *Acta Psycholog.*, 57:69–81.

——— Nitzan, D., & Glicksohn, J. (1983), The influence of task difficulty and external tempo on subjective time estimation. *Percept. & Psychophysics*, 34:451–456.

Zelaznik, H. N., Shapiro, D. C., & Carter, M. C. (1982), The specification of digit and duration during motor programming: A new method of precueing. *J. Motor Behav.*, 14:57–68.

Perspectives

"What in music is not *process?" asked one reviewer. The following set of three papers on process in music will reinforce his point, demonstrating that one can approach the general issue of musical process from several viewpoints. Hasty focuses on a single, vital aspect of the musical process—the projection of duration by means of reciprocal conditioning between prior and subsequent musical events. For Brodhead, process is dual—the formation of deep, hierarchic structures as a result of the integration of opposites, the oppositions between linear sequence (time's arrow) and sequential recurrence (time's cycle). For Rowell, process in music is multiple, and his typical response (I know him well) is to develop an elaborate taxonomy of categories of processes. But these descriptions will be inadequate to convey to nonspecialist readers what aspects of music are addressed in each of the articles. Hasty is concerned with such things as musical duration, timing, simultaneity, beginnings and endings, and the qualities of rhythm; Brodhead, with apparent linearity, the importance of recurrence in music, hierarchy, and the temporal metaphors with which we capture our experience of music; Rowell, with the values and meanings of musical process, the temporal scale of music, and with composite musical structures and strategies.*

The reader will find several resonances among this set of papers: all three are specifically addressed to nonspecialist readers and pointedly eschew technical jargon; all three share a vital concern for musical cognition—how one hears process in music and what meanings are assigned thereto; and all three, for practical reasons, focus primarily upon familiar works from the traditional Western concert repertoire. Most important of all, all three articles challenge readers to test these explanations through their own listening experience. Passive listening, the kind of listening most of us do most of the time, can provide—at best—a subliminal sensation that some vague sort of process is at work; active listening, the kind that each of us ought to do at least some of time, will convince most listeners that music may well be defined as "the art of audible process."

L.R.

7

Music as Process*

Lewis Rowell

Abstract In music, the art of audible forms in motion, we hear a number of identifiable processes in several dimensions, including (1) the process by which a musical work comes into being, that is, composition or improvisation, (2) the internal process manifested in an individual work or genre—repetition, variation, interaction, and the like, (3) the process by which a work becomes actualized, that is, performance, and, conceivably, (4) we may regard the entire idea and experience of music as a process that encompasses all of the preceding. The present article addresses the second of these issues, *internal process*, by developing, analyzing, and illustrating a taxonomy of musical processes: repetitive, elaborative, commentarial, imitative, emergent, interactive, figurative, rhetorical, analogical, disjunctive, syntactical, and parallel. I argue that musical process embodies significant human intuitions of process in general and is instructive in demonstrating how the human mind is wont to function when imagining some type of sustained creative activity based on a limited set of materials. Musical processes are further instructive in that they are scaled to units of human temporal experience, controllable, replicable, and—most important of all—that they succeed in projecting graspable personas with which we can identify and respond with appropriate affect, drawing by subconscious analogies upon our deepest ideas of self and context.

Process is a word that makes a musician's heart leap up—particularly those of us who take a professional interest in explaining how music works. For music must be the quintessential art of process, an art in which we construct and perceive audible forms in motion. Indeed, if music is anything, it is a process: the process of relating sounds to one another in a manner that brings, or is intended to bring, pleasure.

Why then do we have such a poor set of tools with which to describe process in music? One reason may be that we in the West have traditionally regarded musical works more as objects than as processes, and hence our state descriptions have generally been more successful than our process descriptions. If this reason is correct, it is

*Presidential Address, July 10, 1989.

entirely consistent with the well-known human tendency to convert time into space, and to regard the chains of events that pass through our experiences as structures that can be frozen and preserved for our inspection, diagrammed, graphed, and reified in any number of other ways—perhaps in an unconscious echo of Faust's plea to the passing moment: "Verweile doch, du bist so schön!"

Process is a beastly word to define, a conclusion that I suspect many of my fellow authors will endorse. We all have a reasonably good idea what we mean by the word, and it is not hard to cite obvious examples, and yet a precise definition remains elusive. (It is worth recalling that Augustine had a similar problem with the idea of time [*Confessions,* 11.14].) The 1966 unabridged edition of the *Random House Dictionary of the English Language* sets forth more than fifteen definitions of process—eight as a noun and seven as a verb, plus a number of adjectival meanings—which perhaps the best is the following: "a series of progressive and interdependent steps by which an end is attained" (p. 1147). Despite (or, rather, because of) its level of abstraction, this definition is an excellent description of what happens in most musical compositions, and also in large numbers of other things. But are all processes "progressive . . . interdependent . . . teleological"? Clearly not. It may not be possible to formulate a definition that makes sense in the case of music, but it is at least worth a try.

With respect to music, the idea of process may be sought in at least four dimensions, which are surely related to each other but which display certain unique features: first, the process by which a musical work comes into being, that is, composition or improvisation; second, the internal process manifested in a musical work, a process that may occupy the whole or only some part thereof (see below for further clarification); third, the process by which a work becomes actualized, that is, performance; and fourth, I must consider the possibility that music itself, in all its manifestations, may be regarded as a process, although such a view requires us to do no less than to explain all of music—a task that I shall not presume to attempt.

This article addresses the second of these four aspects of musical process: the *internal process* by which a musical work is organized, that which can be taken away and yet that which we call "music" remains. If this sounds impossible, readers may find it convenient to think of "music" as process (singular) and the various categories I shall present in the following pages as processes (plural). Throughout the discussion I shall attempt to state the problem in language that avoids the technical jargon of music theory.

I am not suggesting that musicians have entirely ignored process, only that there has been a certain reluctance to focus upon it. One notable exception in recent years has been Leonard B. Meyer, much of whose work has been directed toward implicative processes in music and both the technical means and meanings of their realizations (see, in particular, Meyer [1956, 1973, 1982]). But if theorists of music have not concentrated upon process, what have we done? We have constructed appropriate models, designs, and schemata; and we have become skillful in segmenting a musical work, isolating and naming its components, identifying the relationships among its parts, dividing it into hierarchical levels—seemingly everything, in fact, except the vital question of how it works. It is ironic that music is often regarded as an intricate clock, whose motions are savored but whose mechanism is usually inspected when at rest. Even Heinrich Schenker's method of analysis (see below, p. 134), arguably the most successful of recent approaches to the theory and analysis of tonal music, seems to me to focus on only one limited aspect of musical process—the tonal process (i.e., the pitch construction) by which an individual work actualizes one of a small set of models (Forte, 1959; Bent, 1987, pp. 38–41, 49–52, 81–85; Cook, 1987, pp. 27–66; Dunsby and Whittall, 1988, pp. 23–61).

To study process in music goes against much of our training and habits: it requires us to step back and let our hearing and vision blur; it requires us to listen at a level of generalization that we often find uncomfortable. It is true, as arms negotiators are fond of saying, that "the devil is in the details," but, vital as they may be, details have a way of inducing myopia (the music theorist's occupational hazard, according to Kerman [1985, p. 73]).[1] The traditional approach to the great issues of pitch- and time-organization in music has indeed been to begin with the details, recapitulating our first childhood steps in learning what music is all about. As a result, if we can make statements about music that are both true and detailed, these will automatically become the preferred statements. In abandoning our obsession with the microscopic in music, we may also have to abandon some of our technical jargon, a jargon that often conceals trade secrets from the uninitiated. It is a subversive thought that we may, perhaps, have to resort to describing music in language that we use for the rest of the arts, and to realize thereby that what we call "music" may be only a subset of a larger entity that we might label as "artistic process." But, whatever the reasons, we are not in the habit of thinking about music in terms of process. Let us try.

What can be learned from such an inquiry into process in music? I am arguing that this particular subset of artistic process can tell us much about our attitudes toward process in general. what kinds of processes human beings like to make, participate in, and take pleasure from; and, in an even larger sense, how the human mind is wont to function when imagining some type of sustained creative activity, based on a limited set of materials. (I offer the preceding as a "minimalist" definition of art.) Artistic processes are man-made processes, working models that reveal intuitions of the world of process that we inhabit. Plato once wrote that "when the modes of music change, the fundamental laws of the state change along with them" (*Republic,* 4.424c). In a recent article (Rowell, 1987) I attempted to interpret this provocative assertion by reformulating Plato's thesis, with the cause-and-effect reversed: "When fundamental human intuitions of their world begin to change, as is now happening with respect to their concept of time, then their music must inevitably begin to reflect that change in one way or another" (p. 194). If this is true, then the processes embedded in musical works may be read as evidence of changing attitudes toward, and/or new understandings of, our world.

Musical processes are useful to investigate for a number of other reasons: so many of the physical processes of the universe are on either a macro- or a micro-scale

[1]Lest historical musicologists be tempted to agree, let it be remembered that on the same page Kerman identifies superficiality as the characteristic failure of musicologists.

that effectively removes them from our direct perception, and we require complicated means to get at them. The temporal scale of a process must be among its most vital properties, and, at the same time, one of its most concealing features. But not so with music! Music, like most artistic processes (with the possible exception of architecture), is constructed to scale with human beings, and for the most logical of reasons. We are able to savor its smallest details, and yet see it through to completion—usually within the span of thirty minutes or less, occasionally to the duration of an entire evening. And, further, the musical components are usually "chunked" in a way that permits and invites communication within the span of the units of our attention, short- and long-term memory, and the other thresholds of human perception. And why should this not be so? When humans make, it seems normal that they should do so in a manner that takes advantage of the normal scaling of their temporal experience. In sum, to infer the existence of a process (such as the evolution of a species or the formation of a canyon) is one thing, to *perceive* it is another.

Musical process is also a tightly controlled and replicable process: it can be started and stopped on demand, and it can be repeated on demand—although concert audiences are not usually given the choice.[2] Music embodies these properties to such an extent that its trademark beginnings, endings, and repetitions invade every dimension and stratum of musical experience and have become standard features of the musical language. Music is *always* beginning, ending, and repeating (for further discussion see Kramer [1988, pp. 137–169]; Rowell [1981]). There is a moral to be drawn: We cannot control the universe, nor seemingly even our own small corner thereof, but we *can* control our arts, and they provide thereby that evidence for regular continuity that we find so viscerally satisfying—even when we know better! The experience of musical process brings reassurance that a part of our world is knowable, predictable, retrievable, and controllable; and therein may lie much of music's appeal.

Musical process is also, as I shall argue in the concluding section of this article, a profound analogy. It arises from an individual human being with a strong sense of personal identity, and functions by projecting a virtual identity against a background of musical activity. This is the core of my idea of musical process, and a major strand in my argument is that the aesthetics of musical process center around the recognition of musical personas or identities (taking the form of individual tones, keys, themes, "gestures," instruments, and semes of meaning), coupled with a sense of their context and destiny—the process in which these identities are caught up, transformed, and brought to completion. We are led subconsciously to identify with these personas, suffer their fate, and respond with appropriate affect. It will be clear here that I am endorsing Susanne K. Langer's oft-quoted statement that "music is a tonal analogue of emotive life" (1953, p. 27). The foregoing is in no way a new argument. It has been eloquently expounded by Edward T. Cone (1974), and I shall return to his contributions in a later section.

[2]With the exception of some recent experimental compositions in which audience voting determines the actual course of the piece. It is also interesting that a recent Broadway production based on Charles Dickens's unfinished mystery novel *Edwin Drood* was prepared with multiple endings, one of which was chosen for performance by audience vote during the final intermission.

What has been said so far? That musical process is scaled to human size; that it is reassuringly controllable and replicable; that some of the deepest human values are embedded therein and serve as standards for our judgments of valuation; and that we receive from it that sense of identity-caught-up-in-process with which we tend to characterize our own experience. What else can be said? Let us go beyond music.

Any artistic process is embedded in a medium: it is audible and/or visible; it is concrete. Not only limitations of time but limitations of space prevent us from perceiving the totality of many other processes—especially those crucial ones that may threaten our very existence. But in a musical work, a poem, or a painting, we are presented with a graspable whole. Everything is there for our inspection and savoring. It may be related to other, less knowable parts of our world—to other works, other people, other times, other places—but it is essentially complete in itself, and thus permits us the extraordinary satisfaction of experiencing completeness.

Process in Music

Just what is a musical process? Two simple examples may help to clarify. Most musicians and listeners will readily agree that the following four measures, from one of W. A. Mozart's rondos for piano (K. 494), illustrate the idea of process—perhaps several processes—and that this short excerpt may be accurately described as "a series of progressive and interdependent steps by which an end is attained." To elaborate: *a series,* in that the music falls naturally into repetitive patterns (both to the ear and to the eye); *progressive,* in the obvious sense of moving forward in time, but also in the sense of descent (as the following analyses will demonstrate) and approaching a tonal goal (*f,* the keynote); *interdependent,* not only in the equally obvious sense that the various notes are linked in meaningful, consonant relationships with one another, but also in the more technical sense that (1) ascent in one line is countered by descent in another, and (2) stasis in one part is countered by activity in another[3]; and *an end is attained,* in that the entire pitch construction of the four measures departs from, and unmistakably signals a return to, the chord built on the keynote of the scale, *f*.

Two simplifications may help to point out the repetitive, progressive, interdependent, and teleological properties of Mozart's phrase. The first of these is a

[3]The foregoing is also a good general description of *counterpoint*.

harmonic reduction that eliminates the ornamental notes and reveals the underlying chords; the second is a graphic representation of only the part for the left hand, in effect ''straightening out'' its up-and-down contour and thereby revealing that it is a complete cycle through all the basic notes of the scale by taking each fifth note.[4] The original excerpt is obviously much more musically satisfying than either of the simplifications, and the notes I have eliminated play just as important a role in the processive qualities of the music as do the underlying tonal relationships revealed in the two reductions.

1

2

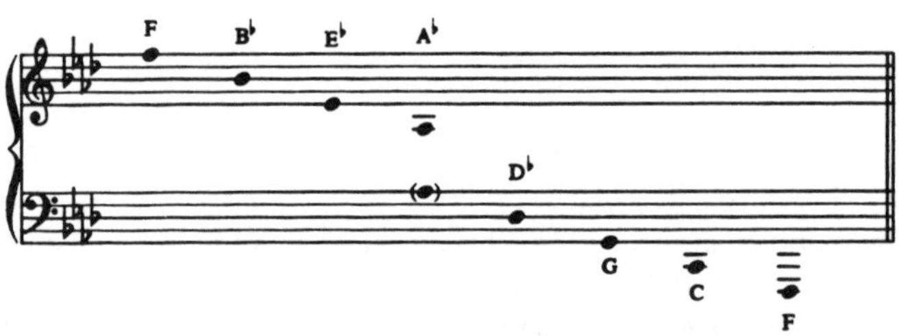

A few bars of ''Chopsticks,'' the archetypal children's game at the piano, will remind readers in a more homely way of the elements of musical process:

A learned musician—especially one who lacks a sense of humor—might provide the following technical account of ''Chopsticks'' and its traditional accompaniment (by a second player): an isorhythmic process of alternately diverging and converging consonant intervals in C major, with interludes of parallel diatonic thirds, supported by a chord progression of alternating dominant and tonic triads, in triple meter! A child's instructions might go like this: First find middle *c* and then find the same key (that is, in the pattern of white and black keys) to the right; use only the white keys; move your two index fingers together until they are as close as they can be without playing the same key; now play these two notes in waltz time (six notes to each group), choosing only pairs of keys that sound good together, and gradually moving your two fingers farther and farther apart until they both get to *c;* then reverse the process until you reach your original two notes, and

[4]That is, if the reader imagines a continuous downward series of notes, taking only those indicated by the key signature of four flats (*f, e* flat, *d* flat, *c, b* flat, *a* flat, *g,* and *f*).

start over again![5] But no matter which set of instructions one follows, ''Chopsticks'' is clearly a process by any definition.

In the Mozart excerpt we have an example of what I shall subsequently refer to as ''local process''; while the same progression could easily be expanded and elaborated so as to occupy a much longer span of time, it would be absurd to think of it as a gestalt, a complete composition. ''Chopsticks,'' on the other hand, *is* a complete piece, albeit on a very small scale; one can think of it as a single process, although two distinct subprocesses or local processes can be identified in its two sections.

Musical Processes

With the foregoing as an introduction, let us proceed to investigate the various types of musical process. The purpose of this section is to propose and explain a set of categories (Table 7.1), with a highly selective set of recommendations for listening. I am hoping to provide some new thinking on the subject of music as process, but the novelty, if any, will lie more in the overall scheme than in what I have to say about any of the individual processes. The aim is to suggest a framework for classifying musical processes, and most of my suggestions and illustrations will (indeed, *should*) be immediately obvious to my musical colleagues. I have deliberately approached the subject in a naive way, because I believe that a certain degree of innocence is necessary if we are to appreciate the global role that processes play in music.

The display in Table 7.1 is an attempt to group the standard forms, models, and yes, processes of music in a new way. All of the following depend upon the way in which the components of a musical work are disposed in time; and, in a similar manner, also depend on the essential relation between musical ''identities,'' their destiny, and their context—bodies, motions, and background. There are admittedly problems in this line of thinking, among them (1) that musical identities (melodies, instrumental roles, and the like) are revealed to us in motion, and it is therefore impossible to separate the ''bodies'' from the ''motions,'' the nouns from the verbs, any more than we can separate the dancer from the dance. And (2) that various musical processes exist on many levels and interpenetrate one another, so that a controlling master process may be a composite of, or even be contradicted by, many local processes that cannot be separated from the overall framework. This sounds more complex than it really is, and I suspect it is among the things that we relish most in music. Process is a messy business.

In the remainder of this article I shall distinguish between two terms when such a distinction is called for: *process,* with reference to a musical whole of any size and scale—a complete composition, a self-sufficient movement (of a sonata,

[5]This describes only the first and the last parts of ''Chopsticks''; the middle section has been omitted for the sake of brevity.

Table 7.1
Categories of Musical Process

Type of Process	As in
1. Repetitive	Strophic forms, primitive dance music, ostinato and other isorhythmic devices, cyclical rhythm, various manifestations of stasis in music
2. Elaborative	Themes with variations (i.e., successive elaboration), heterophonic textures (i.e., simultaneous elaboration)
3. Commentarial	Compositions on a *cantus firmus* (i.e., simultaneous commentary), troping (interpolation as a commentary), parody and paraphrasing (posthumous commentary), quotation and collage (allusion)
4. Imitative	Contrapuntal forms and genres such as canon, fugue, motet, and the like
5. Emergent	Music as "becoming," especially as found in the free/strict archetype in Asia and the West (Rowell, 1981)
6. Interactive	Musical interaction between solo/group (e.g., the primitive call/response model, the concerto), music with mingled subprocesses or combined themes, theatrical "happenings," and music constructed upon various models of social interaction
7. Figurative	Music as pure, abstract patterning
8. Rhetorical	Musical organization following the classical oratorical models, formulaic musical constructions
9. Analogical	Narrative and descriptive music, texted compositions, dramatic accompaniment (melodrama and opera), music forged from leitmotifs, music constructed on any model of process (e.g., organic, inorganic, physical, mental, emotional)
10. Disjunctive	Mosaic forms such as rondo and the so-called part forms, reprise forms, moment forms (Kramer, 1988), and other musical forms with prominent interruptive procedures
11. Syntactical	Music in which cross-references are drawn between separated sections and events (as in sonata form, the standard "first-movement" form of the eighteenth century)
12. Parallel	Music constructed in independent streams of activity (each of which may display a different process)

quartet, or symphony), or an individual number in a long series (as in a suite or a cycle of songs); in contrast, *local process* will be used to refer to any identifiable process within one of these gestalts. Indeed, all the processes listed in Table 7.1 appear frequently, pervasively, at the local level, and the number of types of local process must be much larger than the number of processes that are capable of sustaining an entire composition. They divide, in fact, into a virtually infinite array of subcategories, and the greatest objection I can imagine to the present line of thinking is that it fails to address the much more challenging issue of local process in music. As we continue, however, I am hopeful that the categories proposed above may prove useful in expanding and helping to organize our thinking about the nature of local process.

Some other caveats and explanations are in order. Music, in all but the simplest and single-minded types of compositions, is by nature resistant to organization by a single process, and many of the categories are prone to hybridize successfully with one another. As examples, many commentarial processes are also elaborative (as in grounds, passacaglias, chaconnes, and other *cantus-firmus* forms), and interactive processes can mingle effectively with almost any of the others. Therefore, it would be an error to think of these processes as mutually exclusive. And further, any directional process (as, for example, elaborative and emergent processes) offers the possibility that it may be reversed, although statistics are clearly on the side of the direction indicated in Table 7.1.[6]

The order in which the twelve categories have been listed in the table resembles, in a rough sort of way, the order of their appearance in world music: the first six are all present in primitive and ancient art music, and in the traditional musics of many non-Western civilizations; the second group of six includes more sophisticated, recent approaches to musical organization in the European tradition of art music since 1600, perhaps the most important watershed in the history of Western music before the twentieth century. Still other processes are developing today, and I do not foresee a point at which the possibilities will have been exhausted.

Some of the categories are more overtly processive than others. Here I am thinking especially of repetitive, elaborative, and imitative processes: repetitive, as in much recent minimalist music; elaborative, as in themes with variations; and imitative, as in fugues and the other forms listed in this category in the table. In these approaches to musical organization the processive element dominates, often to the virtual obliteration of the overall sense of formal design. To use a cliché, they are more process than product, more action than object.

I do not regard all of the categories as equally important. Figurative process, for example, merits recognition as a separate and independent process, but more

[6]That is, most themes with variations proceed in the direction from simple to complex, and, similarly, the general trend in music is toward greater, not lesser, organization, with emergent processes preferred to recessive ones. Two well-known exceptions are (a) the *Istar* symphonic variations by Vincent d'Indy, and (b) the poignant, thematic disintegration at the conclusion of the funeral march from Beethoven's *Eroica* Symphony.

often appears at the local level.[7] Rhetorical processes were among the major formal models during the seventeenth and eighteenth centuries and may, in a somewhat broader sense, influence the organization of music in many non-Western traditions, but I would not suggest that they are as widespread as most of the others. Parallel processes are a fairly recent development, although simultaneous local processes that involve different musical dimensions are musical commonplaces.[8] The nine that remain, however, must be recognized as major categories that are not restricted by such things as time, place, ethnic group, or style.

Musicians will notice some conspicuous exclusions from the list, and explanations for three of these may help to clarify the general issue of process in music. I have not mentioned *tonality,* the network of relationships and probabilities by which the priority and finality of one pitch is established in a musical work.[9] I find it difficult to conceive of tonality as a process in itself: composers of tonal music can choose from many local processes in delineating their particular version of tonality. In the writings of the Austrian music theorist Heinrich Schenker (1867–1935), it seems clear that he thought of tonality more as a model (of musical structure) than a process per se, but the manner in which a particular musical work implements this model is clearly a process—a complex process that is part elaborative, part analogical, part commentarial, and perhaps others included in my list: elaborative, in that the surface of a musical work includes many decorative elements not present in the underlying fundamental structure; analogical, in that the compositional unfolding of the work may be heard as a metaphorical demonstration of the biological urges that Schenker believed to be embodied in musical tones; and commentarial, perhaps, in that the various structural levels (e.g., the foreground and middleground) may contain motivic references to each other.

Nor have I mentioned implicative processes, except for a brief reference to the work of Meyer. Most music is implicative by nature, most if not all of the time, with both short- and long-range implications. Implication followed by realization is surely one of the most vital and pervasive aspects of local process, but something more is required to sustain an entire composition.

I have also not mentioned referential processes, for many of the same reasons. Again, most music is referential, most if not all of the time, but to say this is to say no more than that music consists of relationships—both audible and inaudible. And indeed it does: relationships of all kinds, in all dimensions, and on all levels! Most of the categories of process listed in the table are referential in one way or another, in that a particular component of the music will refer to what has preceded, what is to follow, or to something farther removed in time—or even something

[7]The Mozart excerpt analyzed above is an excellent example of figuration at the local level, whereas the opening Prelude in C major from the first book of J. S. Bach's *Well-Tempered Clavier* may be heard as a continuous process of figuration that extends throughout the piece.

[8]That is, a transition passage in a classical sonata-allegro movement will often combine counterpoint (an interactive process), a modulation (change of key), rhythmic repetition, and dynamic intensification—four distinct local processes.

[9]This is admittedly an extremely broad definition, and tonality is often defined in the narrow sense of a major or minor key (see Rowell [1983, pp. 234–236]).

extrinsic to the music. It is the way in which these references are disposed that determines the particular process that they manifest. Tonality, implication, reference—these three properties belong to the explanation of "music" itself (at least until recently) as a process, the fourth aspect of musical process to which I referred in the opening pages of this article.

To label a musical work or genre as a particular type of process is to give priority to one particular mode of attention, one way of hearing. In the normal experience of music, numerous processes and local processes compete for our attention, interpenetrating with one another. It is conceivable that an entirely different set of categories could be devised if one selected a different set of aspects or approached the subject of musical organization from a different perspective. The present scheme is admittedly untidy, but so, it seems to me, is art. Is there, as a perceptive reviewer suggested, a much smaller number of processes "lurking" behind the set I have proposed? I shall give this suggestion further consideration at the conclusion of the following section. The test of the set presented here will be their usefulness in helping to organize one's own listening experience.

Listening for Process[10]

In this section I shall suggest illustrations of several of the major categories of musical process. Many will be familiar to readers, most are drawn from the standard repertoire, and all are readily available on discs and tape. The task will be not only to demonstrate and explain the categories, but also to show how multiple processes combine to intensify the sense of musical action. As always, the best test is to listen.

Bolero, by Maurice Ravel, is an almost inevitable choice as a demonstration of process in music; indeed, the work is virtually nothing else! Repetitive process dominates the twenty-minute duration of the *Bolero,* organized into three nested cycles of 6, 54, and 108 beats—or, as a musician would count them, 2, 18, and 36 measures of 3/4 time. But the composition is at the same time an example of elaborative process, in that the tonal colors and mixtures of Ravel's large orchestra become progressively enlarged, more complex, and more brilliant over the course of the piece.

The opening of the first movement of Beethoven's Quartet in C-sharp minor, Op. 131, is an extremely simple and effective illustration of imitative process in its step-by-step presentation of the main theme in each of the four instruments. The movement is a fugue, a form that is perhaps the most relentlessly processive of all the types of musical organization. Listening to any of G. P. Telemann's canonic sonatas for flute and violin will provide an even clearer illustration of imitative process for only two parts.

[10]This section replaces the set of recorded illustrations and informal discussion presented in the original address.

Emergent processes are commonplace in many of the world's musics, especially as a tactic of beginning, as when, for example, an Indian musician gradually explores the melodic contours of the *rāga* in which he or she has chosen to perform, or in the free opening (*Jo*) section of a piece of eighteenth-century Japanese chamber music.[11] The initial section of the first movement of Mozart's so-called "Dissonant" Quartet in C major, K. 465, provides an example from the Western tradition, a section in which the music becomes progressively less ambiguous, more orderly, clearer in tonality, and directing the listener's expectation toward the subsequent faster section (the main part of the movement). The impression is not unlike the sensation of order emerging from primal chaos.

Interactive process is featured in concertos and other musical genres that exploit the solo/group relationship or the contrast between several soloists or groups. As examples, J. S. Bach's Concerto in D minor for two violins and orchestra or Brahms' Double Concerto (for violin, 'cello, and orchestra). A compelling example of interactive process in one of the major non-Western traditions may be heard in a percussion improvisation for South Indian drum (*mrdangam*) and tambourine (*kanjira*) near the end of an hour-long performance recorded on HB-72023 (in the Nonesuch Explorer series): in this "rhythmic dissertation" (*tani avartam*) the two performers literally take dictation from each other, trading patterns back-and-forth without prearrangement and gradually reducing the length of their patterns until they arrive at an exact unison, thereby signaling to the singer the precise point at which he is to enter and bring the performance to a conclusion—all of this taking place within the strict metric control of the *tāla* cycle chosen for the whole performance. In this stunning feat of virtuosity and spontaneous communication the performance is shaped as we listen.

A famous example of interactive process from the Western tradition occurs in the finale of act 1 in Mozart's opera *Don Giovanni*. It is the passage beginning with the well-known minuet, in which the main orchestra and two small on-stage bands play in three different meters to accompany the dancing of three sets of characters on stage. The general effect is pleasantly complex but, again, within the controlling meter of the minuet.

Syntactical processes have become increasingly important since ca. 1700, especially in the sonata-allegro—the formal structure heard in the first movements of virtually any quartet, sonata, or symphony by Haydn, Mozart, or Beethoven. In this case the form is one thing, and the process is another. Syntactical process requires the listener to make comparisons between noncontiguous, often distant events and thereby demands much of the memory: the practiced listener registers the original presentation of the musical materials in the exposition, savors their fragmentation and disarray in the following development, and mentally reassembles them in the recapitulation. This level of listening is clearly not for everybody, and listeners who fail to make these comparisons are more likely to receive a subliminal sensation of some unidentified process working itself out. In the standard first

[11]For further background and a discussion of the so-called free/strict archetype see Rowell (1981, pp. 204–208).

movement of the classical concerto (as in Mozart's clarinet concerto or Beethoven's violin concerto), syntactical and interactive processes combine in one of the most popular and successful hybrids.

The type I have labeled as analogical process covers a lot of territory and requires some extended commentary. As a category it may seem suspiciously like a convenient catch-all, but in many cases it provides a valid description of the musical organization when no other single category applies. In recognizing musical analogies, the comparison may be to another piece of music, or (more problematic) to something external. In the latter case, we are drawn into the slippery question of musical meaning and the many related issues that have arisen in the developing discipline of musical semiotics.

Analogical process is not new in music. Descriptive music has been a special category in the history of Western music from early times, and music devised to accompany a sung text has often incorporated analogies to the content of the text—by helping to paint a mood, to enhance the qualities of motion and action, or even by graphic puns. The range of expression runs therefore from naive simplicity (cuckoo calls, rippling brooks, and turning millwheels) to complex analogies that are far from obvious and communicate only at subtle, deeper levels. Music can be conceived as analogous to virtually any type of process—physical, biological, mental, societal, artistic—and deep-seated cultural meanings underlie many of our understandings of process and have thereby been influential in shaping our artistic preferences. To cite but one case, cultural intuitions of the nature of the creation process and of beginnings in general have helped to mold the standard beginning gambits and strategies in many of the world's musics (Rowell, 1981). I suspect that there has always been a strongly analogous component in our thinking about music, in our instinctive attempts to understand music's transitive and intangible organization by comparing it to other, more familiar processes and structures. What else has attracted so many musical scholars to the study of time, if not to gain some deeper understanding of the models by which we shape and arrange our experience of audible time?

The central problem, as I see it, is one of verifiability, and the new field of musical semiotics has not so far distinguished itself by providing objective means for verifying their analogies. Music that embodies a valid analogy is by no means assured of automatic communication along the chain from composer to performer to listener, and these analogies may be implanted unintentionally, passed along unknowingly, and/or perceived subconsciously. But, despite all these problems and the undoubted fact that the analogies may often be trivial or lend themselves to overly facile interpretation, we cannot afford to dismiss such a powerful means of communicating the sense of musical process. It may not matter, in the end, just what musical analogy is being asserted, so long as familiar and appropriate affective responses are triggered in the listener.

One illustration must suffice: the orchestral prelude to Richard Strauss's opera *Der Rosenkavalier* has been heard and enjoyed by thousands of operagoers, many (perhaps most) of whom have not fully realized the explicit sexual depiction in the music. Many listeners will prefer not to recognize such an analogy in music and will instead take it in on a symbolic or stylized level, but anyone who listens with knowledge of the dramatic situation behind the closed curtain and empathy for the emotional content of the opera can scarcely fail to follow the erotic narrative of the prelude as it moves through a dizzying series of climaxes to postcoital languor. I am not suggesting that sexual analogies are common in music—they are not—but they do constitute a distinct seme of musical meaning, especially in descriptive interludes from operas as diverse as William Walton's *Troilus and Cressida* and Hector Berlioz's *Les Troyens*—in each case the familiar dramatic trope of sexual intercourse during a storm, with the lovers' passion counterpointed against the fury of the elements. A related and effective dramatic trope in opera is the combination of onstage action set against offstage process—a formal church service (as in Barber's *Vanessa* and many other operas), triumphal music (as in Verdi's *Aida* and Bizet's *Carmen*), or even a piano recital of salon music (as in Giordano's *Fedora)*.

I continue with some illustrations of compound musical processes, in many cases functioning as virtual allegories in that they offer the possibility of reception and understanding upon multiple levels of meaning. The final coda to the third movement of Paul Hindemith's *Mathis der Maler* is a striking example of process layered upon process: imitative, in that the opening perpetual-motion theme is passed through each group of the string section in turn; repetitive, with reference to the short ostinato figure heard first in the clarinets and then in the horns, continuing almost to the end of the work; commentarial, in that a Gregorian chant melody (which introduces new connotations because of the associated text) is spun out in the high woodwinds as a *cantus firmus;* and interactive, in that the listener's attention is drawn first to one and then to another of these components as the music reaches its powerful climax in the final brass chorale. The section will repay repeated listening.

In a somewhat different combination, the final movement of Gustav Holst's Suite No. 2 (for military band) may be described as both interactive and commentarial in its brilliant counterpoint of two Hampshire folksongs: interactive, in that the two tunes ("The Dargason" and "Greensleeves") require the listener to switch attention from one distinct musical level to another—one fast and one slow; and commentarial, in that the work is constructed above and around the two preexistent melodies.

My final illustration is a highly stylized process originating in, and limited to, the music of the nineteenth-century opera composer Gioachino Rossini—the so-called Rossini crescendo. Examples of this process abound in his many opera overtures, but a superb instance occurs twice in the overture to *Il Turco in Italia*. Like most processes, it is easier to recognize aurally than to describe: the music begins with vigorous patterned activity at a very soft level and at rapid speed, then becomes gradually louder and more active as instruments are added and the texture thickens, finally reaching a high plane of dynamic intensity with tremolo strings and brilliant trombone scales. The process is highly repetitive, figurative, and perhaps also emergent with respect to the steady accumulation of instruments and the growing intensity.

In all of the preceding examples, we must acknowledge the problem presented by the figure/ground phenomenon in music. Process need not, and often does not, involve all the musical elements. At times, process is perceived in the form of musical figures against a background; at other times, process *is* the background or context within which unique musical figures are displayed. Part of the composer's skill lies in his or her ability to direct our attention toward, or away from, the processive components set forth in the music.

As I reflect on the general question of process in music and the specific types I have identified and illustrated, it seems to me that four components stand out from the rest: repetition, progression (in the sense of implied motion towards), figuration, and interaction. This is the stuff of which musical process is made, and perhaps we have come back to our original, tentative definition—"a series of progressive and interdependent steps by which an end is attained." The composer who wishes to suggest a strong sense of process, whatever its particular meaning, must introduce and maintain some element of regular continuity, imply forward motion, engage the listener's attention with graspable patterning, and provide the means by which perception is drawn back and forth among the musical actors in the unfolding drama.

I have said very little about recent music, not out of disinterest but unreadiness. It is clear that composers of new music have been successful in devising an array of new processes, as well as new twists on the traditional ones, and it is safe to say that a large quantity of recent music is more "process-minded" than "form-minded." But new means of organization do not guarantee new processes. To make but one point, serial music—in the sense that a twelve-tone row or an ordered series of durations is causal for a musical work—may be regarded as a particularly sophisticated version of what I have termed a commentarial process, in addition to whatever other processes arise in the construction of an individual work. When scholars have managed to sort out the present musical scene, I am convinced that we will be able to identify a number of effective and attractive processes that will continue to add to our understanding of how the musical mind works and of the cultural analogies that have influenced musical thought in the latter half of this century.

Music, Process, Meaning, and Value

I have avoided questions of value in the preceding sections, and it is time to return to some of the issues I raised in my introductory remarks. To be specific, what can be said about the aesthetics of musical process and the temporal structure of music? Two words of warning: first, my suggestions will do no more than scratch the surface of an enormously important and complex issue; and second, I can offer no authority nor means of verification for these proposals—like most statements of value, they can be tested only by means of individual experience. I shall propose a short list of what might be called, rather awkwardly, "process values"—predicates of the musical process to which we may respond with savor, with recognition of their special role or effect, perhaps even with the judgment of "beautiful."

I shall suggest first two values that seem to me to be inherent properties of music itself, and which may thereby set music apart from all the other arts. One is the subliminal perception of music itself as vibration: no matter how successfully the common fictions of music persuade us that we are hearing single tones, sustained durations, and pure sound colors, it can never be far below the surface of our consciousness that musical tone is alive, pulsating, and testifying (correctly or incorrectly) to a totally animate universe of cellular and molecular vibration. Part of a composer's job is to create and maintain the illusions of simplicity, purity, and stasis, and part of the listener's job is to accept these fictions; but in our deeper consciousness lies the subliminal realization that we are responding sympathetically to the imprint of an external vibratory energy that seems to mimic our own inner vibratory life energies.

And second, I suggest that our equally subliminal perception of the web of music as audible, manifest process brings a deep satisfaction of our human need to find continuity, order, and predictability in our experience of the world, for all the reasons mentioned in the introduction; that is, our perception of musical process as something that is controllable, replicable, integral, and scaled to the units of our attention and memory. To receive the sensations of process from a musical work is to receive confirmation that human life continues amidst some semblance of lawful process.

My third proposal is, I believe, the most important one on my list—the recognition of identity or *persona* in music and our consequent identification with it. I have already presented some of the arguments for this line of thinking. These identities—in the form of themes, motifs, gestures, instrumental roles, and the like—are the actors in the unfolding drama. What happens to them, in the form of apparent motion, involves them deeply in process—growth, progress, interaction, interruption, maturation, development, reaffirmation, decline, termination. Quite unconsciously we are led to identify with them, suffer their fate, mimic their qualities with our nervous and emotional being, and project our own human values upon the fabric of sound that we experience. Let us call this value the perception of identity-amidst-change, a sonorous analog of our sense of personal identity—perhaps our most prized and jealously guarded possession.

This line of argument has given me a welcome opportunity to return to a book by Edward T. Cone entitled *The Composer's Voice* (1974). In this wise little book Cone explores questions of voice, role, persona, and identification in performing, analyzing, and composing music. The following passage, assembled from various parts of the book, presents the gist of his argument:

> Any instrumental composition, like the instrumental component of a song, can be interpreted as the symbolic utterance of a virtual persona . . . [a persona] that is the experiencing subject of the entire composition, in whose thought the play, or narrative, or reverie, takes place—whose inner life the music communicates by means of symbolic gesture [p. 94]. . . . One who achieves full identification with the complete persona of any complex work must not only participate in the fortunes of each component persona, character, and leading agent, but also experience,

> vividly and intimately, the course of events produced by their relationships. Above all, to achieve identification a listener must fully comprehend the composite rhythm that constitutes the locus of musical form,[12] and he must synchronize his own psychic movements with the symbolic gestures audibly embodying that rhythm. . . . To hear a composition in this way is, no doubt, an infrequently realized ideal; yet it is an ideal that must implicitly inspire any attempt to comprehend a composition as something more than pure design on the one hand or an opportunity for emotional indulgence on the other [pp. 125–126].

It will be clear that I endorse Cone's eloquently stated position. I am not suggesting anything as simplistic as the notion that music is a discursive language that only requires translation for its story to become manifest. I am suggesting, along (I believe) with Cone, that listener identification and affective response arise at a far deeper, more abstract, and symbolic level—quite beyond the reach of consciousness. This is, I believe, why audiences find music that resists their efforts to discern identities therein so deeply disturbing, and are apt to respond with Kafkaesque analogies and a sense of profound discomfort. I am not claiming that new music has uniformly failed to provide graspable musical personas, but I think one might with some justice contend that the most immediately successful of recent compositions have been those in which the composer—no matter how complex his means and how difficult his style—has been able to facilitate this type of identification.

Cone was probably drawn to his position by his reading of Edmund Gurney's extraordinary book on *The Power of Sound* (1880), on the strength of statements like the following:

> There is one characteristic of melody which attention to its aspect as motion brings out with special clearness; and that is our sense of entire oneness with it, of its being as it were a mode of our own life. We feel in it, indeed, an objective character, inasmuch as we instinctively recognize that it has for others the same permanent possibilities of impression as for ourselves; but our sense of it nevertheless is not as of an external presentation, but of something evolved within ourselves by a special activity of our own [p. 166].

I have propounded a powerful argument: that we regard music or any artistic process as seemly when it invites and rewards our instinctive efforts to identify with its persona and respond with affect to the vital rhythm within which it moves and the symbolic transformations it undergoes. Music is thus heard as a profoundly human analogy, and, yes, Schopenhauer and the other romantic philosophers were right: it is by virtue of music's very generality that we are moved, not its particularity—by the power of the symbolic language of music to summon up abstract personas that can be readily grasped and experienced by a large cross section of humanity. In one way or another, each of the proposed categories of musical process

[12] By this I believe Cone means what I understand as "musical process."

provides for this type of identification with a particular type of musical persona and its destiny, whether by analogy or direct reference.

To continue: having recognized the inherent nature of music as vital vibration and found confirmation that orderly process exists therein, and having identified with its virtual persona(s), to what other aspects of process might one respond with affect? Next, it follows, to the degree of inexorability, the speed, the rate of change, the teleological nature, the duration of the process, and other similar attributes. Or, indeed, to the opposite qualities. Oppositions provide a potent source of affect: processes whose operation resists interpretation, whose rate is not easily discernible, whose end is in doubt, and which suggest stasis are sources of ambiguity precisely because they resist expectation and thwart our attempts to identify, and thereby serve as means by which the composer is able to arouse emotion.[13]

If the general suggestion of identification is correct, we may draw upon the analogy to a drama. Among the qualities to which we instinctively respond must be (1) the feeling of events piling relentlessly upon other events; (2) breathless pace or languid quiet; (3) action moving to an end; and (4) intuitions of aging, ripening, growth, and perhaps simple endurance. Limitations of space will not permit elaboration of these suggestions, other than to point out that affect arises most strongly at the extremes of the oppositions we have constructed. We take some mean as a norm and regard deviation in either direction as expressive—with the pulse rate, as many have suggested, as a norm for speed—and with affect increasing as the music departs from this norm. This is not to contradict Aristotle and claim that moderation is unseemly, merely to point out that a presentation of any musical identity is interpreted both as a norm and a base for subsequent expressive deviation.

I would like to suggest an additional source of special value in music—the perception of intricacy, nexus, figuration, and texturing, whether of a musical surface or a complex multidimensional web of sound. Music that brings this element to the fore is valued in much the same way that we respond to textural values in a Jackson Pollock painting or an elaborately worked surface of any kind. Music has innumerable means of invoking this effect, within a single line or a complex texture, and it is an aspect of music that has not been as thoroughly studied as it deserves to be.

My final proposal deals not with the sensations of continuous process but with the interruptions, conflicts, and challenges to that process—which then become an essential part of the process, heighten affect, increase a sense of restoration when (if) restoration occurs, and participate in a higher level of process that intensifies our sense of identification in the way it mimics our feeling of human identity caught up in ongoing process. I know of no art which is better able to imitate the sense of identity created, maintained, challenged, restored, and, at times, inescapably, painfully lost. And why not indeed? Music may be the art that approaches most closely our perception of self, without the aid of verbal concepts and devoid of visual images—an art that is constantly presented to us as alive, mobile, offering a projection of self-indentity persisting throughout process, whose fate is our fate,

[13] This is an important part of the main argument in Meyer (1956).

and whose context is interpreted with the same range of affect that arises as we interpret our own context.

I have presented a very old argument. If, as Protagoras asserted, "Man is the measure of all things," he must surely remain the measure of his own audible creations. And, in his audible creations, he must just as surely have implanted his own habits of thought and feeling: abstract, external symbols of self, transformations, rhythms, passage, and his own vision of process as culture has taught him to conceive it—whether as a continuous refinement and ripening, a perilous course of conflict leading to an ultimate prize, a serene procession of works and days, an unpredictable nightmare, or a numbing routine of impersonal frustration and boredom. Music provides an ideal medium with which to capture our intuitions of process and frame them in a sensuous, abstract language that draws by subconscious analogies upon our deepest ideas of self and context.

The participants in the conference from which the present volume of essays has been selected heard many details about various processes that operate on a cosmic or microscopic scale, without the agency or intervention of humanity, and devoid of any relationship to human values—which is, perhaps, one of the reasons why we regard them with such awe. One must adopt a theistic position to assert that the physical processes of the earth or the cosmos have any essential relationship to mankind, beyond our continuing attempts to survive their accidents, and find meaning for ourselves. But our relationship to the arts, and to the manifold forms of process that they embody, depends upon all we have learned and all that we feel as human beings. Whatever process they display, we have put it there, and we remain the standard of measure. To hear music as process exposes our cultural attitudes toward process—and when these attitudes change, as Plato affirmed, so does our music.

If you wish to know about process, listen.

References

Bent, I. (1987), *Analysis*. New York: W. W. Norton.

Cone, E. T. (1974), *The Composer's Voice*. Berkeley & Los Angeles: University of California Press.

Cook, N. (1987), *A Guide to Musical Analysis*. New York: Braziller.

Dunsby, J., & Whittall, A. (1988), *Music Analysis in Theory and Practice*. New Haven, CT: Yale University Press.

Forte, A. (1959), Schenker's conception of musical structure. *J. Music Theory*, 3/1:1–30.

Gurney, E. (1880), *The Power of Sound*. New York: Basic Books, 1966.

Kerman, J. (1985), *Contemplating Music: Challenges to Musicology*. Cambridge, MA: Harvard University Press.

Kramer, J. D. (1988), *The Time of Music: New Meanings, New Temporalities, New Listening Strategies*. New York: Schirmer Books.

Langer, S. K. (1953), *Feeling and Form*. New York: Scribners.

Meyer, L. B. (1956), *Emotion and Meaning in Music*. Chicago: University of Chicago Press.

——— (1973), *Explaining Music: Essays and Explorations*. Berkeley & Los Angeles: University of California Press.

——— (1982), Process and morphology in the music of Mozart. *J. Musicology*, 1/1:67–94.

Rowell, L. (1981), The creation of audible time. In: *The Study of Time*, Vol. 4, ed. J. T. Fraser, N. Lawrence, & D. Park. New York: Springer-Verlag, pp. 198–210.

——— (1983), *Thinking about Music*. Amherst: University of Massachusetts Press.

——— (1987), Stasis in music. *Semiotica*, 66/1–3:181–195.

8

Duration and Rhythmic Process in Music

Christopher F. Hasty

Abstract Musical rhythm is often identified with meter and thus regarded as an arrangement of discrete, quantified durations. In this essay it is argued that the reduction of experienced duration to its extrinsic measurement excludes becoming and process. As an alternative to abstracting duration from musical events, duration is regarded as the palpable, qualitative extensiveness of events. Rhythm, released from its ties to quantity, is returned to its broader connotations as the particular manner in which duration comes to be formed. In this way rhythm as active formation replaces the static, atemporal concept of form. The phenomenon of meter is then investigated in light of the views of duration and rhythm that have been presented. Aspects of meter are addressed in sufficient detail to prepare the nonspecialist reader to follow an analytic discussion of portions of Beethoven's First Symphony.

Visual arts create space—not the generalized, abstract space of the mathematician which, in whatever conceptual realm it exists, cannot be touched by art, but rather a particular sort of space, one aimed at communicating to human perception a specific character or quality, something that we say is expressed in the material medium. We have no word for this ''something'' or rather we have too many words—feeling, idea, *quidditas*, sentiment, form, to name just a few. Whatever we may wish to call it, this something is remarkably touchy—a change of color, the displacement of a mark on the canvas, and the character of this space can be significantly altered. The conscientious artist labors to create a space of great particularity (even if this be a particularization of convention) and knows that there is a possibility of failure. Aesthetic judgment bears witness to the creation of space.

To draw a parallel, we might be tempted to say that music creates time, but this formulation fails to ring true. While I may, with some justification, claim to see space or rather a space, I cannot claim to hear time—much less *a* time. Time is for us much more of an abstraction than is space, at least in the aesthetic–phenomenological sense I am using the word *space*. What I can claim to hear is a temporally extended event. The extensive character of an event I will call its duration, and thus I would like to say that music creates duration. From the root *durus* (hard, solid) duration implies a substance or object, something that persists or endures. In common parlance, duration is usually abstracted from event as an extrinsic measurement—thus we speak of an event taking place through or over a duration. In this usage, different events, or classes of events might possess the same duration. This can be a very useful concept in that it permits us to coordinate and compare different events by purging them of content and process. Since this reduction omits precisely those qualities that are germane to an analysis of musical experience, I would like to draw event and duration together and say that, as events are unique, so are their durations unique.

For example, let us consider a week-long conference such as the one in Montana at which this paper was read. Our conference can be viewed as a single, relatively lengthy, and complex event. Like most events, its boundaries are not very sharply drawn. It begins as a continuation and fulfillment of our preparations; but as it proceeds, it will gradually condense into a world apart from the worlds we were in before the conference began. Though we cannot precisely determine when this event begins and ends we may say that it lasts about seven days. These seven days correspond to the duration of the event. Within this duration this patch of earth will seven times be warmed by the sun; but the abstracted measurement of seven days is not, I will say, the duration of our meeting, though we might say that these seven particular days are a part of this duration. By duration I mean rather the unique course of the event that as it transpires acquires for us an almost corporeal quality of massiveness, a feeling that arises not simply from the conglomeration of the event's constituents, but from the way in which they grow together. When we leave this place and look back on our stay, it will seem that we were here a long time. This world will have congealed and to a large extent will have become separated from the worlds we left. Our judgment that this has been a long time will have little to do with a computation of the number of days that have elapsed. Rather, it will issue from a complex feeling of the emergence of the event as a totality—as a relatively autonomous creature, durationally embodied. We feel the extent of this event, its "length," as a certain massiveness, or fullness, or depth. This feeling will be especially vivid as we depart—the week will then have a vastness for us that we shall not later be able to fully recapture.

We tend to discount this feeling of duration because, in thinking of duration as a quantified lapse of time, we find our feeling of quantity highly indeterminate—only within the span of a few seconds and under special conditions that I shall outline later, can we comprehend duration as a determinate quantity. We also tend to discount the feeling of duration I have described because the feeling varies greatly in intensity. Where the event is weakly defined and has little importance for us, our feeling of its duration is vague. Also, since we tend to focus our attention on various components of the event that capture our interest, we easily forget that it is precisely their deep implication in the whole of the event that gives the components the intense particularity that we find attractive.

There is also considerable conceptual difficulty involved with the extensive nature of events. I mentioned earlier that duration implies persistence—something that lasts or endures. But what can be said to endure in the passage of an event? There are two modes of endurance I would like to consider—that of the event in presentation and in memory. Once the event has begun and at any stage in its process of formation, the event is a present for us. As at any moment there are a multitude of events so are there at any moment a multitude of nows. Returning to our conference in midweek, let us say, the event that was our first sight of the hotel is not in itself now, and yet it is now a part of our stay and is present in a manner that our last meal at home is not. In fact, our initial sight of the hotel has become inseparably connected with subsequent events. Our explorations of the interior, the people we have met, the course of all our experiences at the conference keep that first impression present, active, and alive. It is through these countless modifications that the constituent event can gain significance and a presence that extends throughout the whole of the event. This significance and presence I have been calling *particularity*. The word is, I think, appropriate because it expresses one of the paradoxical aspects of events—that something should owe its individuality and uniqueness to assimilation and modification. To say this is to reformulate the familiar observation that what persists can do so only through continuous change; that change can only occur to that which endures. However, this reformulation raises the question not of the event's endurance, but of the endurance of its apparently evanescent components. Through an analysis of the event, we might attempt to differentiate the successive modifications of one of the event's constituents as more or less discrete stages of a process, but these modifications that qualify and particularize the constituent are not in fact *other* than the constituent. In this sense the constituent persists through the course of the event, and in this sense its duration is, I will say, *simultaneous* with the duration of the whole event.

Thus all the parts, all the constituents of an event, while sequentially ordered, are also altogether present at the same time—the unitary time of the event. Simultaneity is not to be confused with instantaneousness. The concept of the instant both annihilates sequential order and relegates our feeling of duration, of extensiveness, and massiveness to the category of illusion.

Evidence of the persistence and simultaneity of the components of an event can be found outside of the event's presentation, in recollection. The memory of an event is, of course, not the event and is not part of the event's duration, though, as I mentioned earlier, we can retain a feeling of this duration as a certain vastness or massiveness. When we later recall our conference it will not take us seven days to perform the act. In the brief span of our recollection, a swarm of more or less distinct features will present itself in a largely nonchronological fashion.

It is only with considerable effort that we will be able to isolate rather gross components of the event and reconstruct their order. Without this effort, our recollection will have something of the quality of a dream. I submit that this nebulous quality of a vivid recollection is not a defect of memory but arises from the irreducible wholeness of the event, from the fact that all the constituent experiences of the event have become involved with one another in a continuous duration with which they are coextensive.

I began this discussion by saying that music creates duration and have gone on to show that this is hardly a unique feature of music. Any activity can create duration to the extent that it forms an event. I would now like to turn our attention to the question of *formation* and to the special witness of musical events. Musicians generally speak of form rather than formation, of the arrangement of settled, durationally embodied components rather than the genetic process through which events become embodied. The traditional term that I think comes closest to expressing the processive formation of duration is *rhythm*. But this term has acquired a technical meaning that is severely limited, precisely as the term *duration* is commonly limited to quantitative measurement. In this narrow sense, rhythm refers to patterns of measured duration; and the study of rhythm is relegated to metrics. Rhythm is regarded as a separable component of music and is distinguished from melody, harmony, and form. Thus, if I ask you what is the rhythm of a musical passage you will understand that I mean for you to clap or to sing on a single pitch or perhaps to write a string of symbols representing the relative lengths of individual impulses. Likewise, if I ask what the duration of the passage is, you will answer in numbers of beats, measures, or seconds. But if I ask you what rhythm in the passage is, how will you reply? A suitable response might be to give a spirited performance of the passage or to dance or perhaps to describe the gestural character of the passage. Such responses would be appropriate for another sense in which we use the word *rhythm*. We say, for instance, that a performer has superb rhythm, meaning that he vividly expresses musical shape or gesture. We speak of rhythmic vitality. We approvingly call rhythmic a person's graceful carriage, an animal in chase, a tennis match. Rhythmic also are buildings, drawings, collages, flower arrangements. Common to all these examples is an aesthetic judgment involving the perception of fluid, comprehensible form. But rhythm in this sense is more than an aesthetic category; our tennis player will lose the match if he does not comprehend the rhythm of his opponent and if his own rhythm is too easily communicated to his opponent—if, as we say, he telegraphs his movement. Likewise, our animal will go hungry if it fails to assimilate the complex maneuvers of its prey, and its prey will likely perish if its rhythmic activity is fully comprehensible to the predator. The consciousness of rhythm requires a sort of participation with the whole of the event and involves a discrimination of the event's constituents as they are in the process of solidifying into duration.

The wholeness of the event, the gradually emerging simultaneity of the event's constituents, forms a tendency or telos which it is the function of rhythmic awareness to comprehend. This orientation toward the future, toward the event's fulfillment, makes it possible for us to speak of the "rhythmic impulse" and of "rhythmic drive." Rhythm entertains (literally, "holds together")—it makes it possible for us to comprehend process and it fastens our attention to the process. Incidentally, it is a pity that recently *entertainment* and *Unterhaltung* have become terms of denigration, for entertainment is precisely the function of music. The arhythmic is the incomprehensible, the senseless, the meaningless. The arhythmic event does not involve us in its process—it is something that *we* must endure; it produces ennui because nothing in the event rises to particularity—nothing in the event endures. Thus, I would like to say that music creates duration by means of rhythm.

The broader connotations of rhythm that I have indicated can be found in the Greek *rhythmos*, which, before Plato allied the term with the concepts of measurement and number, meant "form" in a special sense—the distinctive and characteristic arrangements of the parts of a whole that is in flux. Benveniste (1971) defines the ancient meaning of *rhythmos* as " 'the particular manner of flowing,' used to describe configurations without fixity and arising from an arrangement that is always subject to change" (p. 283). I believe there is something in our understanding of the word that responds to a similar intuition. Benveniste shows that before Plato, there was no indication of periodicity in the use of *rhythmos*. *Rhein*, to flow, could not be said of waves, nor of the sea. And yet it is difficult for us to separate periodicity from rhythm. To do this it is sufficient to regard periodicity as one among many means through which rhythm can be engendered—a powerful but not a necessary component of rhythmic activity. After all, we do speak of such things as the rhythmic flowing of robes or the rhythmic wind-up and throw of a pitcher. And in music there are supremely rhythmic compositions that are largely nonmetrical—for example, certain genres of East African mbira music and Japanese shakuhachi music. Gregorian chant can be performed virtually ametrically; and for many listeners such performance actually enhances this music's rhythmic effect. Much of Karlheinz Stockhausen's electronically generated composition, *Kontakte* is, I would argue, fully rhythmic, although it bears only the faintest traces of meter.

However, the sort of music I would like to consider now, Western classical music of the tonal tradition, is intensely periodic. Thus I have no desire to disparage periodicity and meter. In fact, I hope that by investigating the contribution of meter to rhythm I can (to some extent) disentangle the two concepts. I hope, also, in this discussion to clarify my statement that music creates duration by means of rhythm. Since my view of meter does not conform to customary definitions, I shall have to point out, from time to time, some of the ways in which my understanding of this phenomenon diverges from current thinking.

In Figure 8.1 I have indicated two series of brief sounds or impulses such as could be made by clapping one's hands together. The first series is nonperiodic, and the second series is periodic, the periodicity being approximately 1½ seconds. In listening to the two series, we readily perceive this difference—that the first series is erratic and that the second is regular. But there is another, perhaps more subtle difference that we also perceive. This perception is spontaneous and is not based on judging whether or not the series is periodic. In the second series, the impulses acquire a perceived duration that extends beyond their actual acoustical duration. In the first series, the sounds are sufficiently cut off from one another that the silences between them are in a sense empty—they do not seem to participate in the durations of the impulses. Thus, for example, the duration of the last sound is highly indeterminate and is in a sense shorter than the last sound of the second series which has acquired a relatively determinate duration that extends into the subsequent silence.

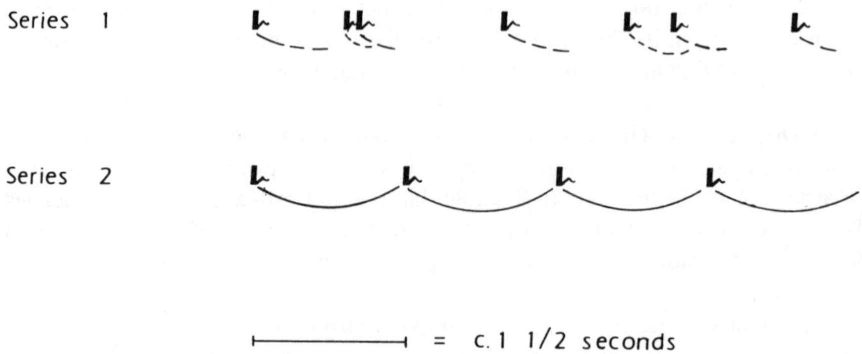

Figure 8.1

In both series the sounds endure in what is called echoic memory, and so it is a special sort of endurance that I call attention to in the second series. In the second series there arises a feeling of duration that is not present in the first series. This novel perception is the special contribution of meter to rhythm. I do not know how this perception comes about, but I do not think it happens because we can predict the next impulse. We may say that the regularity of the previous impulses allows us to predict the next one, provided we understand that such a prediction rests not on imagining the upcoming impulse beforehand but rather on *knowing* the duration of the present event. (Note that the durational character of the last impulse is not altered by the absence of a succeeding impulse.) We can predict because we perceive a determinate duration and not vice versa. If prediction does not account for how this perception arises, it does account for why the perception arises. Any creature must predict in order to act in the process of its world; and if this world includes complex and articulated motion, the creature must find some means of assimilating this motion and comprehending its pattern. Since we must also assimilate events that are not so clearly articulated and that do not display a comprehensible periodicity, what we call meter must be considered only a species of rhythm. Nevertheless, it is an extremely important one, and we instinctively seek it out in our experience. Meter grabs us, or rather we grab after it. We cannot escape a ''beat''—it gets into our bodies, and particularly into our feet.

As a guide to action, our grasp of periodicity is limited. In the second series, the length of the period was between 1 and 2 seconds. If we expand the period to 5 or 6 seconds, our sense of extended duration will vanish and with it will disappear the accuracy of our prediction. Now, 1 or 2 seconds does not give us much time in which to react. To overcome this temporal limitation we instinctively seek a greater periodicity by consolidating events. Conventional theory in fact defines meter as the grouping of pulses and does not recognize the pulses as metrical. But I see no reason to require our second series to be grouped in twos or threes before we may speak of meter. Such grouping produces essentially the same phenomenon we observed emerging in the second series—the creation of palpable, determinate duration. And yet there are concomitant factors that help to explain why theorists

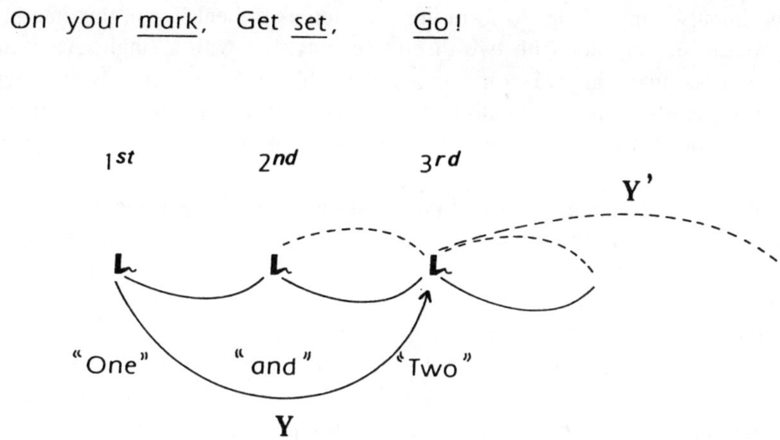

Figure 8.2

have thought grouping necessary for meter. For, as we listen to a series of pulses, particularly if the period is relatively short, we shall in fact find it difficult *not* to group pulses. As I have indicated, this proclivity serves our interests by giving us a greater basis upon which to act. Where the period is longer, say 2 to 3 seconds, we are less inclined spontaneously to group pulses.

Before I consider the question of grouping, I would like to examine in more detail what occurs as we listen to a sequence of pulses. Figure 8.2 represents the beginning of the second series. Until the second impulse there is no determinate duration, in the sense that there would be no determinate duration without the second impulse. This little word *until* camouflages some conceptual difficulties that emerge from the conflict between the notion of an extended present and the notion of discrete, durationless "time points." If we regard the initiations of the two impulses as time points we will be confronted with the problem of understanding how the first of these can instantaneously acquire a continuous duration. Modeling perception on the method of extrinsic measurement, we might simply say that with the second impulse we recall the first, make a comparison, and judge the extent of the duration. However, in perception we cannot so easily distinguish these three acts. If we recall the first impulse at the instant of the second, its duration has vanished. We could return its duration, rather begging the question, by saying that we recall the first impulse together with its duration; but how could we recall the duration since that duration is now? If we claim now to hear the duration there can be no part of the duration that is not now. The only way to save the phenomenon for recollection would be to imagine that the duration is composed of finite parts (what Husserl calls "primal-impressions") that are in the end summed to give us a measure of pastness for the first impulse (Husserl, 1964, pp. 50–57). But where are these parts; how many of them are there? It seems to me preferable to regard this first event (and by event I now mean this particularly extended impulse) as fully present throughout its duration. Its duration is certainly dependent upon the second impulse which helps to form this duration as present. Upon this interpretation, we are dealing not with two discrete events, but with a single event and a single process, involving two components, that while they are successively ordered, are simultaneous with the duration of the event. The principle of simultaneity must be invoked here to account for the coextension or copresence of the event's components.

The second component, or that part of it that is the second impulse, has upon its inception a determinate duration commensurate with the first duration. This duration is, we should note, suddenly present, before it has actually come to pass. It is as if the second impulse, having acted to form the first duration, has somehow acquired that duration as its own. I shall call this phenomenon projected duration, or simply projection, and represent it graphically, as in Figure 8.2, by a dotted line. Again, the projection is not separable from the first duration in that it is dependent upon both impulses. A more complex instance of projected duration is shown in Example 8.1, taken from the ending of the first movement of Beethoven's *First Symphony*, Op. 21. The last sound, at R, is apparently very brief—its notation indicates that it lasts for only about a quarter of a second. But the piece obviously continues beyond this quarter second. Beethoven has notated the ending some 2 seconds later with the double bar at S, a projection I have indicated beneath the score. This is the minimal duration of the last sound. Incidentally, it may be possible to hear a greater projection marked T' above the score. This latter projection is not so clearly defined and requires our perception of the duration marked T, a perception that depends on the performance and our attentiveness. Later I will discuss some of the vagaries of projection.

Returning to our simpler example in Figure 8.2, the second impulse, through its participation in the formation of the first duration, projects a duration commensurate with the already formed first duration. In this respect, the initiation of the second impulse is different from the initiation of the first impulse, whose duration was to be determined. But this novelty does not qualitatively distinguish the two durations; that is to say, the first duration does not stand out from the other members as qualitatively distinct. This novelty does, however, for the first time enable us to predict a future event—the initiation of the third impulse. Again, the projection of the second duration is not in itself prediction—the duration is not produced by imagining a third event, and we will not be disappointed if a third event does not occur.

However, we can predict if we want to, and if a third event does occur we judge its arrival on the basis of the projection. If the new event arrives a certain length of time before the completion of the projected duration, say at three-quarters of this duration, we will feel that it has come "too soon," that it interrupts the course of events, or perhaps that it signals an acceleration. I should mention here that acceleration in this sense is something of a paradox—it is the perception that durations are in some sense equivalent, but that they are getting shorter or speeding up. If the new event arrives a certain length of time after the completion of the projected duration we will sense delay, hiatus, or slowing down. Such possibilities are too complex to discuss here in detail. They are determined by the relative

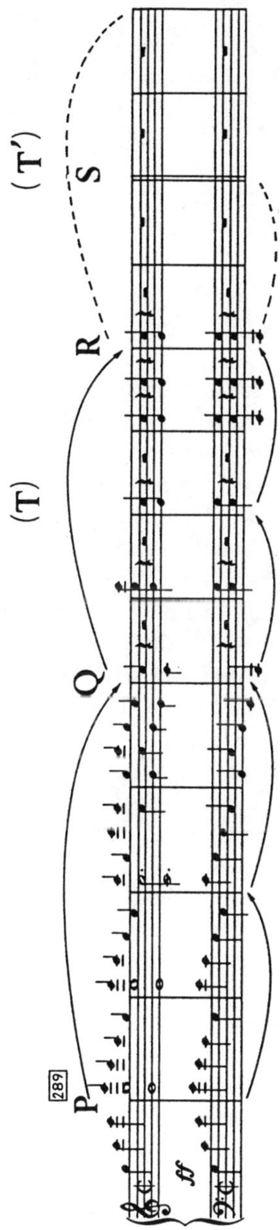

Example 8.1

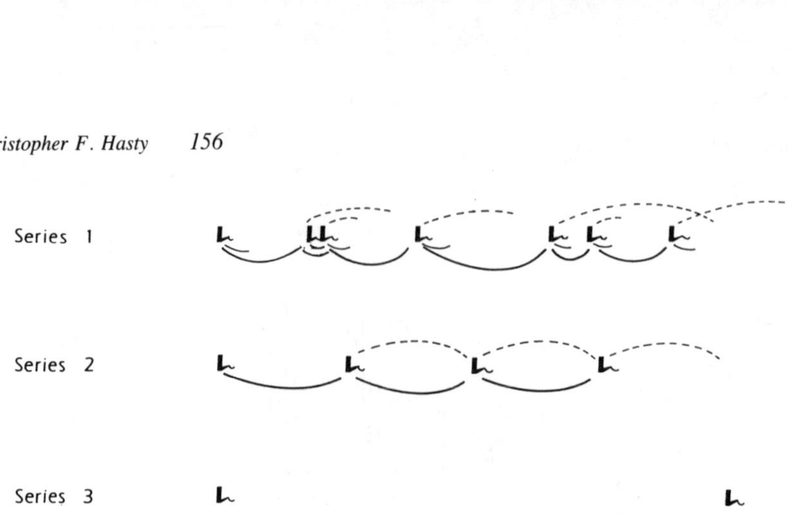

Figure 8.3

lengths of events, the time scale, and the qualities of events, all of which determine grouping. But I should point out that we tolerate considerable inaccuracy in our judgments that durations are equal or equivalent. Performances of series 2 will be heard as fully periodic even if the intervals are far from equal by the clock. This is not simply an unfortunate flaw in our perceptual capability. After all, we can accurately discriminate differences in very small durations. It is rather a considerable practical advantage, for we cannot count on the world being precisely periodic. If we are to act successfully (and have time enough to do it) we must be flexible and able to modify our grosser predictions as the occasion arises.

I conjecture that our fine discrimination of duration can come in handy as we more closely approach the new event. But we would not have time to act on this more detailed information without an earlier approximation. One of the things that would hopelessly confuse and evade us if our projections were absolutely fixed is an object that accelerates or decelerates.

It has taken a long time to get to our third impulse. This is as far as we need to go at this stage. Nothing new to our understanding is added by a fourth impulse. Runners, for example, need only three commands—"on your mark," "get set," "go!"—anything more is superfluous. We did not even need the third impulse to speak of meter; nor would runners need the command "go!" were our sense of projected duration sharper. In fact, this points to a flaw in my initial demonstration and to the reason it may have been difficult to perceive the distinction I was trying to point out, for it is not at all difficult to perceive meter.

As I have indicated in series 1 of Figure 8.3, any two successive impulses will create a determinate duration—that is to say, there is meter also in more erratic series. All I was able to do here was to weaken our perception of meter by composing the series so that projections would not be fulfilled. As a result, sensations of "too early" and "too late," interruption and hiatus, arise to detach the components from one another. The fact that projection can affect the previous and the following duration is again evidence of the simultaneity of components. If I want to eliminate meter altogether, I should have to compose something like series 3 of Figure 8.3.

I have certainly defined a duration, but, it is not, as I say, a palpable, determinate duration, and nothing is projected. We cannot feel this duration. But we will be in serious trouble if events of 6 seconds' duration escape our comprehension. Fortunately, we can comprehend the duration of such an event if it is rhythmicized. Again, meter is not the only means for creating rhythm, but it is the most powerful means of creating relatively long, palpable duration. To accomplish this, we instinctively attempt to consolidate events as components of a greater event, thus creating what may be called a metrical hierarchy. This hierarchy is usually called meter; but, as I have said, meter precedes its hierarchization.

In Figure 8.2 I have indicated the consolidation or grouping of the first two durations as a single duration which becomes determined only with the initiation of the third impulse. If this, in fact, has happened, we will find evidence in the projection of the third impulse, for this impulse must necessarily project a duration commensurate with the duration labeled Y in the figure. This consolidation does not, however, obliterate the determinate duration of the first and second constituents, for the third impulse will also project that smaller duration; that is, we do not lose our finer discrimination in apprehending a larger event. Musicians refer to this latter aspect of the phenomenon as subdivision. If, as in Figure 8.2, a unitary duration Y has been formed, the third impulse will have functioned as a second constituent. The second impulse will have relinquished some of its "secondness." Musicians express this nicely by counting in such situations not "one, two, three," but "one, and, two." The second impulse in a sense recedes as an event; hierarchically speaking, it suffers some degradation and is traditionally called a weak, or bad (*schlecht*) beat. The question of how this distinction of weak and strong, good and bad comes about has been the bane of metric theory, where it has usually been answered, somewhat tautologically, by replacing the terms *strong* and *weak* with *accented* and *unaccented*.

Returning to Figure 8.2 and the runners' instructions, notice that I have supplied a very dramatic accent to the third impulse (we can imagine that a pistol is fired on the command "Go!"), but such an accent in the sense of an objective, supplementary marking is not necessary for the creation of hierarchical grouping. We will spontaneously group a series of objectively undifferentiated pulses, say the ticking of a clock; and with some effort we can freely alter the perceived grouping. Although it is true that some of these pulses will be heard to be distinctly louder than others, one might just as well say that the hierarchy creates accents as say that the accents create the hierarchy. Indeed, in the case of subjective grouping, it would seem that our inclination to perceive larger durations takes precedence over the fact that we mark such durations with an increase of intensity.

It seems to me that certain eighteenth-century theorists expressed an understanding more subtle than ours with the notion of *quantitas intrinsica* (Houle, 1987). In their view, meter (or, as we now would say, metrical hierarchy) arises from an innate proclivity we have for grouping beats. Extrinsic quantities—what we could call various forms of accent—very often correspond to the basic opposition of "good" and "bad" beats, but do not in themselves create this opposition and can lack alignment with intrinsic quantities without thereby compromising the meter.

If this phenomenon is, like meter, mysterious in its origins, it is clear in its effect and purpose. While our runner, on "Go!", will have no use for the perception, an aurally attentive spectator could sense a projected duration of about 3 seconds, and come to perceive a palpably extended duration of the magnitude of about 6 seconds—a duration that is not perceived in series 3.

How long can such durations be extended? It is difficult to say. Psychologists have measured aspects of this perception, generally by asking a subject to reproduce the total duration of a series of pulses, and have gotten a variety of answers. But while effects of an event may endure in memory, the event does not so endure. Also, a series of pulses is for our purposes an artificially simple test. To create duration, music employs, among very many other things, tones. Tones are not simply qualities that are attached to pulses. We quite properly speak of "the duration of a tone" and not of "the tone of a duration." Tones are extraordinarily important for musical rhythm because, within the context of a tonal system, they can endure for a very long time after they have objectively ceased to sound—in a way similar to that in which the sound of the first impulse in Figure 8.2 endures for the third impulse. A highly organized pattern of tones can achieve greater duration than a pattern of toneless impulses. Not just tonal distinctions, but any of the distinctions we are able to make among sounds can jointly and inextricably function to extend palpable, determinate duration.

The limit to this extension is generally called the musical phrase. The phrase, like the spoken sentence, or like any event, is created not simply by articulation or the immediate discontinuities of beginning and end, for there must be a beginning and end of something, some process that coheres as an object for our attention. Even under the most favorable circumstances, phrases cannot exceed a length of about 12 seconds without losing durational coherence. As we approach this upper limit our feeling of determinate duration becomes diffuse; and not, I think, because we find it difficult to retrieve the initial stage in memory. Psychologists have shown that, if we choose, we can recall the initial stage much more clearly than we can the intervening stages. This diffuseness, like that of projection, serves to give us the opportunity to revise our interpretation, possibly to prepare us to understand the present event or some phase of it as part of a larger event. As phrase joins phrase we will rarely be able to feel their combined duration as a definite intrinsic quantity, but this is not to say that meter has ceased to function, for the formation of the second phrase will be in part determined by the metric-durational character of the first phrase. To illustrate this process, I would like to examine another excerpt from the first movement of Beethoven's opus 21. This excerpt, shown in Example 8.2, is in a sense the beginning of the movement proper—an allegro movement preceded by a slow introduction of 12 measures.

In Example 8.2 I have greatly simplified the score, but this reduction should not present difficulties for the reader who wishes to follow the course of the music in a recorded performance. (The reader can consult Example 8.3, constituent A, for a more detailed notation of the first four measures.)

Before entering into an analysis of the passage I will explain some of the symbols that appear in the example. Circled numerals above the score refer to

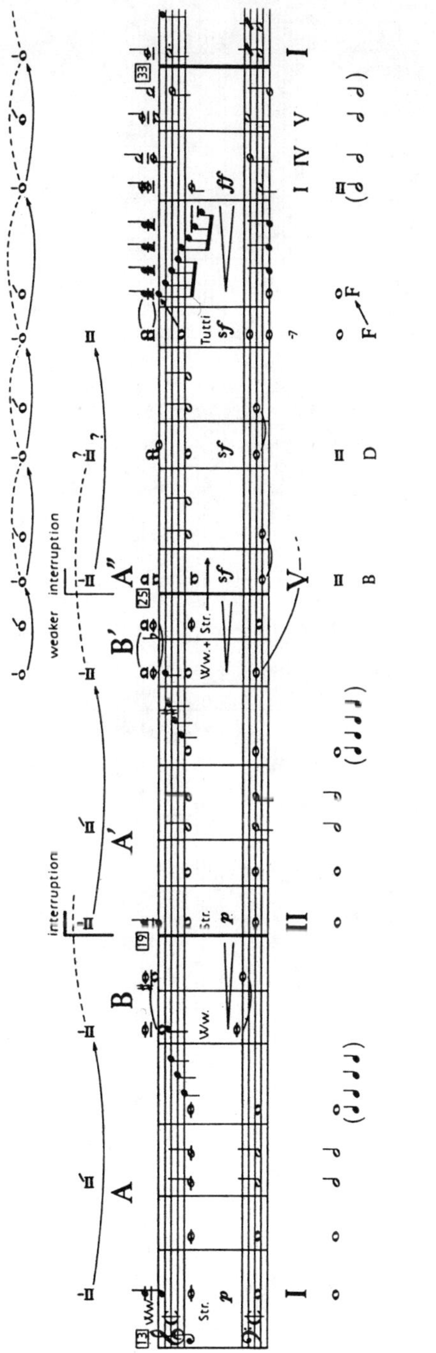

interruption
weaker interruption
A
B
A′
B′
A″
13
19
25
33
Str.
Ww.
Ww.+ Str.
Tutti

Example 8.2

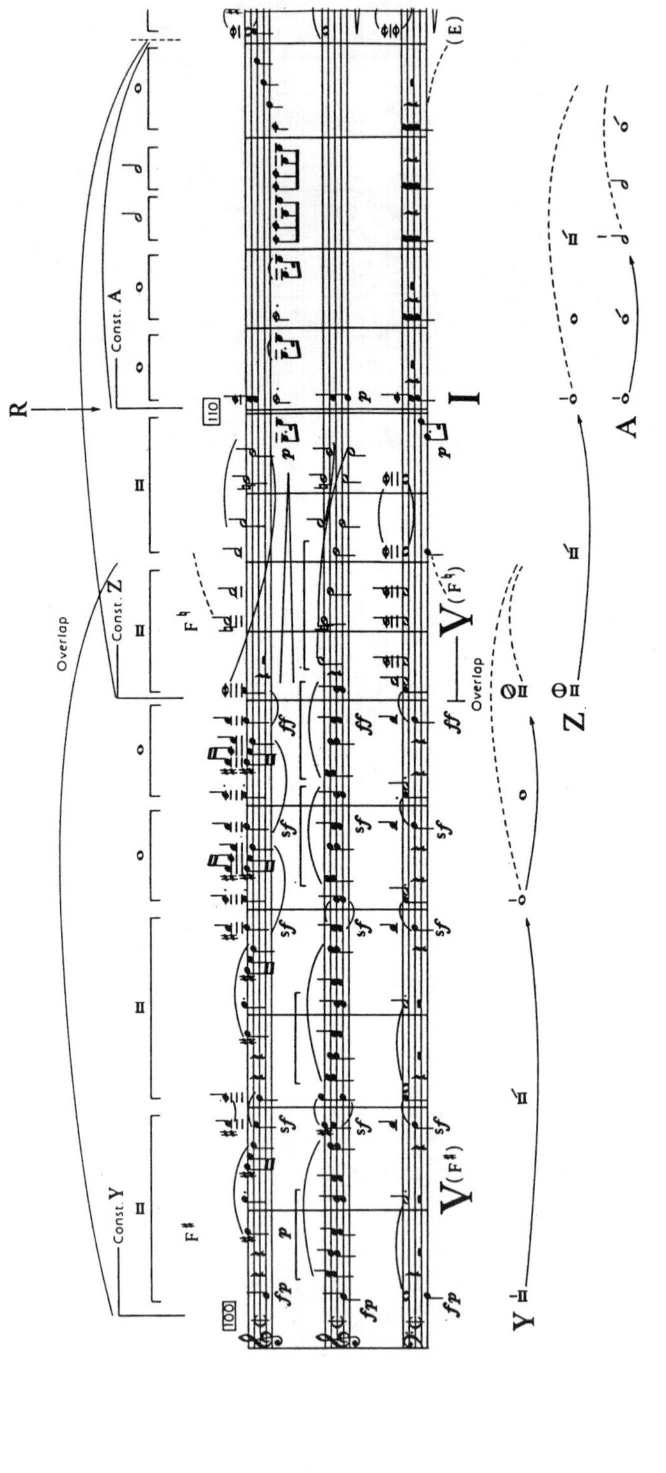

Example 8.3

measure numbers; thus, the first section of the allegro begins with the thirteenth measure, and the next section begins in measure 33. Strings (str.) alone perform measures 17 and 18; winds and strings are combined from measure 23 on; and all the instruments of the orchestra are employed in measures 29 to 32. Immediately above the score, I have proposed a metric interpretation similar to that used in Figure 8.2. Here the square symbols indicate breves and represent a duration twice that of a whole note or, in this case, a measure. The symbols | and / indicate, respectively, strong (or good) beats and weak (or bad) beats.

While it is problematical to isolate components of an event, for purposes of analysis I have identified a first constituent labeled A (measures 13–16), which might be called the first motive or theme of this part of the movement. A is a metrical unit, but it is not closed, for it ends in a projection commensurate with its already elapsed duration; that is, a projection of four measures. Now, as I mentioned before, longish projections are rather weakly defined in their duration, and so (as was the case in our attempt to hear the projection T' in Example 8.1) it will be difficult to precisely feel this full projected duration when nothing is happening. We will generally estimate short for fear of losing precious coherence. This phenomenon is well known to performers in the tendency to rush through rests and sustained sounds.

If we stop with the initiation of measure 17 we may in fact be inclined to settle for a projection of only two measures, a hierarchically lower projection. However, the effect of the larger projection is far from negligible in the present case, for it will give particular urgency to the next statement of the theme, A'. Beethoven chose to fulfill only half of the four-measure projection before initiating the second phrase in measure 19. Because the projection is not fulfilled we come to feel that the second statement of the theme has arrived too soon, that it has interrupted the projected duration. Similarly, the third statement of the theme, beginning in measure 25, can be heard to enter too soon; but here the effect of interruption has been mollified by several factors. The third statement has been modified in such a way as to allow it to be heard both as the initiation of a new phrase and as a completion of the projection.

Unlike the first two statements, the third statement of the theme leads to no interruptions—it plays itself out free of the abortive gestures that have led to it, and yet it has also absorbed something of those gestures in a sense I shall try to explain. The first two thematic statements create a durational unit of four pulses and thus project a duration of four pulses. The projections are cut off after only two pulses and for this reason sound incomplete or "too short." In the third phrase, four-pulse units and four-pulse projections have become more tenuous. It is difficult now to say what constitutes the third thematic statement, A''. Has the theme been condensed into two-measure units (measures 25–26 and again measures 27–28), or expanded to comprise six or eight measures? I believe that both of these interpretations are involved in the complex process that makes the third phrase the final stage in the synthesis of this section.

As this section concludes, two-pulse units that were earlier heard as anomalous have become absorbed into the metric structure and function to accelerate the pace of the music; not, as before, to hinder its progress. This acceleration and intensification culminates in the initiation of the next section (measure 33). The consolidation of measures 25 to 32 as an unbroken durational unit is supported by the adventures of a thematic rhythmic pattern shown at the bottom of Example 8.2. Constituents A and A' each present the pattern, long-long-short-short-long (two whole notes, two half notes, whole note). In the third phrase echoes of the pattern can be heard partly through the agency of contrapuntal–harmonic factors. A'' begins like A and A' but fails to present this pattern of durations. Instead, the pattern is enlarged to cover the entire eight-measure phrase (two breves, two whole notes, breve). Simultaneously with the expanded pattern, there is a telescoping of pulse from breve to whole note to half note—an acceleration that orients us toward the next section.

We have been attending to the form of this music, or better, to the *formation* of its comprehensible duration; that is, to rhythm. Rhythm uses meter, but it is certainly not to be equated with meter. For one thing, there is no metrical extension that encompasses the three constituent phrases we have identified; but it was the whole of these constituents that we have heard, and in fact more than this, because with the completion of this whole we were already projected into the next section of the work. What then is the rhythm of this whole? It is nothing less than the entire process by which a unitary and singular duration comes to be formed. What we have just heard, if we have been fully entertained, is rhythm—not meter, not tones, not instrumental combinations, not themes, though all these have served rhythm and take whatever meaning they have from the process they jointly serve. If we wish to *describe* rhythm our description must take the form of the reconstruction of a genetic process (to borrow Whitehead's phrase). Thus, in attempting to describe the opening of the allegro, I said that there appears an interruption that is mollified and finally assimilated into an expansive gesture that accelerates toward the entire duration's end—the initiation of the second section. This is obviously an inadequate description, though not, I think, wrong.

Certainly, there is much more going on here which could be noted to modify and further particularize my rather crude reconstruction. But questions of adequacy aside, it should be noted that to describe process I needed to refer to quasi-substantial things in the process of becoming; thus, *an* interruption that is mollified and assimilated, or the first, second, and third appearance of *the* theme. This manner of speaking is, I believe, appropriate owing to the simultaneity of the components of a duration. For example, how could the second interruption be mollified were not the first present in it—present not as recollection, for we surely do not recollect the first *as it was* while we attend to the second. We can say simply that we compare the two, but the comparison is numerically one.

I have concentrated on the conditioning of later components by earlier components. Nevertheless, this conditioning is fully reciprocal. The tonal intensification of the second appearance of the theme makes the first more relaxed than it might otherwise have been; for example, had the second phrase moved to A minor instead of D minor. That the second appearance of the theme is a step higher than the first makes our experience of the first appearance one of "below." Earlier components *need* later components—to keep them alive; to extend *their* duration; to explore their latent possibilities; even, I would say, to make them more particular. Components are simultaneous also in the sense that, once they have together formed a duration, they cannot be separated. This attribute of duration again is revealed in *durus*, hard, solid. Thus in the Beethoven, it is from the first section as a whole that the second section issues and not from any part of the first section. For this reason, the first section does in fact function to project the duration of the second section. This projection is not determinate—determinate duration is solely the contribution of meter and we have passed far beyond the reach of meter. Nevertheless, the beginning of the second section (*as* the second section) immediately promises a duration in some sense commensurate with the first section. It happens that the second section has precisely the same number of measures as the first section—20 measures—though the precision here I take to be more a mark of serendipity than necessity.

In saying that duration hardens and solidifies, I must stress the fact that while this happens, as it were, all at once (this once being the extensive present of the whole duration), it does not happen at an instant. There is no single moment at which an event begins or ends. This observation led Bergson to believe that articulation is illusory; but articulation and continuity are not at all incompatible. The first section of the Beethoven Allegro comes into being only with the second section, and we can know that a second section is being initiated only if we have felt the whole of the first section. Like projected duration, any duration must be open-ended. In an unpredictable world, we cannot afford to be inflexible in judging the articulations of events.

To illustrate the openness of beginning and ending, I would now like to consider the conclusion of the first part of the allegro, the part that is called the exposition. A reduced score is provided in Example 8.3. The exposition ends at the place marked R above the score. There immediately follows an exact repetition of the entire exposition which begins with the phrase constituent labeled A. Again, I have chosen an illustration that involves meter because the effect of meter is so tangible and because I can thereby avoid confronting the general reader with a more specialized discussion of harmony and counterpoint.

Constituent Y presents a very clear instance of the thematic pattern (long-long-short-short-long) we noted in Example 8.2. Constituent Z, however, enters before Y has played itself out. When shall we say that Y has begun? We are tempted to say that it begins where I have marked it in the score, at the initiation of measure 106. In a sense, this is true; but it is also true that we first come to perceive a new constituent Z considerably later, somewhere in the neighborhood of measures 108 and 109. I should also point out that constituent Z presents a change of harmony. The F-natural introduced in measure 107 replaces the previous F-sharp to signal a return to tonic harmony in measure 110. And yet we come to feel that the whole of measures 106 to 109 constitutes the new harmony and that the change was effected from the "beginning" of constituent Z.

The phenomenon we are observing, known to musicians as overlapping, is essentially different from interruption; its effect is known only after it has already

Example 8.4

taken place. Thus, there is no constituent Z until it has escaped the duration of constituent Y—as it comes to be, it will have been.

Moving now to measure 110, we can simplify matters by saying that the repetition of the exposition begins at R with the reinitiation of constituent A, but we should keep in mind that beginning is a process that takes time. As it begins, constituent A is to some extent a constituent of an earlier event (notice also the acceleration that joins Z and A). It will take some time for A to become the beginning of the repeated exposition. It is obvious but not trivial to note that the reinitiation of constituent A, while it may promise a repetition of the exposition, must fulfill that promise in order to be a beginning. Thus, (questions of stylistic norms aside) I can drastically alter the situation by changing the third and fourth measures of A as in Example 8.4.

In the music as it stands, constituent A will only gradually become detached from Z, as it emerges that the whole of the exposition leads to its repetition. In the same sense that the first impulse in Figure 8.2 will come to have been from the start metrically extended, constituent A in Example 8.3 will come to have been the beginning of the large-scale repetition; again, not through retrospection, for it will never have been other than the beginning of the repetition owing to the fact that from its start in measure 110 it has always held this possibility.

The two problematical instances of beginning we have encountered in Example 8.3, though they are special cases, point to the complexity of rhythmic segmentation and to the general difficulty of abstracting an instant or moment from temporally extended process. In its beginning, the event is in the process of formation and presents possibilities which are yet unrealized. Only as the event comes to be determined, does the beginning of the event come into existence. To argue otherwise is to separate the notion of beginning from that of event, just as the concept of instant is separated from time.

These difficulties would seem to be minimized in the case of ending, where formation is completed and possibilities are closed to further actualization within the context of the event. Might there not come a moment at the end of a process when duration has hardened as a fixed simultaneity of contents? There are, I think, a number of reasons for answering no. If the components of an event were ever open to possibility and interpretation, that openness must remain a part of their history—they would not have become what they are without that freedom and its exercise. Thus, what was unrealized in an event is also part of that event. In this sense an event is not a finite thing, and it is for this reason that we shall never be able to exhaustively describe a process. Also, to argue for a momentary coming-to-be of an event is to deny our very real feeling of duration's extensivity—a feeling of thickness, massiveness, or vastness. This is not a feeling merely of temporal lapse. Very weakly or ambiguously defined events do not inspire massiveness. The solidness of duration is a mark of richness and intensity of experience, not of fixity.

Processes cannot be broken into moments. If we wish to subject processes to analysis we can at best distinguish phases or stages. Beginning and ending are phases of a process and take time. For my last example, I would like to return to the ending of the first movement, shown in Example 8.1. I have already indicated that the composition does not end with the last note because this sound projects a duration of several seconds. Like all projections of this magnitude, this projection will not be precisely determinate—there will certainly be no instant at which the music will cease. But even if we could locate such a point, this would not be the ending, for the piece has already ended or has been ending for a long time—for about 25 measures at least.

This last phase of the piece is called its coda, or tail (a bit like the tail of a comet). Twenty measures before our illusory point R (shown in Example 8.1) there was a moment O that marked the conclusion of the work; and then another moment Q marked the conclusion; and then S too marked the conclusion. But the piece has not ended three times; it has ended once, all at once, with the entire duration. The reason it had to take so long to end is that the coda was projected by a very substantial section that led to the close. Even in leading to the close, the piece was ending. In fact, we shall find it difficult to say when the piece began to end. Again, we speak of the ending of the piece; we say that the piece ends. When we say good-bye to the last sound we are simultaneously saying good-bye to the movement in its entirety.

Our difficulty in trying to extract an ending is the difficulty of isolating phases of a process. Phases are not events—they are inseparable components of events. The concept of event is our attempt to claim autonomy for what is ultimately a component of boundless process. If this ultimate boundlessness is the claim of the universe, the boundedness of the event is the claim of the world. I have speculated in a variety of connections that our grasp of musical events is aimed at expanding our comprehension of a world of activity and motion. I have also speculated that the special world of music is deeply connected with the greater world from which all our temporal perceptions arise. If this is the case, the study of musical formation and musical rhythm in the largest sense could well serve as evidence in the investigation of fundamental questions of temporality and process.

References

Benveniste, E. (1971), *Problems in General Linguistics,* trans. M. E. Meek. Coral Gables, FL: University of Miami Press.

Houle, G. (1987), *Meter in Music, 1600–1800: Performance, Perception and Notation*. Bloomington: Indiana University Press.

Husserl, E. (1964), *The Phenomenology of Time-Consciousness,* trans. J. S. Churchill. Bloomington: Indiana University Press.

9

Metaphor and Deep Time in Music

Garry L. Brodhead

Abstract In a recent work, Stephen Jay Gould coined the phrase "deep time" to mean the cyclic records of geology, as contrasted with the metaphor of "time's arrow," standing for the human experience of passage. Both metaphors—time's arrow and time's cycle—may be usefully applied to music.

The observable characteristics of music, directionality through time and deep structure, lend themselves to the application of the metaphor of time's arrow. Cyclicity is less applicable. The application of both metaphors, however, makes possible a concept of deep time that can be experienced while listening to music. Recurring events that underpin the structure of a musical work, while they unfold in succession along time's arrow, also create circularity, thereby embodying time's cycle and providing a basis for a view of deep time.

Cyclicity is readily evident in music where the composer uses a circular formation in the layout of the musical score. Two examples of circular score format are works by Goerge Crumb and Baude Cordier. In these works the music we hear is a sonic representation of the circular format displayed by the score. However, music written in the traditional linear score format can also present aural cyclicity. In both formats, circular and traditionally notated, two musical factors that create cyclicity and evoke a sense of deep time for the listener are repetition and recurrence of large-scale musical elements—for example, themes, textures, harmonic and tonal structure—after contrasting material. Repetition and recurrence are critical as factors in the representation of time's arrow and time's cycle in George Crumb's "¿ De dónde vienes, amor, mi niño?" from *Ancient Voices of Children* (1970), J. S. Bach's Prelude in C major, *Das Wohltemperirte Clavier*, Book 1 (1722), and the Allegretto movement from Beethoven's Symphony No. 7, Op. 92 (1812). A careful examination of these works illuminates a number of important aspects of deep time and its relevance to our perception of musical events. Factors that create large-scale repetition and recurrence in musical works (e.g., beginnings, endings and their modifications) and the cognitive processes involved in their perception provide us with some of the tangible manifestations of deep time.

> I looked at Hampton's Throne and I saw Burnet's frontispiece. The two structures are identical in concept; they display the same conflict and resolution between time's arrow of history and time's cycle of immanence [Gould, 1987, p. 184].[1]

In his book *Time's Arrow, Time's Cycle,* Stephen Jay Gould presents a compelling explication of geological deep time in which the dual metaphor—time's arrow and time's cycle—provides a framework that accounts for the temporal properties of geological formation. The monumental scale of geological deep time makes welcome the use of metaphor to render the ideas in question more manageable. As Gould observes, "Deep time is so alien that we can really only comprehend it as metaphor" (p. 3). While the scale on which geological deep time occurs differs from the duration of a musical work, which is infinitesimal by comparison, the dynamic processes observed through the functions of the elements at work in a piece of music (in tonal music, its melody, harmony, tonality, textures, etc.) and experienced during its hearing exhibit similar temporal characteristics and properties in a microcosmic context.

Gould's view of geological deep time, as presented in his theory of punctuated equilibrium, is that "cycles of uplift and erosion" (p. 6) account for a geological process in which there is both a linear succession of causative events and a succession of cycles described by a sequence of events recurring in similar order. Thomas Burnet, a historical geologist upon whom Gould relies for the dual metaphor, saw geological formation as the result of the tension between uplift and erosion, a duality of narrative and vector (1691). Gould observes that Burnet's narrative and vector result in "repetition *with a difference*" (p. 56). From this perspective Gould draws together the traditional metaphors of time's arrow and time's cycle as a basis for explaining the linear and cyclic character of geological formation and change. This dichotomy embodies the tension between linear succession and sequential recurrence. While either may be seen as characteristic of temporal process at a specific moment in the evolution of dynamic phenomena, both characterize the large-scale structure and deep-time properties of these phenomena.

The dual metaphor of time's arrow and time's cycle juxtaposes the divergent views of temporality that each conveys and, as Gould demonstrates, yields a concept of deep time that not only explicates certain temporal features observed in geological formation, but also provides potential linkage with questions relating to structure and deep time in other disciplines. Linear time, the domain of time's arrow, accounts for the properties of succession that are exhibited by events in and through time. Events that constitute time's arrow are discrete and, together with other events in the succession of which they are a part, they create continuity and give direction to the vector of linear time. These events exhibit specific characteristics that limit their structural importance and, as a result, they are situated at the bottom of a hierarchical structure. Cyclic time, on the other hand, is defined by the course of

[1]Hampton's throne is a room-size folkart sculpture which depicts in a circular arrangement of objects a narrative of events in Judeo-Christian history. The sculpture is located in the National Museum of American Art, Washington, DC.

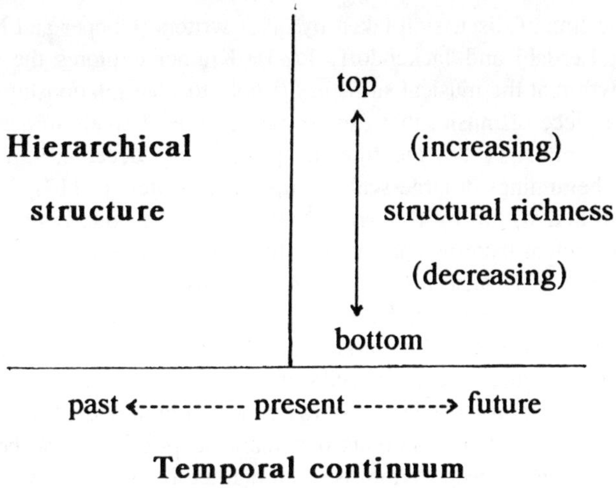

Figure 9.1. Coordinates of hierarchic structure and temporal mode in dynamic systems that exhibit deep time.

large-scale repetition or recurrence that events follow as their path continues to define a unity of succession. Events that constitute a cycle are interdependent and are related through varying degrees of structural richness. Accordingly, cyclical structures are of a higher order within a hierarchical system than linear events. The temporal properties manifested by this duality account for a deep time that is applicable in dynamic systems that exhibit both phenomenal unfolding within a continuum of past, present, and future and structural depth. Figure 9.1 illustrates a dynamic temporal system based on these two properties.

Music, when considered as a temporal art, exhibits just such a dynamic system. Events that constitute a piece of music succeed one another along the temporal continuum bounded by past and future and, in most styles and genres, these events are related also within a hierarchy of structural levels of varying degrees of importance. It is in the leap from a time of static metaphor (McTaggart Series A) (Gale, 1968, p. 9) to the dynamic time of experience (McTaggart Series B) in which the listener associates a musical event being heard in the present through time with related, but widely separated events previously heard in the past, and events expected to be heard in the future, that the duality conveyed by both time's arrow and time's cycle triggers an awareness of deep time for the listener.[2]

Two properties of deep time in music, the time evoked by events that organize and articulate the large-scale structure of a musical work, that are examined by Jonathan Kramer in *The Time of Music*, are deep meter and deep rhythm (1988).

[2]Cyclic time in music is not literally circular. It is an evoked time that relies on the recurrence of events, or the repetition of events which results in cycles that resemble other experienced recurring phenomena (e.g., going to class at a particular hour each day or experiencing a similar event on a regular basis, such as a holiday, anniversary, or birthday).

Following the line of discussion taken by other writers (Cooper and Meyer, 1960; Cone, 1968; Lerdahl and Jackendoff, 1983), Kramer explores the properties of meter and rhythm at the musical surface, the note-to-note relationships heard when listening to a piece of music, that can be translated effectively to deeper levels of organization. Kramer (1988) identifies deep meter as "accented timepoints" that articulate the beginnings of large-scale groups of measures (p. 117). These measure groups are unified within by properties that convey the underlying structure of a musical work such as patterned harmony or the use of a particular register or timbre. As a unit a large-scale measure group is referred to as a "hypermeasure" (Cone, 1968, p. 79). Concomitant with the articulation of a large-scale unit by a metrically accented timepoint is a preceding large-scale "weak beat" or "up beat," a unit of less structural importance (Kramer, 1988, p. 116). At the same time Kramer views deep-level rhythm as large-scale grouping delineated by a wider range of structural features than the accented timepoints that mark deep-level metric beginnings. In contrast to deep meter, deep rhythm is often marked by arrival at a cadence which enunciates the end of a large-scale structural unit and by a beginning which is metrically weak (Kramer, 1988, pp. 116–120).

As a visualization of the dual metaphor in music, the linear arrangement of notes in notated music is an easily perceived example of time's arrow. As an experienced phenomenon, linearity is reckoned as sonic events directed through time. On the other hand, in conventionally notated music cyclicity is not visually evident. However, cyclicity is readily apparent in examples of music where the composer uses a circular formation in the format of the score. In George Crumb's *Ancient Voices of Children (1970),*[3] the movement entitled "¿ De dónde vienes, amor, mi niño?" is written in circular format (Crumb's "circle music"). Reference to the score (Example 9.1) reveals that Crumb attaches significance to the circle, calling it "Dance of the Sacred Life Cycle." It is beyond mere novelty that Crumb chose to employ a circular score format. Even if the listener does not see the score as the piece is being heard, the cyclic recurrence of events A–E produces an aural experience of cyclicity, a most appropriate means for Crumb to demonstrate the recurring musical ideas of this movement. After an introductory statement is sung by the soprano (the line of music notated across the top of the score), the sequence of cyclic ideas is begun. The five event classes, A–B–C–D–E, are performed in a carefully cued manner: A^1, B^1, C^1, D^1, E^1, A^2, B^2, C^2, etc., until the sequence arrives at C^3. With each A-event recurrence a sense of cyclic stability is evoked. The sequence of events has gone full circle and departs again through another rotation. A sense of deep-level metric beginning is perceived articulating a unit of deep time for the listener.

Another aspect of visual representation in "¿ De dónde . . . ?" that makes it an especially appropriate example of the dual metaphor is the repeated accompanying pattern, a bolero rhythm, that is notated across the bottom of the score. Heard in the percussion and continuing throughout the movement, the underlying repeated

[3]George Crumb, *Ancient Voices of Children* (New York: Peters, 1970), p. 4. (c) 1971 by C. F. Peters Corporation. Used by permission.

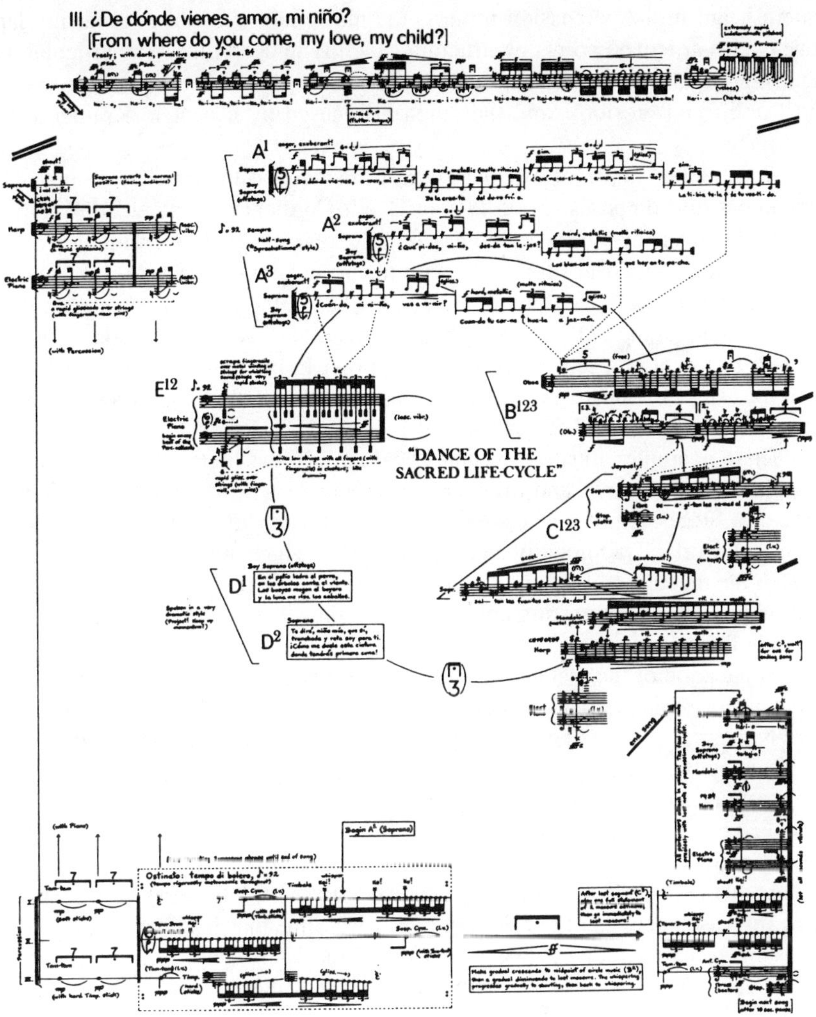

Example 9.1. Ancient Voices of Children. George Crumb (1970). " ¿ De dónde vienes, amor, mi niño?" (c) 1971 by C. F. Peters Corporation. Used by permission.

rhythm pattern is a literal musical representation of time's arrow, projecting the piece to its conclusion. In comparison to the succession of contrasting events A–E that constitute the cycles of deep time, the limited features of the brief, two-measure bolero pattern restrict it to the musical surface causing its underlying linearity. As a whole it is articulated by singular beginning and ending events. Reinforcing its directness are the *crescendo* and *diminuendo* which shape its entirety.

Crumb's music weds linear and cyclic elements into an aural form that embodies both time's arrow and time's cycle: while the continuous two-measure rhythm pattern heard in the percussion projects the music into the future, recurring large-scale events articulate points of structural stability in deep time. The articulation of recurring large segments of music is reminiscent of Gould's ''episodic renewal'' and ''uplift and erosion'' and the emphasis caused by accent in Kramer's deep-level meter.

A second example of circular score format is the three-voice canon, ''Tout par compas suy composes'' (c. 1400), by Baude Cordier (Example 9.2)[4] A perpetual canon,[5] the music may be repeated indefinitely, making its circular notation a literal representation of its heard form. As with the articulation of each cyclic recurrence in George Crumb's ''¿De dónde . . . ?'', the beginning of each restatement of the Cordier canon (at the ''O'' and ''3'' below ''M. Baude Cordier'') marks a structural juncture that evokes a sense of cyclic deep time. In both pieces, large-scale recurrence of a unit of music as well as the articulation of the beginning of that unit characterize the sense of deep time that is evoked.

These examples illustrate visually and aurally the literal application of both metaphors, time's arrow and time's cycle, to music and, through the cyclic recurrence of a succession of musical events, a basis for the articulation of deep time in a musical work. Traditionally notated music in which recurrence occurs, either through the repetition of a large section of music or the restatement of a large-scale musical unit after a contrasting unit(s), also exhibits features of cyclicity and evokes a sense of deep time.

Application of the metaphor of time's arrow to represent linearity in traditionally notated music is easily supported. As the music progresses forward in time from its beginning to its ending, it presents an aural representation of the notated events embodying a Series A succession. J. T. Fraser visualizes the process as the movement of a beam of light across a musical score much in the same way a score is tracked by its reader whose eyes follow deliberately the succession of notes from the beginning to the end of a musical work (Fraser, 1985, p. 183).

At this point it is helpful to distinguish between the linearity of time's arrow being discussed here and the category of musical time that Jonathan Kramer identifies as *linear time*. Kramer defines linearity as ''the determination of some characteristic(s) of music in accordance with implications that arise from earlier events of the piece'' (p. 20). Other characteristics that Kramer ascribes to linearity include: ''teleological listening,'' ''horizontal,'' ''motion,'' ''change,'' ''progression,'' and ''becoming'' (p. 63). These descriptors are appropriate for the quality of linearity in music being discussed here. However, Kramer also lists implication and goal-directed activity as characteristics of listening and musical structure, respectively, as important qualities that distinguish linear time from the other types of time he identifies.[6] Succession and direction in time, whether or not they are characterized

[4]Baude Cordier, ''Tout par compas suy composes,'' Musée Condé, Chantilly Ms.

[5]In a perpetual canon (also a circle canon) the voices enter at a designated interval in imitation of the leading voice and return to the beginning at the conclusion of the music, which may be repeated indefinitely. A round is a type of perpetual canon.

[6]Kramer identifies five types of musical time: directed linear time, nondirected linear time, multiply-directed linear time, moment time, and vertical time. Goal directedness is an important distinguishing characteristic of Kramer's taxonomy. In addition to explicating its unique qualities in defining the types

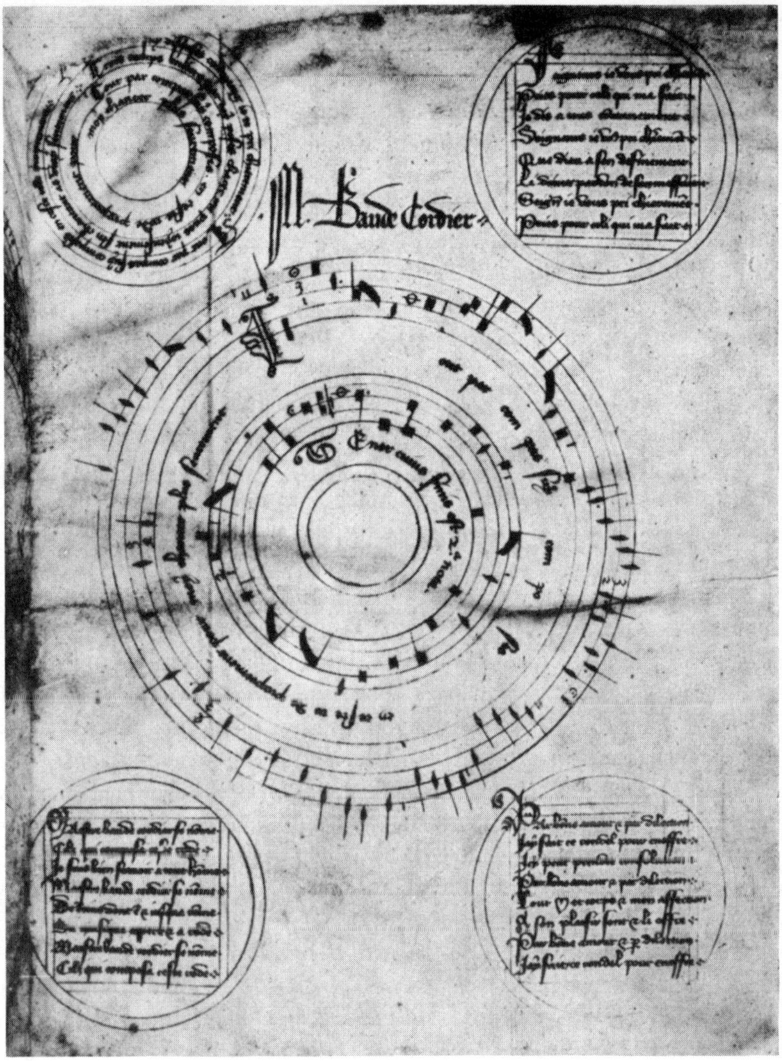

Example 9.2. "Tout par compas suy composes," Musée Codé, Chantilly Ms.

by implication or goal directedness, are primary characteristics of linearity as it is being considered here. Succession through repetition of surface events and dynamic change such as is heard in the accompanying bolero pattern in the percussion instruments in "¿De dónde . . . ?" embody linearity and evoke for the listener a

of time in music, Kramer's discussion points up the role that goal directedness plays in the evolution of musical styles.

sense of directionality that is not based on functional implication and goal directedness.

However, linearity as a function of ordered succession and cyclicity as a function of large-scale recurrence and repetition may be observed at deep levels which reinforce both the distinction among Kramer's types of musical time and his views on deep-level meter and rhythm. J. S. Bach's Prelude in C major, *Das Wohltemperirte Clavier*, Vol. 1, Example 9.3, is given by Kramer as an example of nonlinearity which he contrasts extensively with linearity (Kramer, 1988, p. 30). Kramer's definition of nonlinearity is "the determination of some characteristic(s) of music in accordance with implications that arise from principles or tendencies governing an entire piece or section" (p. 20). He cites the unchanging surface of Bach's Prelude—its consecutive arpeggios—as the basis for its nonlinearity (Kramer, 1988, p. 40). This consistency of texture and figuration highlights the importance that Kramer places on all of the elements of a piece in determining its structural and temporal properties. The continuous sixteenth-notes from the beginning may also be viewed, however, as a surface-level motoric feature which directs the music forward in time creating a sense of linearity.

Pitch function through the progression of chords also reinforces the sense of direction and linearity, but it can be viewed as cyclic as well. Measures 1–4 constitute a small cycle: the triad, C–E–G, the primary (tonic) triad upon which the harmony of the piece focuses, initiates the harmonic activity that progresses through measures 2–3 and arrives back at the same triad in measure 4. The first three measures cannot be considered superfluous based on the recurrence in measure 4 of the opening triad so soon after its first occurrence. Bach's mastery is not in question. Instead, the listener who is concerned with deep-level phenomena will perceive the cycle identified in measures 1–4 as a miniature of that generated by the harmony of the complete piece, as well as that heard on an intermediate level in measures 4–19. Bach returns to the C-major triad at measure 19 as well as at measure 32, but at both points the triad is an octave lower. Such a return is not unusual; in fact, it might even be anticipated[7] (Schenker, 1969, pp. 36–37).

Figure 9.2 illustrates the cycles described by this piece. In this analysis, deep time is articulated by the recurring C-major triad first heard in measure 1, then in measures 4 and 19, and finally in measure 32. Measures 33–35 extend the arrival at measure 32 by prolonging the tonic, C major, through its repetition in the bass.

The tonic triad occurrences that articulate these cycles also reinforce the large-scale, deep meter of the Prelude. Using Kramer's definition that meter is a "patterned succession of accented timepoints" (p. 83) that may be irregular on deeper levels (i.e., resulting from hypermeasures of differing lengths), the tonic triads mark variously the metric characteristics of the hypermeasures of which they are a part. The tonic triad in measure 1 is the local level metric accent that begins the piece. But the deep-level accent for the first hypermeasure, measures 1–4, occurs

[7]In Schenker's analysis the positioning of the C-major triad an octave lower is a distinction of tonal space not significant from the standpoint of structure defined solely by pitch class.

Example 9.3. Prelude in C major, *Das Wohltemperirte Clavier,* Volume 1, by J. S. Bach (1722).

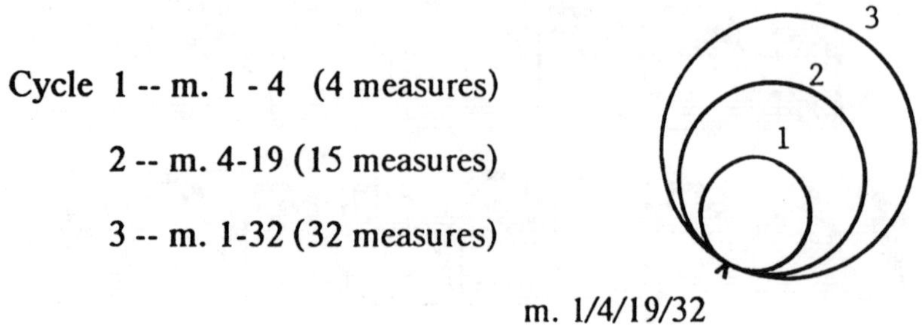

Figure 9.2. Cycles in Bach's Prelude in C major.

with the arrival of the tonic triad in measure 4. These four measures follow the common four-beat metric pattern, weak–strong–weak–strong, respectively.

The tonic triad in measure 19 also marks the end of a hypermeasure, that of measures 12–19. (An intervening hypermeasure, measures 5–11, is not part of the cyclic plan being discussed here.) Here the triad is a weak beat ending, but for a hypermeasure eight measures in length. Measure 20 articulates the beginning of the final hypermeasure which lasts to measure 32, followed by three measures of extension. This 13-measure hypermeasure illustrates another traditional metric plan, that in which both the beginning and ending are accented.

The sense of deep time evoked by the articulation of large-scale repetition and recurrence and the recognition by the listener that a section of music heard previously has recurred, translates the metaphors of time's arrow and time's cycle to the cognitive domain. An underlying criterion for the application of metaphor is that it is useful only when it conveys the meaning of the subject it exemplifies (Goodman, 1976, pp. 52–71; Greene, 1984, pp. 15–17). Or as Greene states: "Seeing [hearing] an exemplification is a matter of recognizing the aptness of the metaphor" (p. 16). The dual metaphor, time's arrow and time's cycle, is representative of linearity and cyclicity in music, not only as an intellectual observation of morphological organization as has been shown, but also as perceived aural form.

Music is perceived as a succession of ordered events that unfold in time. This linearity is perceived within the breadth of the psychological present or, as Fraser has identified it, the musical present, normally a duration of about 8 seconds (Miller, 1956, pp. 81–97; Jones, 1976, p. 344). The music constituting this unit of time is perceived by the listener as a psychological construct, a reality which is processed in its entirety in short-term memory. Because the listener organizes this perception based on the syntactic properties of the music being heard, that is, listening from the beginning of a musical unit—motive, phrase, period, section—to its end, the perceived unit occupies a psychological present for the listener, a "musical 'now' " using Fraser's terminology (1985, p. 183). The attentive listener will usually join together successive units that constitute the music being heard. For this listener, perception is therefore a succession of psychological or musical presents which recede from short- to long-term memory as each present is replaced by the next.

This process describes the activity of time's arrow as it is borne ahead by the succession of events and musical units that constitute a piece. At this level of perception the surface structure of the music evokes in the listener a sense of linear time.

At the same time, time's cycle is a metaphor that is realized aurally as the listener draws together structurally related events and sections that may be widely separated in a piece of music. At this deep level of cognition long-term memory is invoked. An event succession that constitutes a psychological present previously heard is drawn out of long-term memory when the succession is repeated or when it recurs (Klatzky, 1980, pp. 180–186; Gjerdingen, 1988, pp. 59–63). If this association with previous music were not made, as listeners we would not appreciate the formal and aesthetic value of repetition as a significant factor in providing unity in the music we hear. The perception brought about by the association of events of the musical present and those heard previously is a result of retrospective cognition in which deep-time perspective momentarily supersedes the linearity of surface-level tracking; the sense of having heard previously, a musical *déjà vu*, occurs. The deep-level perception evoked reinforces the sense of cyclicity and time's arrow, of the music going around again.

Integral to the perception of surface structure as linear and of certain deep structures as cyclic is the cognitive processing done by the left- and right-brain hemispheres. Recent research by cognitive psychologists has begun to provide evidence of the perceptual functions of the hemispheres. Although this information is not conclusive, it has opened the way for musicians to begin to consider important questions relating to the perception of musical time. In a discussion of left- and right-brain hemisphere functions vis-à-vis their roles in processing musical perception, Kramer identifies the left hemisphere as the processor of serial information. According to Kramer, the left hemisphere "understands time as containing a sequence of events" (p. 10). He also identifies the left hemisphere capability as "analytic," "deductive," "discrete" and "sequential" (p. 9). In processing music the left hemisphere tracks the succession of events heard in the music and orders them in short-term memory, providing continuity to the succession. This analysis suggests that the function of the left hemisphere is linear.

In contrast, Kramer identifies the analytic processing accomplished by the right hemisphere as "holistic." The right hemisphere "looks at wholes," "sees correspondences and resemblances," and "understands time as containing a complex of events" (p. 10). In processing music the right hemisphere organizes the large-scale features of a musical work. Repetitions that exceed the duration of the psychological present and large-scale recurrences that invoke the cyclic metaphor and evoke deep time are right-hemisphere functions. From this discussion, meaningful application of the dual metaphor can be extended beyond its use as a literary device to an effective means for identifying properties of deep time in music, one that is supported by psychological and neurological functions.

In pieces where sectional repetition is prominent, a balance is experienced between perceptions of linearity and cyclicity, reinforcing the dual metaphor as a property of deep time. One such piece in which the dual metaphor can be applied effectively is the second movement of Beethoven's Symphony No. 7, Op. 92 (1812), an Allegretto. This well-known work is a uniquely treated theme and variation movement in which a contrasting large-scale section is interposed between two framing sections in which the variations occur (Radice, 1988). For the purposes of this study, the discussion will focus on the first large section of the movement—the theme and the first three variations—although the ideas presented can be applied to the third large section of the movement as well. Section two employs other structural and temporal processes. Refer to Example 9.4a. Repetition of the second phrase of the theme (measures 11–18) in measures 19–26, is the listener's first experience of large-scale repetition and cyclicity which results in a sense of deep time. Here measures 11–18 are heard in a distinctive, low register voicing leading to what sounds like the end of a two-phrase musical unit—a period. (Phrase 1 consists of measures 3–10, phrase 2, of measures 11–18.) However, repetition of these measures in 19–26 draws the listener back in time, retracing the path of the previous phrase hearing. The time sense evoked during the second hearing is unique compared to the first. The listener realizes that this music has just been heard, but listening through the repetition evokes the sense of temporal compression, a recycling of the music measure by measure from its beginning to end, but seemingly in a different time. Here we are reminded of Gould's "repetition with a difference." The experience is not just hearing the music again, but instead coming around again through the unique harmonic progression and registral treatment that was. Each variation affords the listener a similar experience of cyclicity when the repeated second phrase occurs, dressed in a figural expansion of the theme. After variation 1, the experience is almost predictable, except for Beethoven's skillful use of a change of tonality from C major back to A minor, from the beginning to the end of the repeated phrase, which moderates the listener's expectation for repetition in each variation.

An average real-time performance of each phrase of the theme is approximately 20 seconds. Because there is little change of material within any of the phrases, the attentive listener is able to track the progression of chords and simple melody of each phrase as a perception constituting a single psychological or musical present, despite its length.

On another perceptual level the listener is able to recognize the recurring music of the theme as each of the variations begins. This 24-measure unit becomes the basis for a cycle of greater duration than that of its repeated second phrase. With the beginning of each variation, another cycle is articulated and a sense of large-scale deep time is evoked. Examples 9.4b–4d show the beginnings of variations 1, 2, and 3, which are identified here as cycles 2–4.

According to this explanation, two levels of deep time result: that evoked by the repeated phrase within each section, and that evoked by the section itself. Figures 9.3a and 9.3b illustrate this discussion. On level 1, that created by phrases, there is a balance between linear and cyclic time. Phrase 1 is not repeated and as a result is linear. Direction toward the end of the phrase is projected by the repeated surface rhythm, 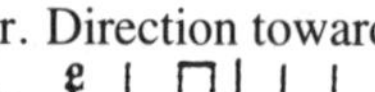 , which will be discussed later, and by the simple harmonic progression from a minor in measure 3 to C major in measure 20. As we

Example 9.4a. Symphony No. 7, Beethoven, Allegretto, m. 1-26 (Cycle 1).

Example 9.4b–9.4d. Symphony No. 7, Beethoven, Allegretto, m. 27–42 (Cycle 2-beginning); 4c. m. 51–58 (Cycle 3-beginning); 4d. m. 75–82 (Cycle 4-beginning).

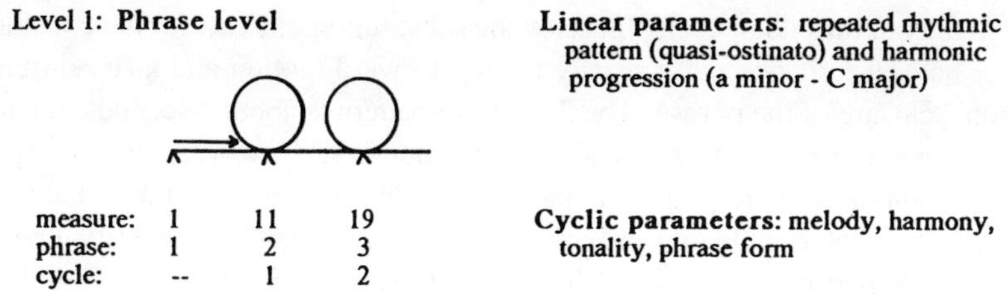

Figure 9.3a. Linear and cyclic design for the theme and each variation. Symphony No. 7, Beethoven, Allegretto, m. 1–27.

have noted, phrases 2 and 3, measures 11–18 and 19–26, are cyclic based on repetition.

On level 2, that constituted by the theme and its succeeding variations, cyclicity is created by the recurring 24-measure melody, the progression of chords and the tonality scheme which accompanies it, and the three-phrase form they support. On this largest level, there are four cycles. Each section, that is, the theme and each variation, is a cycle defined by the elements that delineate its beginning and ending. These features encode the theme for storage in long-term memory. As a listening experience, each restatement of the melody and its tightly aligned harmonic supporting component draws from long-term memory the recollection of previous statements of similar music and marks the beginning of another span of deep time at that level. Melody, harmony, tonality, and the formal sectioning that their patterned restatements articulate, function in this work as parameters that evoke a sense of cyclic time.

Time's arrow is conveyed in this same music by the elements that create linearity through each of the sections. The repeated 2-measure rhythmic pattern identified above, 𝄽 ♩ ♫ | ♩ ♩ , heard first in the violas, cellos, and basses in measure 3—a quasi-ostinato—articulates the surface of the theme and provides continuity and forward direction throughout this large section. The pattern is altered

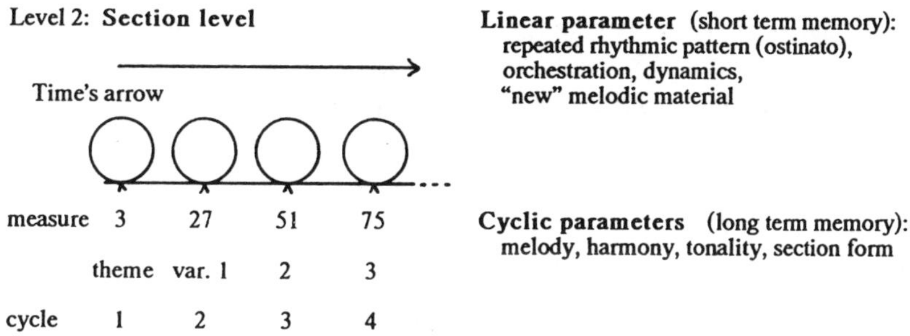

Figure 9.3b. Linear and cyclic design of the theme and variations 1–3, Symphony No. 7, Beethoven, Allegretto, m. 3–98.

only slightly at the end of each phrase by the substitution of a rest for the concluding quarter note, the silence momentarily halting forward motion and also reinforcing harmonic closure of the phrase. The 2-measure pattern is about 4 seconds in length. As it is heard, it is encoded in short-term memory. However, its brevity and subordination to the more prominent parameters identified above—those that constitute the fullness of a single musical present that cycles constitute—identify it perceptually as only part of such a unit. Without the larger ordering brought about by the melody and harmony organized by phrase and section, the elements evoking time's cycles, the repeated rhythmic pattern would result in a predominantly linear time experience for the listener, one based on a succession of short-term memory encodings.

Two other elements that create linearity through their consistent and unvaried use are orchestration and dynamics. Each section employs an unchanging group of instruments. Cycle 1—the theme, measures 3–26, uses violas, cellos and basses; cycle 2—variation 1, measures 27–50, adds the second violins. Cycle 3—variation 2, measures 51–74, increases the orchestration through the addition of more violins and woodwinds in measure 67. At measure 75, cycle 4—variation 3, the full orchestra is employed. Through increments the orchestration is expanded, each section establishing a plateau on which homogeneous instrumentation results in a sense of linearity through time for the listener. At the same time, the change of orchestration at the beginning of each variation marks the juncture which reinforces the beginning of a new cycle.

Within the same music, the use of dynamics complements the treatment of orchestration and reinforces the listener's sense of linearity within the section. Beginning softly at the level of *piano* (p) in measure 3, each section maintains the dynamic established at its beginning. As is the case with orchestration, dynamic change occurs at the beginning of each variation, further reinforcing the sense of cyclic time. Each cycle brings with it an increase in dynamic intensity both through the addition of instruments and the deliberate marking of a louder dynamic. Cycle 4 culminates this activity with a *fortissimo* (ff) dynamic.

Finally, the linearity of time's arrow is projected by the inclusion of "new" melodic material in each variation: measure 27—viola and cello; measure 51—viola and cello; and measure 75—viola, cello and bass.

The unchanging identity of the 2-measure rhythmic pattern, as well as the incremental treatment of orchestration and dynamics by section, cause a deliberate, forward pacing throughout this large section of the movement. At the same time, the recurring melody, harmony, tonal scheme and resultant form detail the series of four cycles which bind together this first large section. The delicate balance between linearity and cyclicity and the temporal senses they evoke create the "magic" which projects the listener through time—first hearing the clear, forward thrust of succession in phrases 1 and 2, then the "rehearing" (but differently) of phrase 2 through repetition—first linear time, then cyclic time, and then, as the music continues through the succeeding sections, at times both marvelously mixed.

Finally, there is possible one experience of cyclicity and deep time on the largest level—that of the entire movement.[8] This largest level of deep time is brought about by the recurrence of the woodwind chord that opens the movement (measures 1–2) as the concluding event the listener hears at the end of the movement. In addition to the two levels of cyclic time discussed earlier, the recurring woodwind chord identifies a third level. Such opening sonorities have a historical precedent and are seemingly superfluous where the structure of the movement is concerned. Similarly, Beethoven uses a repeated opening chord to begin the *Eroica* Symphony, No. 3, composed in 1803, approximately nine years earlier than the Seventh Symphony.[9] Beethoven's predecessors used this technique primarily for nonmusical reasons—to silence a talking audience and to signal the beginning of the music. For Beethoven, however, especially in the Seventh Symphony, the opening sonority is more a sonic veil that is lifted to reveal the beauty of the music which follows. The delicate quality of this woodwind chord indicates that its role is more than that of an attention getter. Where previous uses of an introductory chord announced a first movement opening, in the Seventh Symphony the energy and drama of the lengthy first movement are seemingly assuaged by the woodwind chord which begins the second movement. It is an effective separation for the sombre and quiet tone of the opening of the theme in measure 3. At the end of the movement the chord reappears in its exact scoring and signals closure to the listener—replacing the veil—and a final return to the starting point. This macrolevel recurrence brings the listener full circle inscribing the entire movement as a cycle on the largest scale.

In summary, although deep time in music is a subject in need of extensive investigation, application of the dual metaphor, time's arrow and time's cycle, to linearity and cyclicity in music is promising as a source for exploring the properties of deep time. Linearity, as directedness through time, and cyclicity, as repetition and recurrence of large-scale units of a musical work, together in certain contexts, may be experienced by the listener as concomitant qualities of deep time. The musical richness of works that exhibit both time's arrow and time's cycle is profoundly experienced by the listener as structure and time of great depth. Especially in works in which cyclicity is a characteristic of structure and time does the listener recognize the limitations of music that is only linear. As the study of time in music progresses, further examination of these and other properties in the relationship between structure and deep time offers direction for serious inquiry.

References

Burnet T. (1691), *Sacred Theory of the Earth*. London: R. Norton.
Cone, E. T. (1968), *Musical Form and Musical Performance*. New York: W. W. Norton.

[8]This view reflects thinking presented in a number of current analytical approaches that account for structure on the level of a musical work as a whole (e.g., Schenker [1979, pp. 3–21], Lerdahl and Jackendoff [1983, pp. 36–39]).

[9]A historic precedent to this opening is the ''hammerstroke'' beginning often heard in symphonies of the middle-eighteenth century. Hammerstroke beginnings are used by Mozart in three symphonies:

Cooper, G., & Meyer, L. B. (1960), *The Rhythmic Structure of Music*. Chicago: University of Chicago Press.

Fraser, J. T. (1985), The art of the audible "now." *Music Theory Spectrum*, 7:181–184.

Gale, R. M. (1968), *The Language of Time*. New York: Humanities Press.

Gjerdinger, R. O. (1988), *A Classic Turn of Phrase*. Philadelphia: University of Pennsylvania Press.

Goodman, N. (1976), *Languages of Art*. Indianapolis: Hackett.

Gould, S. J. (1987), *Time's Arrow, Time's Cycle*. Cambridge, MA: Harvard University Press.

Greene, D. (1984), *Mahler, Consciousness and Temporality*. New York: Gordon & Breach.

Jones, M. R. (1976), Time, our lost dimension: Toward a new theory of perception, attention and memory. *Psycholog. Rev.*, 83:323–441.

Klatzky, R. (1980), *Human Memory*. New York: W. H. Freeman.

Kramer, J. D. (1988), *The Time of Music*. New York: Schirmer Books.

Lerdahl, F., & Jackendoff, R. (1983), *A Generative Theory of Tonal Music*. Cambridge, MA: MIT Press.

Miller, G. A. (1956), The magical number seven, plus or minus two: Some limits on our capacity for processing information. *Psycholog. Rev.*, 83:81–97.

Morris, R. (1984), *Time's Arrows*. Aurora, CO: Schwartz.

Radice, M. (1988), Dual function and composite form in middle period Beethoven: The Allegretto of the Seventh Symphony. Paper presented at the meeting of the American Musicological Society, St. Lawrence Chapter.

Schenker, H. (1969), *Five Graphic Music Analyses*, ed. F. Salzer. New York: Dover Publications.

——— (1979), *Free Composition*, trans. E. Oster. New York: Longman.

No. 7, K. 45 (1768); No. 20, K. 133 (1722); and No. 27, K. 199 (1773). Haydn used hammerstroke beginnings in only two symphonies, No. 9 (1762?) and No. 65 (c. 1771–1773).

Perspectives

The literary arts are the foremost depositories of articulated views about time felt and time understood. But writers and poets can do something more than write about *time. Through the magic of the word they can create a variety of temporal experiences. The two papers that follow deal with both of these methods—the articulate thought about, and the creation of temporality—but they have two very different foci. Hasty writes of time and process in the works of the Russian poet Marina Tsvetaeva, while Soulsby explores the creation of temporality in literary experience.*

Marina Tsvetaeva was one of the great twentieth-century poets in the Russian language, yet her work is hardly known in the West. This fact, remarks Hasty, is clearly due to the difficulty in translating her rich and complex poetry. One of the external readers of Hasty's paper spoke of

> "The electrifying magnificence and lacerating pathos of Marina Tsvetaeva's song. I say *song* deliberately because it is sound and the shattering of sound into sharp and brilliant shards that distinguishes Tsvetaeva from all other Russian poets of her time. It is those shards that wound and excite the reader."

These shards are the "acoustic properties of words" discussed by Hasty where she speaks of Tsvetaeva as a mediator, through the sounds of words, between different modes of perception.

The metaphor of shards of glass may be extended with uncanny ease to the life of the poet. Considering how intimately her poetry and her views of time match her life, and how unknown she is in the West, it is appropriate to quote here a few lines from her extensive body of poetic works not related to her views on time. The quotes are woven around her biography as sketched in Hasty's paper. I shall add to that biography only marginally, where it is necessary for continuity.

Tsvetaeva was born into a family of Moscow intellectuals of the turn-of-the-century upper class, into a milieu where music and art, the speaking of foreign languages, and travel abroad were considered the obligations of the privileged. She was a strong, independent, and fiercely romantic woman who began writing poetry at the age of six and traveled to Paris alone when she was seventeen. In a poem she wrote during that period, called A Prayer *(Markov and Sparks, 1967, p. 429) she caught the mood of her adolescent spirit on the wing.*

I thirst for all roads—in one cup.

I want all things: the souls of gypsies—
To walk with songs and rob someone.
To hurt for all midst organ-playing.
To rush to war—an amazon.

She met Sergei Efron when she was eighteen, he seventeen, a cadet in the Officer's Academy. She resolved to marry him and so she did.

The revolution of 1917, as Hasty tells us, found Tsvetaeva with her two small daughters trapped in Moscow. Efron, a White Army officer, escaped to Prague. During the famine of 1919 she gave up her younger daughter to a state-operated orphanage, where she died of starvation. In a cycle of poems called Insomnia *(Tsvetaeva, 1986, p. 14) written during that period, she spoke about a world very different from her fantasy of a gypsy life.*

Who sleeps at night? No one is sleeping.
In a cradle a child is screaming.

She was of the wrong class, had the wrong sympathies, and had little love for the dictatorship of the proletariat (Tsvetaeva, 1980, p. 125).

So, mother and daughter—two lone wanderers.
The churlish mob surge, chiding, on at us. . . .
Turn homeward, daughter—tread submissively
The path of plain folk, of simplicity:
Back to the land of Dreams and Loneliness—
Where *we* are Majesties and Highnesses.

After her reunion with her husband and three happy years in Prague she, her daughter, and infant son traveled to Paris. Though, as Hasty relates, she was welcomed by the Russian literary community, her lack of interest in the endless intrigues of a rootless emigré community made her an outcast.

In a poem written in 1934, the young girl who thirsted "for all roads—in one cup" was homeless, homesick, terribly lonely, as sensitive as ever, but now sadly rather than romantically (Markov and Sparks, 1967, p. 447).

Alone, and with my bag from the market
To plod home on who cares *what* stone—
To some house. . . .

Each house is strange; each church means nothing.
It's all one and the same to me.
But if a bush looms by my pathway
Especially a rowan tree. . . .

Her husband followed her to Paris but by 1935 they led separate existences. Efron underwent a political change of heart, began working for the Russian Secret Police, and in 1937 he and their daughter Alja returned to Moscow.

By early 1939 Tsvetaeva's voice was that of despair. She ended one of her poems (Karlinsky, 1966, p. 98) with an apostrophe to God:

To Thy insane world
[There is only] one answer: a refusal

In 1939 mother and son followed Efron. Hasty's paper recounts the events from then on to Tsvetaeva's suicide. After her death her son volunteered for action against the invading Germans; soon thereafter he was reported missing. Her daughter Alja was released from labor camp in the early nineteen-fifties and undertook the tracing and editing of her mother's extensive literary heritage of poetry, prose, drama, essays, and memoires. She also placed a cross at her presumed burial place.

Tsvetaeva's real life story is that of her land, as is the story of Pasternak's fictional Doctor Zhivago. *In an earlier age, we find Dickens's Sidney Carton in* A Tale of Two Cities, *witness to the French revolution. Marina, having to give up her infant daughter in real life differs from Sophie's fate in Styron's fictional* Sophie's Choice *only in the species of the evil implicated.*

As we look further into the past, the distinction between fiction and nonfiction tends to blur. The poems and narratives, those great continuities of the cultural process remain. They acquire lives of their own, interpreted and reinterpreted according to the changing values of the historical environments. Wherein resides the power of literature and its ability to be endlessly reinterpreted?

In terms of what we learn from Soulsby, it may have its roots in the ability of the literary arts to create a spectrum of temporal experiences by taking inner journeys to moods of different temporal coherence. She speaks of the tendency of the mind to map experience into a coherent order, and do so against the pull of passion and instinct toward disorder, toward the less coherent experience of time felt. The creation of time in literature involves the skillful manipulation of qualitatively different temporalities always present in the human mind and among human feelings.

The conflict between order and chaos is a part of the dynamics of the human mind. It drives a multitude of its civilized and uncivilized expressions without determining the form of those expressions. Or, as Soulsby puts it, "Writers and poets of various ages and cultures have been engaged in the same process [of creating order out of chaos] but using different modes of expression."

Soulsby offers illustrations from the works of Franz Kafka, Milan Kundera, D. N. Thomas, Margaret Atwood, and Wang Wei, an eighth-century Chinese painter and poet. These and other literary artists create the nature of time that informs their material. The temporalities so created may be free of the iron rules of passage, or remain powerless slaves to them, or represent any and all stations and combinations in between.

When Tsvetaeva was thirty-two she published excerpts from the diaries that she had kept when she was twenty (Karlinsky, 1966, p. 289). She wrote that "The two things I love most in the world: song and formula. (This is my note of 1912: chaos and the victory over it!)"

This quote from the young Tsvetaeva helps us circle back to Soulsby, who sees in the victory over chaos the victory of the mind's need for temporal order, over the pull of emotions toward the chaotic and incoherent. This closure is a pleasing feature of the two essays that follow.

J.T.F.

References

Karlinsky, S. (1966), *Marina Cvetaeva: Her Life and Art*. Berkeley: University of California Press.

Markov, V., & Sparks, M. eds. (1967), *Modern Russian Poetry*. Indianapolis: Bobbs-Merill, 1967.

Tsvetaeva, M. (1980), *The Desmesne of the Swans*. Ann Arbor: Ardis.

——— (1986), *Selected Poems of Marina Tsvetaeva*, trans. E. Feinstein. New York: Dutton.

IO

Time and Process in Marina Tsvetaeva's Definition of Lyric Verse

Olga Peters Hasty

Abstract "Every poet is an emigrant. . . . An emigrant from Immortality into time," the Russian poet Marina Tsvetaeva observed in her 1932 essay "The Poet and Time." The works of this major avant-garde poet reverberate with an issue that had become a central concern to her Russian (and many European) contemporaries—that of probing the question of temporality and of developing a lyric style to reflect this quest of the poetic imagination.

In her highly expressive lyric verse and in her penetrating essays on poets and poetry, Tsvetaeva demonstrated and described her art as a means of transcending commonly accepted divisions that impede a wholeness of vision. Poetry was to heal the essentially illusory rifts in our perception of the world. It was to replace existing boundaries with new connections that expand perceptual possibilities and thus also meaning. To this end she denied a limited view of measured temporality and insisted instead on the far broader concept of process which became central to her definition of lyric verse. Indeed, lyric poetry is, according to Tsvetaeva, distinct from all other genres precisely in its *demonstration* of process. The emphasis on process establishes a relationship of interpenetration between part and whole and serves to obviate the apparent fragmentation effected by the formal characteristics of the genre of lyric poetry.

In this essay I seek to describe the background and development of Tsvetaeva's definition of lyric poetry as process and to examine the purpose served by this designation. To this end, I explore the artistic concerns to which Tsvetaeva responded in her writing and consider how they shaped and were in turn shaped by her definition.

The Russian poet, essayist, and critic Marina Tsvetaeva (1892–1941) defined lyric verse as process and maintained that it was this feature that distinguished it from all other literary genres. The theme of this volume—time and process—is thus an ideal context in which to examine the writings of this remarkable poet. Even as it deepens our understanding of Tsvetaeva's poetry, an examination of the purpose served by her designation illuminates many aspects of the problem of process. I propose here to outline the artistic concerns Tsvetaeva sought to respond to in her writing and to explore what her account of poetry as process entails.

The difficulty, not to say impossibility, of translating Marina Tsvetaeva's richly complex poetry has resulted in the fact that she is not as well known to Western readers as are, for example, her contemporaries Vladimir Maiakovskii, Osip Mandelshtam, Anna Akhmatova, and Boris Pasternak. Yet Tsvetaeva's contribution to the development of the genres of both lyric poetry and the longer verse form, designated in Russian by the term *poema*, is of great significance, and Joseph Brodsky (1978) is not alone in regarding Tsvetaeva as a leading poet of the age. Tsvetaeva's essays and correspondence mark her importance as a prose writer and thinker as well.

Tsvetaeva's artistic development coincided with the dynamic, fertile period of Russian letters of the first two decades of this century. Devoting herself early in life to the calling of poet, she remained apart from, but by no means deaf to, the many poetic schools and circles that blossomed during this period. Tsvetaeva's art is one of synthesis, and it is more fruitful to consider her poetic heritage in terms not of influence, but of confluence, as she herself suggests. She is an avant-garde poet whose dynamic and highly original style is organically linked with Russian, French, and German culture of the eighteenth and nineteenth centuries.

Tsvetaeva's first volume of lyric verse, *Vechernii al'bom* (Evening Album, 1910), was noted by critics for the remarkable intimacy and specificity of its thematic detail. In subsequent collections she developed rhythmic, semantic, and acoustic constructs of highly expressive mobility that took her (and those readers able enough to follow) into uncharted poetic territory.

Tsvetaeva's personal life was one of deprivation and tragedy. Her biography illustrates dramatically the extent to which her poetic life superseded the mundane. At the outbreak of the 1917 revolution, Tsvetaeva's husband joined the White Army and she was left alone in Moscow with her two small daughters, the younger of whom died of starvation. In 1922, upon learning that her husband was still alive, Tsvetaeva emigrated via Berlin to Prague to rejoin him. She was by this time a recognized poet. Three years later she moved with her surviving daughter and newly born son to Paris. There she was greeted enthusiastically by the Russian literary community, but much of the enthusiasm soon changed to hostility when it became evident that Tsvetaeva had no intention—in either her personal or her artistic life—of joining or supporting any of the literary or political circles that expected her allegiance. Her fierce independence of spirit and critical attitude toward influential critics resulted in a self-imposed isolation. As her verse became more difficult, the initially lionized poet encountered increasing hostility and a diminishing readership, a situation made all the more difficult by the fact that her publications were the major source of income for the entire family. In mid-1939, Tsvetaeva, giving way to the demands of her husband, son, and daughter, agreed to return with them to the Soviet Union. The fact that she left a large portion of her works in the West for safekeeping suggests that she knew there could be no place in Stalinist Russia for a poet like herself. Shortly after the family's return to Moscow, Tsvetaeva's husband and her daughter were arrested; her husband was later shot.

Her sister, as she now learned, had been arrested two years earlier. Tsvetaeva's situation was dangerous politically and impossible artistically. The demands of Socialist Realism, proclaimed in the 1930s as the only acceptable form for Soviet arts, had nothing in common with the demands she had herself determined. After war broke out with Germany in 1941, Tsvetaeva and her son were evacuated to a small town in the Tartar Republic where Tsvetaeva hanged herself on August 31 of that year. No one attended the funeral. Her grave was not marked.

In the course of the forty-nine years of her extremely difficult life, Tsvetaeva evolved an understanding of the poet's task based on an uncompromisingly exacting set of criteria. Poetry, according to Tsvetaeva, aspires to indicate from within the limitations of language, the infinite and uncharted realm of creative potential. The poet's task is thus a difficult one and can perhaps be compared to that of a cartographer who must depict a three-dimensional earth on a flat piece of paper. In the designation of lyric poetry as process Tsvetaeva found a solution to the many paradoxes and apparent contradictions attendant to this situation.

Marina Tsvetaeva postulates two separate realms of existence—the mundane world which is inimical to poetic concerns, and another world far removed from day-to-day cares, an ''elsewhere'' for which the poet longs. (See Vitins [1977, pp. 644–657] for a penetrating study of Tsvetaeva's other world.) What is essentially a German Romantic ideal becomes for Tsvetaeva the basis of the controlled, clearly articulated, and highly consistent poetics that she evolved in her writings. Tsvetaeva describes the ideal ''other world'' as a realm of infinite creative possibility and the source of poetry. It is a noncorporeal realm defined by an absence of commonly accepted boundaries, limitations, and relationships, which, according to Tsvetaeva, fragment and impede the wholeness of vision the poet seeks to recover.

Here we must observe that Tsvetaeva's ''other world'' is not simply an imaginary poetic paradise. It is perhaps best understood as Tsvetaeva's metaphor for a different mode of perception, free from predetermined, prefabricated constructs. It is the individually discovered, supple, and subtle perception of advance. However unarguable the various delimitations that govern conventional perception may appear, and however necessary these conventions may be in day-to-day life, they remain nevertheless oversimplifications that jeopardize the individual and idiosyncratic perceptual potentialities on which the poet must draw.

Tsvetaeva most frequently describes her ''other world'' as spatially removed from—usually above—the world manufactured by convention (c.f., e.g., *Novogdnee, Sivilla,* interalia [1980]). This distancing need not be conceived of exclusively as separation, for its primary function is to present a new perspective. In the Russian designation for the ''other world'' (*tot svet*), the world *svet* means both ''world'' and ''light.'' Thus the newness of perception that Tsvetaeva's other world represents is borne out semantically in the suggestion within the same word of a different light in which to see things. In this new light, perception dominated by linearity and causality is replaced by a more complex nonlinear and noncausal, creative one.[1]

[1]Tsvetaeva deploys this double meaning of the word *svet* in a number of works. Perhaps the most notable example of this occurs in her poem titled ''Novogodnee'' (''New Years'') (1980–1983, Vol. 1, p. 263), written as a letter to Rainer Maria Rilke after his death, in which Tsvetaeva asks him about the

In her writings (and her life) Tsvetaeva adamantly opposed definitions based on delimitation, and, in this connection, frequently expressed her hostility to the demarcation of space and time. She regarded the boundaries imposed by the material world as impediments to perception and believed that they had to be overcome before a genuine understanding and wholeness of vision could be attained. Thus, for example, Tsvetaeva speaks with animosity of her own body because it can present only an apparent and illusory definition of her being. Her oft-expressed preference for correspondence over physical encounters is based on a desire to circumvent this limitation.

Ideally, space and time are to be regarded as unlimited and unlimiting. They have the power to separate, to divide only in the ready-made world of the mundane. Distance can prevent only physical contact and bears no significance in the context of more profound unions. Thus, for example, in her 1932 essay titled "The Poet and Time," Tsvetaeva writes of herself in France and Boris Pasternak in the Soviet Union: "Boris Pasternak—there, I here, across all expanses and prohibitions, external and internal . . . , Pasternak and I, without conferring, think about the same thing and say the same thing" (Tsvetaeva, 1979, Vol. 1, p. 380).[2] Temporal designations, like spatial ones, also come under frequent attack in Tsvetaeva's writings. Tsvetaeva's last collection of lyric verse, *After Russia* (1980–1983, Vol. 3, pp. 11–128), which was published in 1928, provides several prominent examples of her attitude toward time. In a 1923 poem that she titles sarcastically "In Praise of Time," Tsvetaeva punctuates each stanza with an exclamatory expression of her antagonism: "Time! I cannot keep up." "Time, you will deceive me!" "Time, you will give me short measure!" "Time, you will betray me!"

Время! Я не поспеваю. **Время, ты меня обманешь!**

Время, ты меня обмеришь! **Время, ты меня предашь!**

She concludes the poem with a triumphant stanza addressing time: "For I was born past/Time! To no purpose and in vain/You struggle! King for a day: /Time! I pass you by."

Ибо *мимо* родилась
Времени! Вотще и всуе
Ратуешь! Калиф на час:
Время! Я тебя миную.

10 мая.

[1980–1983, Vol. 3, pp. 74–75]

---

other world and the new light in which he now sees things. All references to Tsvetaeva's poetry will rely on this edition. The volume and page numbers will be indicated in parentheses in the text. The translations are mine.

[2]Unless otherwise noted all references to Tsvetaeva's prose essays will be based on this edition. The volume and page numbers will be indicated in parentheses within the text; the translations are mine.

In another poem Tsvetaeva contemplates how victory over time can best be achieved (1980–1983, Vol. 3, p. 76). In the concluding poem of a lyric cycle titled *Trees* (1922) she juxtaposes the "injury of time" to the "coolness of eternity" (1980–1983, Vol. 3, p. 35). In the final poem of the cycle *Sibyl* (1922), the Cumaean Sibyl addresses an infant whom she encourages to cry, for as she explains, the infant has by means of its birth fallen from a far preferable undelimited state of existence into flesh and blood and into days and hours. She consoles it with the promise that death, described as a rebirth, will readmit the child to the flowing, unparceled realm of eternity.

In Tsvetaeva's juxtapositions of time and eternity we are able to discern more precisely the nature of her animosity. It is not time itself that Tsvetaeva objects to, but rather (as is also the case with her objections to spatiality) to its measurement—the ticking off of minutes and seconds, the designation of a space of time isolated from the temporal context of what preceded and followed it. Tsvetaeva regards the measurement of time as the creation of arbitrary and artificial boundaries that fragment and interfere with a genuine comprehension of the whole. Temporal divisions detract from the uniqueness of any given event by placing it in a series of identical constituent parts and by shifting the focus from the all-important, signifying links and relationships to the artificially isolated solitary units themselves as they stand meaningless, torn from their meaning-endowing context. Quantification of time, according to Tsvetaeva, shifts our attention to the unit of measurement. Perception is lulled by the sameness born of the predictability and arbitrariness of the divisions. A second, minute, or hour falls at a designated place with no regard for the ongoing and unaltered by it. This entails the acceptance of linearity, and it is here that causality slips in to rule and predictability displaces prophecy. The resulting fragmentation is unacceptable to the poet who must seek significance in dynamic interactions rather than in static classification or divisions.

Tsvetaeva's frequently expressed hostility to the mundane world and her postulation of an ideal other world represents her refusal to accept quantification, fragmentation, linearity, and causality—all of which interact to dictate a severely constricted view of existence. Nonrecognition of traditional boundaries becomes for Tsvetaeva an important aspect of combatting linear perception. She refuses to acknowledge clear-cut beginnings and endings and designates the source of meaning as complex interaction rather than delineation of things or events. The past endures in that it has contributed to the shaping of the present and thus remains always a part of it. The future too is present in the anticipation of its formation by all that leads up to it. Thus present and future do not simply succeed the past; they coexist with it. There occurs not a replacement of one with the other—a situation that would suggest interchangeability and thus also a false sense of equivalence—but rather a synchrony. This type of synchronic perception supersedes linear temporality and does away with boundaries separating one event from another. In the emphasis on connections among events the poet creates a coherence that endows each event with a qualitative identity that permits it to endure. Indeed, for Tsvetaeva poetry is in essence the creation of duration, made possible by the transcendence of linearity.

This holds in all areas of existence. Thus, to cite but one of many possible examples, Tsvetaeva writes in her essay "The Epic and the Lyric of Contemporary Russia" (1932):

> When at some French literary gathering I hear all names except Proust's, and, in response to my naive surprise: "Et Proust?" "Mais Proust est mort, nous parlons des vivants." Each time I feel as if I have fallen from the sky: by what sign is the aliveness and deadness of a poet determined? Can it be that X is alive, contemporary and active because he can come to this meeting, and Marcel Proust—because he will never be able to get anywhere on foot anymore—is dead? Only runners can be judged in this way [Prose, Vol. 2, p. 8].

The poet struggles against our all too passive and unquestioning acceptance of what is in fact not "things as they are," but of "things as we have agreed to see them." This entails a return to the unschooled perceptual realm of a child. Herein lies the freshness, the "for-the-first-timedness," and the truth of artistic vision. The youthful expressions in Tsvetaeva's early poems of reluctance to join the adult world and of her desire to remain a child develop into a statement of what it is to retain a childlike, unprejudiced quality of perception.

The two worlds Tsvetaeva postulates are essentially incompatible, and it appears that the poet who insists so vehemently on a wholeness of vision has, in fact, created yet another division. The situation is indeed a difficult one. The two very different modes of perception cannot be held simultaneously, yet neither can the poet choose one to the exclusion of the other. Tsvetaeva's works are informed by her hostility toward the mundane and her intense longing for the other world. Yet it is the healing of this rift between the two realms that she defines ultimately as the role of poetry. Her goal can be summarized in E. M. Forster's deceptively simple-sounding formula, "Only connect." The poet seeks not simply to escape into the ideal realm of infinite possibility, but rather to move between the two worlds, to act as translator between them. The untrammelled other world of infinite creative potential is essentially the nothingness out of which the poet must create. In an essay about the painter Natalia Goncharova (1929), Tsvetaeva describes paradise as empty, spacious, and full of light. (Here she again deploys the two separate meanings of the word *svet.*) She explains her reason for this description: "Only emptiness does not impose, does not displace, does not exclude. In order for everything to be possible, it is necessary that there be nothing" (Prose, Vol. 1, p. 286). Tsvetaeva's other world—the "empty Aeolian tower"—as she describes it in one of her poems (Poetry, Vol. 1, p. 263) is a perceptual slate wiped clear of all preconceived divisions and relationships. It is the source of unlimited creative potential, but cannot serve as an end in itself for the poet. Poetry becomes the language of communication between the two worlds and in her various definitions Tsvetaeva stresses its role of mediation. "Art is not an end in itself," she asserts. "It is a bridge, not a goal" (Prose, Vol. 1, p. 106). Poetry indicates a different perceptual realm but cannot be equated with it and Tsvetaeva insists that her "other world" is "a realm to which poetry is but a path" (Prose, Vol. 1, p. 178). In a

January 1927 letter to a friend, Tsvetaeva explained, "And the most *heavenly* of inspirations is nothing, if it is not transformed into an *earthly* deed" (Letters, 1969, p. 199). The poet's responsibility is, as Tsvetaeva succinctly formulates it in the 1925 essay "A Hero of Labor," "to create the other world in these" (Prose, Vol. 1, p. 184). (Tsvetaeva uses the plural here to suggest the fragmentation inherent in the mundane perspective.) In the essay "A Poet on Criticism" (1926) she elaborates on this role of the poet and explains the impossibility of a total rejection of the mundane in favor of the ideal world. "It is impossible to speak of weightlessness weightlessly. My goal is to anchor, to give an object weight. And in order for my 'weightlessness' (the soul, for instance) to have weight, something of the vocabulary and custom of this world is needed, some measure of weight already known to this world and confirmed in it" (Prose, Vol. 1, p. 229). Thus each object of the material world can be transformed by the poet into an indicator of the other.

The poet's "translations" require movement in both directions. In the same essay ("A Poet on Criticism") Tsvetaeva writes, "To subject the visible to serve the invisible—this is the life of a poet. . . . And what exertion of outward vision is needed in order to translate the invisible into the visible" (Prose, Vol. 1, p. 229). In the essay "The Poet and Time" (1932) Tsvetaeva expresses this idea in temporal terms as the poet's responsibility "to defend from time that which is eternal in it, or to immortalize that which is temporary in it—however you turn it: to time—the era of this world is juxtaposed the era of the other (Prose, Vol. 1, p. 379).

Poetry comes about as the result of constant movement between two different modes of perception. Both postulated realms are necessary to the poet. Even as the eye must oscillate constantly to prevent the loss of an image from perception, so too must the poet move between two forms of seeing to retain artistic vision. From the two different perceptual realms the poet must bring together from one, the ability to perceive the whole in its unfixedness of boundaries and, from the other, the ability to create connections that can elicit meaning from that chaos of potential. Having postulated the division, the poet brings the two realms together into a relationship of interdependence and interpenetration by means of the poetry that draws on both for its existence.

It is interesting to observe that, as Tsvetaeva herself fully understood, the poet's situation is mirrored also in the material of her art—words. The word, like the poet, mediates between two realms. On the one hand it is linked to the mundane world by its designation of objects in it, on the other, the word is an independent acoustic entity that is free from the many constraints of this world. There exists a clear parallel between the two perceptual realms posited by Tsvetaeva and the two facets of a word. In the course of her creative process Tsvetaeva shifts the focus from what words designate to their acoustic properties. These acoustic properties become the source of new connections and associations that remain outside the laws of linearity, causality, and logic that dominate mundane perception. Once the acoustic links among semantically unrelated words are demonstrated, their meanings rejoin them and reveal a new truth, a new way of seeing. We can say that the range of possibilities that exists between a word and its meaning is the realm of poetry.

Tsvetaeva's emphasis on the "in between" is an important strategy in her intent to erode the hold that temporal and spatial boundaries and divisions have on determining perception. At the same time it enables her to depict the highly improbable but necessary role of the poet as mediator. In her poems Tsvetaeva frequently depicts the "no man's land" between borders, the gray areas that do not fit neatly and entirely into one category. Dawn and dusk (which in Russian are designated by the same word, *zaria*) are endowed with importance in many of her poems. Thus, for example, in an untitled poem written in 1920, Tsvetaeva expresses her certainty that she will die either at dawn or dusk.

> **Знаю, умру на заре! На которой из двух,**
> **Вместе с которой из двух — не решить по заказу!**

[1980–1983, Vol. 2, p. 26].

> I know I will die when the sky breaks! At which break,
> With which break—is not my choice! [White, 1988, p. 89][3]

In another untitled poem dating to 1919, she designates her birth as having taken place between Saturday and Sunday and insists that this determines her dual personality. She sees as significant and appropriate the fact that Rainer Maria Rilke, a poet she revered and with whom she carried on a brief, but intense correspondence shortly before his death, died at the end of December (1926), that is to say, at the boundary between two years.

Trees, Orpheus, and the Cumaean Sibyl appear in Tsvetaeva's works as important symbols closely aligned with poetry by virtue of their links to both the visible and the invisible. The tree draws its sustenance from invisible sources at its roots and reaches upward toward the sun and sky. The Orphic myth is important to Tsvetaeva in its demonstration of the power of poetry over the material world and of the sway it holds even in the underworld. By means of his song Orpheus is able to travel and gain entrance to the realm of the Shades. Moreover, the song continues even after Orpheus has been dismembered; in transcending the material world, poetry transcends also its creator. In poems, essays, and private correspondence Tsvetaeva draws on the Orphic myth to illustrate and support her own insistence that poetry does not succumb to spatial or temporal delimitations. Thus, for example, she describes two beloved poets—Aleksandr Blok and Rainer Maria Rilke—as incarnations of Orpheus. She describes Rilke in a letter to a friend of January 1927, "A Germanic Orpheus, that is Orpheus, who *this time* appeared in Germany. Not Dichter (Rilke)—Geist der Dichtung." (Tsvetaeva, 1982, p. 48).[4]

[3]This book of Mary Jane White's poetry contains a section devoted to translations of some of Tsvetaeva's early poetry.

[4]Rilke had sent Tsvetaeva a copy of his *Sonnets to Orpheus* and in them Tsvetaeva discerned an echo and confirmation of her own ideals, for Orpheus had already appeared as an important symbol of the poet in her own works. Among passages of Rilke's *Sonnets* to which she responded specifically in her correspondence with him are the lines of the Fifth Sonnet "Wir wollen uns nicht muehn//um andre Namen. Ein fuer all Male/ists Orpheus, wenn es singt"; and the opening of the Sixth Sonnet, "Ist er ein Hiesiger? Nein, aus beiden/Reichen erwuchs seine weite Natur."

The legend of the Cumaean Sibyl provides Tsvetaeva with another compelling metaphor for the role of poetry and the poet. Like Orpheus, the Sibyl too travels to the nether realm. In the image of her disembodied prophetic voice is Tsvetaeva's own ideal of poetry: measured time is transcended by the eternal voice, and temporal divisions are done away with as past, present, and future converge in the enigmatic prophecies uttered by the Sibyl.

Central to Tsvetaeva's war on measuring time is the expansion of a given moment beyond itself. Thus not completion, but anticipation is also underscored thematically in her poems and is discussed in prose as a vital aspect of her artistic philosophy. Prophecy creates such anticipation, for the moment of prophecy points beyond itself. Thus, for example, it was not Christmas, the birth of Christ, but rather the Annunciation, its heralding, that was Tsvetaeva's favorite holiday. Renunciation too expands time. The longing that it engenders surpasses the moment of fulfillment which is denied. Ariadne left behind on Naxos becomes forever a part of Theseus.

To provide an exhaustive account of specific poetic devices that reflect Tsvetaeva's artistic goals outlined thus far would necessitate far more citations of her poetry in Russian than would be appropriate here. It is important to observe, however, that Tsvetaeva's complex and innovative rhythms serve to subvert fixed metrical boundaries. Enjambment—both between lines and stanzas—working in conjunction with caesuras, frequently leads to a disruption of anticipated linear divisions or to a disjunction between such divisions and semantic units. By a constant shifting and regrouping of poetic devices that link—assonance, rhyme, enjambment—with aspects of the poem that divide—caesuras, line and stanza division—Tsvetaeva creates a series of dynamic, ever-changing configurations that override the divisions inherent in the genre, and prevent such divisions from becoming rigid or predictable. This is augmented by her use of meter, which carries a pivotal function in its ability to join *or* separate. Semantic and longer grammatical units appear often out of synchrony or, more accurately, out of predictable synchrony with other elements of the poem. The various boundaries do not coincide—there is always something continuing even as something else is ending—and the divisions are consequently weakened.

Here it is interesting to note that the definition of the poet's task and the role of poetry that Tsvetaeva evolves is remarkably consistent and holds for the entire hierarchy of poetic units, ranging from details within individual poems to the entire body of not only her own works, but of all lyric poetry. It holds true also in Tsvetaeva's biography—a fact not surprising in light of her insistence that her life and art were inextricably bound. Aware of the insidiousness of conventional divisions and designations, Tsvetaeva went to great lengths to assert her independence of them. For one thing, she never became a member of any of the literary schools or movements that were so numerous at the time she wrote. In 1914, when war broke out with Germany and anti-German sentiment reached an apogee in Russia, Tsvetaeva recited verse in which she praised German culture, which, as she maintained, remained unaffected by current events. In the years between 1917 and 1922, in Bolshevik-controlled Moscow, she declaimed poetry praising the White resistance to an audience comprised largely of Red soldiers. These gestures were not political statements, but poetic ones, protests against divisions and categorizations, whatever their source, as impediments to perception and comprehension of the whole.

The dissolution of conventional boundaries, the focus on the "in-between," and the description of poetry as translation between two perceptual realms that I have outlined thus far, are important steps toward the accomplishment of the poet's task as defined by Tsvetaeva. It is necessary now to return to the paradox that is central to it and remains as yet unresolved.

As has been described, the poet must leave the world of established, predetermined definitions and relationships to enter the perceptual realm entirely devoid of such preconceptions. Yet boundaries, divisions, and distinctions are necessary to meaning of any kind. The poet cannot remain in what is essentially the perceptual condition of an infant who is not yet able to distinguish among the various stimuli that flood its senses. She must use this chaos of creative potential to shape new relationships, new structures of significance, yet in so doing must herself create divisions and boundaries, isolating aspects of the infinite possibilities presented by the other realm. Tsvetaeva's equation of infinite creative potential with nothingness in the "Natalia Goncharova" essay indicates clearly her recognition of this problem. Here we must also note that parallel to this paradoxical dilemma of the poet is the fact that the very nature of the genre of lyric poetry itself entails divisions. The brevity of the individual poems themselves, and the stanzaic, linear, and metrical divisions within them would make it appear that poetry is a medium very ill-suited to eliminating fragmentation and healing rifts.

There remains also the problem that the "translations" between realms can themselves become as rigid and fixed as the commonly accepted perceptual norms of the mundane world, exacerbating rather than healing the fragmentation of wholeness that the lyric genre on the surface appears to result in any way. Whatever the poet can do to draw attention away from the boundaries that define her genre, the formal divisions persist even as the poems themselves remain discrete texts that—for all the struggles of the poet—present only fleeting glimpses into that infinite perceptual realm that Tsvetaeva designates as the other world. Even the tension and the movement of the poet back and forth between the two realms can be regarded as suggestive of the stasis of equilibrium.

Here we arrive at the focal point of Tsvetaeva's definition of poetry, one that resolves the apparent contradictions and paradoxes that are entailed in her demands for what poetry must accomplish. This is Tsvetaeva's designation of poetry as *process*. Infinite potential can be translated into our limited world only by the demonstration of process.

In the essay "A Hero of Labor" (1925), Tsvetaeva wrote: "I would replace perfection (a state) with the act of accomplishing (a continuity)" (Prose, Vol. 1, p. 179). This concern becomes the center of her definition of lyric verse, whose capacity to demonstrate process is, according to Tsvetaeva, the feature that distinguishes it from other literary genres. This demonstration of process is predicated on the elimination of spatial boundaries, of linear temporality and of causality that has already been discussed.

While Tsvetaeva's metaphor for artistic perception is couched in spatial terms, her description of poetic process draws on temporal comparisons. Art is, according to Tsvetaeva, based on continued and continual development. In her 1932 essay "The Poet and Time," she insists that "art itself, whatever may sustain it, and whatever it may be trying to reestablish, is itself already forward movement. [There is] no return in art: it is unceasing, that is, irrevocable" (Prose, Vol. 1, p. 369). Thus an individual artist who ceases to continue to develop regresses and falls from the process of art, as Tsvetaeva elaborates later in the same essay: "Not to move forward (in poetry as in everything) means to move *backwards*, that is to quit" (Prose, Vol. 1, p. 370). She then continues: "X is not contemporary not because he did not accept contemporaneity, but because he stopped on his artistic path, the only thing a creator has no right to do. Art goes forward, artists remain behind" (Prose, Vol. 1, p. 370).

"Every poet is an emigrant. . . . An emigrant from Immortality into time" (Prose, "The Poet and Time," Vol. 1, p. 372). Tsvetaeva's hostility to measured time has already been noted; underlying *this* statement is also the rejection of linear temporality, which is already entailed in her denial of causality. Yet how is this linearity to be distinguished from the ever-forward movement of poetic process as Tsvetaeva describes it? Tsvetaeva's answer lies in her insistence on the constant and highly complex interaction of past, present, and future in any given moment. "Every contemporaneity is actually the coexistence of times, ends, and beginnings, a Gordian knot which can only be severed" (Prose, Vol. 1, p. 370). Tsvetaeva insists on process in order to draw attention to this temporal simultaneity. To perceive process it is necessary to recognize that any given moment is not only shaped by what preceded it, but that it also demands reinterpretation of what came before it, even as it serves as an indicator of what is to come and will, in turn, be reshaped by it. There emerges a constantly shifting structure of relationships that distinguishes process from causality and linearity, static in their absence of such reciprocal influences.

As one thing ever becomes another and changes everything around it, the significance of completion recedes before the process of becoming. Just as what Tsvetaeva calls "earthly signs" must be perceived as indicators of a beyond, so too must each poem be read as pointing beyond itself. Tsvetaeva maintains that even after completion each lyric poem demands the continuation of the larger interrupted poetic gesture that underlies the individual poem and extends beyond it. She provides an interesting description of this in a letter to Boris Pasternak of February 11, 1923:

> Lyric poems (so called) are separate moments of *one* gesture: an interrupted gesture. Do you remember those turning kaleidoscopes from your childhood? Or didn't you have one? The same gesture, but slightly protracted: a hand, let us say. To the right, a bit more to the right, a bit more, and so on. As you turn it, it moves [Tsvetaeva, 1972, pp. 278–279].

The image of the kaleidoscope is a particularly instructive one. The bits of colored glass in it are made to group and regroup constantly by the gesture of the hand. The fragments are made to participate in the patterns they shape together by means of perceptual ordering. The individual shard surrenders its discreteness to the larger configuration and loses its determinacy—for even though it itself remains unaltered, its position and structural function becomes different in each newly emerging pattern. The shifting relationships of the pieces in the kaleidoscope demonstrate vividly that there can be no one fixed definition, that the configurations—that is to say perceptual possibilities—are endless. We witness here also the overcoming fragmentation that is achieved in the focus on the dynamic relationships among the pieces. In the structural complexity of the interaction of the pieces comprising each configuration we witness a simultaneity that obviates linearity. Their unfixedness carries an insistence on expanding rather than limiting perceptual possibilities, even within the finite space of the kaleidoscope with its unvarying quantity and quality of pieces. In creating it is not that the poet invents something new—the concept of novelty is after all an ephemeral one—but that she insists on ever new modes of structuring and perceiving what already exists. It is this that Tsvetaeva has in mind when in her essay "The Poet and Time" where she states, "In essence all poets of all times say one and the same thing" (Prose, Vol. 1, p. 372).

Even as one kaleidoscopic configuration falls into place, new regroupings are anticipated. Each kaleidoscopic pattern is discrete and complete; yet not one of these patterns can be considered the final one. Indeed, in this context, the concept of finality is not applicable. In her definition of lyric poetry Tsvetaeva focuses on the absence of completion. Not only the entire body of poetry, but also each individual poem comprising it is always in the process of becoming. In her essay "A Hero of Labor" Tsvetaeva criticizes the lyrics of the Russian Symbolist poet Valerii Briusov as being too final in that they do not leave the reader with the desire and expectation of more to come. She complains that the words "the end" appear invisibly, but palpably at the end of each of his poems, and goes on to address the poet, "Bring a work to its end, struggle with all your might to bring it to its end, but if I feel this end, then that is the end for you" (Prose, Vol. 1, p. 179). Each lyric poem represents a renewed, eternal, and unrealizable attempt to suggest what Tsvetaeva calls "the all presence of everything." Poetry must go on because the poet continues her struggle to achieve the unrealizable. Also in "A Hero of Labor" Tsvetaeva writes, "The most perfect creation—ask an artist—is only an intention: that which I wanted—and could not" (Prose, Vol. 1, p. 179).

The goal of poetry, as we can now expand our understanding of Tsvetaeva's account, is not simply to bridge or transcend divisions, but to *demonstrate how* they can be overcome. The individual poems comprising the body of poetry and the formal divisions within them become for the poet a sort of grid or framework on which to demonstrate this. This makes possible the *indication* rather than the *recreation* of the other perceptual realm Tsvetaeva designates, and here too is the dynamic interaction of parts that prevents the ossification of boundaries. Above all, a poem must not be regarded as serving a specific end. Thus, for example, in her essay "My Pushkin" (1937), Tsvetaeva explains the inappropriateness of questions and answers in poems.

> [A] question in poems is an irritating device, if for no other reason than that every *why* demands and promises a *because* and thereby weakens the intrinsic value of the entire process, transforms the entire poem into a gap, riveting our attention to the final external goal, which poems must not have. A persistent question transforms poetry into a riddle and a problem and if each poem is itself a riddle and a problem, then not *that* riddle for which there is a ready response, and not that problem for which there is an answer in a book of problems [Prose, Vol. 2, p. 267].

We must also consider what occurs *between* the kaleidoscopic configurations Tsvetaeva describes in the February 11, 1923 letter to Pasternak. In this letter Tsvetaeva continues her description of lyric verse in the following way: "Lyric poetry is a dotted line that looks solid, black from a distance, but when you look closely: it is continuous gaps between the dots—a vacuum: death. And you die from poem to poem. (Thence the 'lastness' of each poem!)" (1972, p. 279). In another letter to Pasternak written in mid-July of 1927, Tsvetaeva once again returns to the designation of a poem as a fragment to suggest the overwhelming difficulty of the poet's task. In this letter she writes about her collection *After Russia* which she describes as her farewell to lyric verse. "I am tired of tearing myself, of breaking myself into the pieces of Osiris. Every book of poems—is a book of partings and tearings asunder, with the finger of Thomas in the wound between one poem and the next. Who of us has not placed the final stroke without a sinking of the heart: and further?" (Tsvetaeva, 1969, p. 196). Tsvetaeva identifies here with the Egyptian deity Osiris, who was murdered and torn into fourteen pieces by his brother Set. She designates the spaces between poems as wounds—tears in that organic whole that they comprise. Each completion of a poem creates at the same time a gap, a wound that must be healed by the opening of the following poem. In that "death," that "vacuum" between them is the tension of doubt and fear, but also of expectation of continued growth and development. This foreshadowing of a future reopening renders the closure of a completed poem temporary. If the poet is the sundered Osiris, she is also the Isis who gathers the fragments, makes Osiris whole again, and once more imbues him with life.

If we examine what Tsvetaeva seeks to achieve by designating poetry as "process," we can arrive at a clearer understanding of what process entails. The poet seeks both to avoid closure and to create a completed poem; to create poems that are complete yet do not fragment that larger whole which they designate. Here it emerges that Tsvetaeva's understanding of process is based on a special relationship she establishes between part and whole, a relationship that elucidates a central aspect of her artistic vision. Tsvetaeva maintains that the interpenetration of part and whole informs every level of lyric poetry, from the interaction of parts and phases creating one poem to the totality of lyric verse. Thus for Tsvetaeva it is not only that all poems and poets taken together constitute poetry, but also that each individual poem is a reflection of the whole. Her position does not deny significance to the individual poem in the face of the greater phenomenon of poetry. It presents a relationship of part to whole in which equal importance is ascribed to each, for although every poem participates in a larger whole, it remains nevertheless complete in itself. Indeed, it is this very completedness that enables it to reflect the larger whole.

Tsvetaeva's view that the whole is found in each of its parts extends to the entire hierarchy of poetic units. In her essay "Art in the Light of Conscience" (1932), she explains:

> The only goal of a work of art at the time of its creation is its completion, and not even as a whole, but of every discrete particle, every molecule. Even it itself, as a whole, recedes before the realization of this molecule, more precisely: every molecule *is* this whole, the goal of this whole is everywhere, for its entire duration—it is all-pervasive, all-present and it as a whole is the end in itself [Prose, Vol. 1, p. 388].

Each element is complete in itself yet becomes simultaneously a fragment of a larger whole which it encapsulates. As a poem unfolds, each of its parts is seen to alter. The discrete quality of each individual "molecule" is subsumed by the larger relationship into which it is drawn as a poem unfolds. The discrete units are perceived not in isolation, but as participants in ever larger configurations. The emergence of each successive part necessitates a reinterpretation of the entire structure of interactions and thereby also of each of its components. Thus each successive "molecule" of the poem endures beyond its appearance as it participates in reshaping what preceded it and preparing what is yet to come. Here again it is possible to speak not only of interaction, but of an interpenetration of parts and whole. It is in the creation of this interpenetration that the poet can transcend the linearity and fragmentation that appears inherent to the genre of lyric verse. Even as each "molecule" of a poem points beyond itself despite its completion, so too does each lyric poem indicate further development of which it becomes a part. Thus the relationship of part to whole is a constantly expanding one.

The constant development and forward movement of poetry which yet retains an organic link with its own past; the intimation that more is yet to come even as a poem is completed; the refusal to recognize the rule of quantity—spatial or temporal; the elimination of linearity and causality even within the sequential aspect of poetry—are all contained within Tsvetaeva's designation of poetry as process. It is a designation that enables the poet to indicate the infinite realm of creative possibilities that emerges when conventional perceptual norms and boundaries are overlooked.

In an article (1988) that examined Tsvetaeva's genre distinctions (and I draw here on some of my conclusions from that study) I pointed out that Tsvetaeva's means of defining lyric poetry are pertinent to twentieth-century arts in general and that they bear a striking similarity to the philosophical implications of modern science as described by the philosopher Errol Harris in the conclusion of his book *The Foundations of Metaphysics in Science* (1965). Harris summarizes:

> [E]very process is the process of realization of a whole, and . . . the totality being generated is immanent in each successive phase. . . . Moreover each phase is a phase of a particular process, only by virtue of the fact that it is a state in the unfolding of a particular whole. If it were otherwise the process would be something quite different. Its special character consists in its being the progressive unfolding of just that totality which gives it its special character. The whole is thus immanent in the part so characterized, and this immanence of the whole in the part gives the part the status of a special manifestation of the general principal of order. . . . This is the essential character of a system—a scale of forms progressively realizing a structural principle [p. 465].

In the development of her formulations, Tsvetaeva stressed the poet's solitude and the overwhelming difficulties she had to overcome in the course of writing. At the same time, Tsvetaeva stressed the importance of the role of the reader—not the reading public whose praise and censure she scorned—but the ideal individual and individualistic reader who would rise to the demands of her verse. Inasmuch as a poem is not an end in itself, and does not suggest or provide a key, an answer, or a solution to anything, its significance to the reader lies, as Tsvetaeva insisted, in the retracing of the poet's path—participation in the process of a poem's creation and its continual reshaping by subsequent poems. "He who has not read every line of a poet has no right to judge him. Creation is continuity and gradualness," she maintains in "A Poet on Criticism" (Prose, Vol. 1, p. 223) and it is clear that underlying this assertion is her understanding of lyric poetry as process. Tsvetaeva devotes considerable attention to describing the role and responsibility of the reader. In "A Poet on Criticism" she writes, "And what is reading but deciphering, interpretation, the extraction of the secrets that remain beyond the lines, beyond the limits of words. (Not to mention 'difficulties of syntax!') Reading is—above all—co-creation. If a reader is lacking in imagination, no book will survive" (Prose, Vol. 1, p. 237).

At the time this essay was written (1926) Tsvetaeva's poetry was already beginning to receive the designation of "difficult" and even "too difficult." Tsvetaeva scoffed at those who expected her to continue writing in exactly the same way she had written a decade earlier, and refused adamantly to make things easier for the reader—a position of considerable integrity for someone who relied on publication and therefore a readership for her livelihood. Hostile critics regarded this as posturing. Tsvetaeva explained it in terms of respect for her readers and her art. "The reader's exhaustion is not devastating, but creative. Co-creative. It does honor both to the reader and to me" (Prose, Vol. 1, p. 238). The reader must recognize that a poem points beyond itself and must avoid the temptation of reducing it to its component parts. Otherwise there remains the danger that features of the restricted and restrictive mundane perception that a poet strives to overcome in verse can be reimposed by the reader.

> When in response to my given, where by means of drafts form has been overcome, banished, I hear: ten "a's," eighteen "o's," assonance (I do not know any professional terminology,) I think that all my rough drafts have been in vain, that is,

> have again resurfaced, that is, what has been created is again destroyed. Dissection, but dissection not of a corpse, but of something living. Murder [Prose, Vol. 1, p. 238].

Tsvetaeva insists that the reader refrain from imposing limits from the perceptual realm the poet has already done battle with. The individual poem and all of poetry must be respected as a living, developing organism, for, as Tsvetaeva insists in her essay "The Epic and the Lyric in Contemporary Russia":

> [P]oetry is not divided into poets or among poets. In all its manifestations it is one, the same, complete in each—just as there are in essence not poets but only *the* poet—one and the same, world beginning to world end, a force that takes on the hues of given times, tribes, countries, dialects, individuals, flowing through those who carry it, this force, like a river between these or other banks, under these or other skies over one river bed or another [Vol. 2, p. 7].

References

Brodsky, J. (1978), Questions and answers after Brodsky's reading, 21 February 1978. *Iowa Review*.

Harris, E. (1965), *The Foundations of Metaphysics in Science*. London: George Allen & Unwin.

Hasty, O. P. (1988), "*Poema* vs. cycle in Cvetaeva's definition of lyric verse. *Slavic East Eur. J.*, 32/3:390–398.

Tsvetaeva, M. (1910), *Vechernii al'bom* [*Evening Album*]. Moscow: Tipografiia Mamonta.

——— (1969), Pis'ma Mariny Tsvetaevoi [Letters of Marina Tsvetaeva]. *Novyi mir*, No. 4.

——— (1972), *Neizdannye pis'ma* [*Unpublished Letters*], eds. G. Struve & N. Struve. Paris: YMCA Press.

——— (1979), *Izbrannaia proza v dvukh tomakh 1917–1937* [*Selected Prose in Two Volumes 1917–1937*]. New York: Russica Publishers, Inc.

——— (1980–1983), *Stikhotovreniia i poemy v piati tomakh* [*Poems and Poemy in Five Volumes*]. New York: Russica Publishers, Inc.

——— (1982), *Pis'ma k A. Teskovoi* [*Letters to A. Teskova*]. Jerusalem: Izdatel'stvo Versty.

Vitins, I. (1977), Escape from Earth: A study of Tsvetaeva's elsewheres. *Slavic Review*, December:644–657.

White, M. J. (1988), *From Starry Sky to Starry Sky*. Stevens Point, WI: Holy Cow! Press.

II

Order and Disorder: Creating Temporality in Literary Experience

Marlene P. Soulsby

Abstract This paper is concerned with the contribution that the study of literature can make to the knowledge and understanding of time. One may well argue that the concern of literature is not to impart information *about* time, and yet literary works exist as temporal experiences that also create a sense of temporality in the reader. Understanding the process whereby the literary experience of temporality is created provides a key for gaining more knowledge about the nature of time itself. It also provides a link and opportunity for dialogue between literary endeavors and other disciplines seeking to further knowledge of human temporal experience. J. T. Fraser's model of time as a hierarchy of temporal umwelts gives us the means to explore the rich dynamic of this process and opens the door for an interdisciplinary dialogue.

In an age where time has become not only a scholarly but also a popular preoccupation, texts abound which talk to us about time or temporal matters. Literary texts also reflect this heightened interest, and yet literature is certainly not the arena for imparting theories or philosophies about time. Even though well-known "time-writers," such as Mann, Joyce, Proust, or Woolf, explore many facets of temporality, the value and impact of their work result from the experience of temporality that is evoked during the reading process rather than from information about time that is conveyed to the reader.

One might then ask what contribution literature makes to the understanding and study of time. If the primary concern of a literary work were to deliver a message or provide information, we would not be drawn to reread it once we had understood the point. In fact, a condensed summary might serve the purpose just as well. A reader will return to a literary text, however, and reexperience its moods, recreate the flow of its time, and relive the tensions that shape its existence. Why? Because literature can *do* to us what other forms of discourse talk *about*. Here, in the *doing,* or rather, in the process of generating temporality, lies much to be learned about the nature of time itself.

Some critics have questioned whether our focus should be the analysis of literary works as objects, or whether it is more relevant to examine the process of understanding and communicating that occurs through encounters with literary texts. In his book, *The Noise of Culture*, Paulson (1988) argues that literature participates in the very same process of invention by which all new ideas and constructions of reality are formed, but it uses language in a different way. Unlike other modes of discourse whose purpose is to facilitate understanding of information, a literary work may say things that are known in a way that is difficult to understand. The reader then must learn how to read its particular constructs if communication and understanding are to occur. In other words, the reader must become an active participant in the process of invention and the creation of order. According to Paulson, this process is itself a form of knowledge—not knowledge of an unchanging truth or idea waiting to be uncovered in the text, but rather knowledge gained from bringing order to reality and the vast realm of human experience. The process of creating order inherently involves time.

Ricoeur (1983, 1984, 1985), in his extensive study on time and narrative, also places literature in a broader context. He sees human temporality and narrative engaged in a dual process involving order as well as disorder. He explains that our *understanding* of human time derives from organizing, in orderly progression, apparently disconnected happenings and confused experiences "after the manner of a narrative." A plot thus forms a new congruence by creating order. And yet, our *experience* of time is not at all orderly or linear. According to Augustine, it is rather discordant, even enigmatic. To be meaningful, a narrative must also incorporate this experiential feature of human temporality. Thus Ricoeur confronts the paradox of time, only to assert that the acts of narrating and following a narrative allow the reconciliation of these two tendencies in a poetic mode. The reader becomes involved in a "living dialectic" of order and disorder, linearity and unity, necessity and chance.

To pursue this point further, we can define "temporality" in a literary context as the experience that arises from two tendencies: the tendency of mind to map experience in a coherent, linear order, to remember and anticipate events, and to recognize sequence and causality, and the tendency or pull of feelings toward less coherent, nonlinear experiences of time. Clearly, some texts might generate greater levels of tension between these two tendencies than others, and even within the same text tensions may vary. The temporality of a text, or more precisely of a literary experience, depends on the nature of these tensions and the levels of their intensity during the process of reading and constructing a new reality.

A model of time which best corresponds to this concept of temporality in the literary experience is that developed by J. T. Fraser (1975, 1978, 1987). According to his theory, time exists as a nested hierarchy of organizational levels, mirroring the evolutionary order of nature. Each level exhibits increasingly greater coherence and complexity, ranging from the primordial chaos of space–time to the ordering capabilities of the human mind. Significantly, each level builds a distinct reality or *umwelt*, complete in itself with its own boundaries, laws, conflicts, and temporal moods. Mind, as the highest organizational level, however, embraces all others and carries the entire evolutionary structure of time within itself.

Reading a text engages the organizational tendencies of the mind to create and order a new reality. Fraser refers to this level of activity as *nootemporality*. Here the arrow of time is well-defined, allowing cognitive consciousness to draw on memory of a long-range past and anticipation of a projected future in relationship to the present to give coherence and meaning to events and perceptions.

The experience of the reading, however, may engage other levels of time consciousness, each with an increasing tendency toward disorder. At the *biotemporal* level, the point and tail of time's arrow blur, focusing on a sense of "nowness" and the immediate needs and desires of a creative awareness. The *eotemporal* realm has no directional quality, as though only the shaft of time's arrow remained and events could occur in reversible succession. At the level of *prototemporality*, events are evident but disjointed and not orderable in any coherent pattern. Finally, in the last level of *atemporal* chaos, all order, including the entire arrow of time, is absent.

The descent into lower organizational levels implies a loosening of definition and directionality in time and also in the experiencing self. The integrity of selfhood depends on maintaining the nootemporal order with all of the burdens and responsibilities, regrets and anxieties associated with the consciousness of past and future. If we lose ourselves in increasingly disordered temporalities, the noetic mind reacts to this disintegration of order in various ways, evoking feelings of ecstasy and stillness as well as of fear, emptiness, or psychic pain. Fraser refers to these reactions or feelings as "temporal moods." Since such moods do not easily lend themselves to expression in logical discourse, other uses of language, other artistic techniques must serve to convey the experience. A literary text can generate these moods in a reader by creating tensions between the noetic tendency to establish order and the experiential pull of less organized levels of coherence.

The following poem, a *shan shui shih* or landscape lyric by the eighth-century poet Wang Wei, illustrates how a temporal mood is created through the juxtaposition of images. The translator, Wai-lim Yip (1972), has maintained as much of the brevity and concreteness of the original as possible.

Deer Enclosure

Empty mountain: no man is seen,
But soft voices of men are heard.
Sun's reflection reaches into the woods
And shines upon the green moss
[p. 101].

At the nooetic level of comprehension, a reader organizes the images into a coherent picture encompassing mountain, voices, sunlight, woods, and green moss, as well as the deer enclosure from the title. The picture allows all images to coexist in the memory, giving the impression of a beautiful moment, a timeless moment in which each element has its place; yet the picture has been created instant by instant during the process of reading. Visually, attention moves from confinement to expansion, from the vast to the small. Temporally, the reader progresses from lesser to greater specificity and directionality. Each image evokes a temporality which results not so much from the image itself as from the juxtaposition of images as they strike awareness during the process of ordering the experience.

The title suggests a focal point, a confined, enclosed area whose boundaries are sprung by the first image of the poem. "Empty mountain" evokes vastness, the sense of absence, a state of no-time, "no-thingness," no self. The Buddhist connotation of "void" heightens the impression of stillness in the external landscape as well as in the mind. The pause created by the natural caesura after mountain allows the mood of calm, the dissolution of time and space, to settle in the reader's awareness. The next image is negative: "no man is seen." This reference to absence, instead of enhancing the feeling of emptiness, simultaneously evokes consciousness of its opposite. What is not seen only makes sense in relation to what could be seen, as light is apprehendable in contrast to darkness. This image necessitates a shift in the level of temporal organization from atemporality to proto-temporality: distinguishable contrasts exist, albeit without context and form. The sound of "voices" that are then heard bring greater clarity to the experience, indicating the presence of something perceptable and continuous, but without the directionality of intelligible speech, without any overt connection to the deer enclosure. The last two lines, which depict the movement of light toward a specific object, the green moss, provide a sequential order to the perceptions, thereby creating a sense of passage.

The temporal mood of the poem involves the noetic process of ordering all perceptions into a unified whole as well as these subtle shifts through increasingly greater levels of organization. The reader moves out of the confines of self (the deer enclosure), and experiences an inner journey from the depths of timeless being into time, then holds the entire experience together in memory as a single timeless moment—an experience which reconciles the paradox of becoming and being. This process creates a new reality by evoking levels of time consciousness which are inherent in human experience.

An entirely different sort of experience and mood emerges during the reading of the novel *The White Hotel*, by D. M. Thomas (1981), even though the same levels of temporality come into play. Here the mood shifts drastically from one section to another as the reader constructs, then reconstructs, the same reality at different levels of temporal organization. Let us focus on the first four sections where the effect is most intense.

The prologue, consisting of a series of letters, introduces the main character as a patient of Freud's, a woman, we are told, with an extreme libido and an extreme of morbidity. The letters give the illusion of real correspondence in style and tone, forming a striking contrast to the writings of the woman herself. Her poemlike narrative entitled "Don Giovanni" follows the letters and thrusts the reader into a totally different reality—an eotemporal umwelt where sexual images and destructive events appear as a flood of impressions without logical or causal connections. Temporal order must be constructed from sentences that are loosely strung together and from the periodic repetition of images and sounds. The narrative lacks temporal definition or direction, except for these cyclic repetitions.

> One evening they rescued a cat, its black fur
> had been almost lost against the dark green
>   fir,
> we stood naked by the window as a hand
> searched among the foliage, it scratched,
> it had been up there two days since the flood,
> that was the night I felt a trickle of blood, . . .
> [pp. 19–20].

The effect on the reader is disturbing because the noetic desire to create order is engaged in strained tension with the pull of the text towards disintegration. The resulting mood is chaotic and painful. The world does not hold together. Yet, we cannot say that order does not exist. There is indeed order, but of a more primitive temporal level. Here things swim together in a kind of oceanic or dreamlike oneness, and relationships can only be intuited through association not cause and effect. The fluid language, the breakdown of causal order, the jarring associations, and the merging of raw passions with death and despair create the reality of a self without integration. In a letter preceding the "Don Giovanni," we read: "Her phantasy strikes me as like Eden before the Fall—not that love and death did not happen there, but there was no time in which they could have meaning" (p. 12). In Eden before the Fall, in an eotemporal umwelt without sequentiality or the tension of good and evil, no judgmental value can be attached to events. In like fashion, the reader becomes aware of narrated events but derives no *understood* meaning from them, and can, therefore, pass no judgment on them. There is certainly *felt* meaning, however. In the attempt to construct coherence and in the tension that ensues, is the psychic pain of fragmentation. The noetic mind confronts an indefinable succession of events and reacts with the plea: "can you do anything for me can you understand" (p. 18).

The section that follows, entitled "The Gastein Journal," narrates the events again, this time filling in gaps and presenting a more linear continuity. Yet despite its increased coherence, the narrative shifts easily from one situation to another, juxtaposing events in an unusual, surrealistic way. At this higher organizational level, the impressions we remember from the "Don Giovanni" piece fall together as a dreamlike fantasy, where the bizarre appears and is accepted as commonplace. The narrating self is still elusive, but we encounter her in a different temporal world and reconstruct her reality on a higher integrative level.

The novel then makes another unexpected shift in its temporal perspective. A fictional Freud interprets these writings of his patient in a case history. Suddenly the events and impressions are given a critical and analytical structure. The arrow of time is constructed with precise nootemporal understanding. The self of the previous sections is given historical context, and from her "irrational flood of images" and the "untidy fantasy" (p. 133), an objective voice discerns meaning and guides us in the structuring of a coherent order. The objects and events are converted into symbols, creating a language that allows communication *about* the irrational workings of the subconscious and the more primitive levels of temporality. Thus the white hotel, that image that struck with such force at a lower temporal level, is now that place known to all of us, the body of the mother, the place without sin. It has become an abstraction. At this level, reason and understanding replace the feeling of oceanic oneness and autoeroticism generated by the earlier sections.

If we were to reread the "Don Giovanni" after constructing the linear history of Freud's case study, the images would build associations on several levels of integration simultaneously. Each section exists complete in itself, with its own language and its own tensions, and yet each fits nested within the whole. There is overall order, but it is an order built from layers of disorder.

Margaret Atwood's *The Handmaid's Tale* (1985) is narrated from a perspective which lacks the dynamics of creative tension. The protagonist deliberately attempts to suppress noetic consciousness with its attendant linking of past memories and future expectations to the present. She struggles to remain locked in the present for fear that memories of the past would drive her to total mental collapse. She also neutralizes the passions and spontaneity of the present. The reader, constructing a reality from this point of view, is denied the historical context that would make events understandable. Therefore, their horror is intensified.

The emotionless language suggests a mental state of biotemporal continuity and awareness without connection, but as the following passage reveals, there is great struggle beneath the flat tone of the protagonist's voice. In this scene, executed doctors are hanging from hooks on the Wall, the place in town to display those who have threatened the social and political order. The heads of the bodies are covered with white hoods. One of the hoods shows a red circle of blood where the victim's tongue has been cut out:

> I look at the one red smile. The red of the smile is the same as the red of the tulips in Serena Joy's garden, towards the base of the flowers where they are beginning to heal. The red is the same but there is no connection. The tulips are not tulips of blood, the red smiles are not flowers, neither thing makes a comment on the other. The tulip is not a reason for disbelief in the hanged man, or vice versa. Each thing is valid and really there. It is through a field of such valid objects that I must pick my way, every day and in every way. I put a lot of effort into making such distinctions. I need to make them. I need to be very clear in my own mind [pp. 44–45].

The protagonist's attempt to maintain a position between the pain of nootemporal self-integration and the pain of eotemporal collapse into insanity is tenuous. The sight of these men threatens the detachment that she must maintain to insure sanity in an insane world. The red of blood, two lips, death, and horror appear the same as the red of tulips, life, beauty, serenity, joy. But as she insists, "there is no connection." One experience cannot, *must* not comment on the other or cancel the other out. Associations that spring up in her mind threaten to obliterate the distinctions, threaten to make the insane ordinary. Each thing must remain in its own place and time. If she would allow her mind to integrate these impressions and events on a higher temporal level, she would have to cry out in rage against injustice and inhumanity—an act that would spell her own demise. This is a matter of survival, both physical and psychological. The use of the present tense affirms in her mind a sense of self that is disengaged, devoid of a personal past, and oblivious of the future.

The reader, on the other hand, struggles against the suppression and lack of tension which forms the work's point of view. Reading the text compels us to construct the levels of temporal coherence and chaos which the protagonist seeks to avoid. We work to reconstruct its history, we are outraged by the reconstruction, and we look into the possibility of psychic breakdown when the artificial holds we have on reality are removed. The dynamics and temporal mood of the text rely on the reader's refusal to participate in the protagonist's struggle and the deadening lack of creative tension.

In contrast to *The Handmaid's Tale*, Franz Kafka's short piece, ''Before the Law'' (1946), relies on the reader's inability to maintain a creative tension between efforts to construct a reliable arrow of time and the feeling of a rapid dissolution of that arrow. This is the story of a man who leaves his home in the country to seek a goal—admittance to the law. He arrives at the gate only to hear from the doorkeeper that he may not enter ''now.'' This impediment paralyzes him to the extent that he spends the rest of his days sitting on a stool in front of the door, waiting for permission to enter. Within this condensed narrative, a man's life dissolves, his body diminishes, his senses deteriorate, and the expanse of his world narrows to that of this door and the doorkeeper. Essentially, he is reduced from a human being with freedom of choice to a pitiable creature begging fleas to help him. His past and future blur into the immediacy of a present without direction.

When the man arrives at the gate, he is presented with the necessity of making a choice. He can enter the gate and confront the possibility of risk that awaits him, or he can abandon this particular goal and seek out another, or he can simply go home. Instead, he escapes from the burden of making a choice and waits. Because of fear, fear of risk, fear of death, fear of unpredictability and the consequences of his actions, he cannot function at the level of noetic consciousness. The waiting reduces his temporal perspective to the biotemporal umwelt of a child, where, free of the burden of choice, he sits suspended in the present making meaningless attempts to change the doorkeeper's mind. Here he waits for a magic connection that will grant him permission to act. The level of temporality at which the story functions descends even further: ''There he sits for days and years'' (p. 3). With these few words, it appears that time collapses. Actually, the words indicate a further reduction in the temporal coherence of the man's situation—nothing happens! As in an eotemporal umwelt, time loses its directionality and events are merely an unconnected succession. Even his speech becomes incoherent grumbling to *himself*. Finally, he is no longer able to distinguish between inner and outer reality; he sees only darkness and light, and is not sure where or whether either one exists. At these lowest levels of temporal consciousness, at the moment of death, all the experiences of the past point to one question: why no one else has come to ask for admittance to the law. In all this time, he only asks two questions. One deals with the moment, whether or not he can be admitted now; the other deals with the fullness of time. The enigmatic answer to time's question is that the gate was always meant for him alone. The availability was always there, but no longer. The story ends with the image of final dissolution—the gate is shut. The man has not been able to integrate the levels of his experience, and thus he perishes with the bitter knowledge of his failure.

Despite its apparently linear structure from point A, arrival at the gate, to point B, death, the temporal moods that are evoked by the man's degenerating condition disrupt a clear, connected ordering of events in time. It is senseless to ponder when the light begins to shine through the gate, if it has always been shining, or if it is shining even now. In an umwelt defined by lower levels of temporality, all of these are simultaneously possible and the search for causal sequence is meaningless. It is only when we move out of these lower levels into higher umwelts of integration that such factors shape a meaningful structure. As we have seen, the man does not pull himself together, and therefore as readers, we are left with the ambiguity of his existential plight.

As is typical with Kafka stories, they involve us in an endless process of constructing and reconstructing their meaning, of trying to find an order that explains the sequence of events, of looking for causes, and examining other paths not taken. This is the noetic tendency of the reading process to construct order. Even so, the content of the story pulls us to an ever-increasing disorder, where the rules of logic and causation break down, and the arrow of time rapidly loses its definition. In this tension lives the story—no matter what significance one attributes to the law, the gate, its doorkeeper, or the light. Experientially we are caught in a conflict that arises from the impossibility of applying noetic logic to a lower level of experience.

According to Fraser, the mature experience of nootemporality depends on the maintenance of conflicts that sustain our sense of individuation and self. From these conflicts arise our freedom, our creativity, our ability to deal with the unpredictable and new. Kafka's seeker succumbs to the unbearable conflict between desire and fear by removing himself from it. As a result, the integrative levels that define his identity collapse into one another.

Kobo Abé in his novel, *The Woman in the Dunes* (1964), creates similar existential conflicts, but the pull toward dissolution and death ultimately yields new life and new energy. His protagonist comes from the city to a world of sand. The sand is perhaps one of the most masterful literary images of atemporality. It is an all-pervasive presence of absence. Its form is at once antiform, in that it threatens form with dissolution. It appears as both solid and fluid, ceaselessly changing yet unchanged. It is space–time. It is barren, the opposite of life, and yet it is like a living thing, never at rest, shifting, pulsing, swirling, avalanching, rippling, creeping everywhere. As such, it threatens to swallow all that lies in its wake, destroying things, civilizations, history.

The novel tells a story of a man trapped in a sixty-foot sand pit, held prisoner by a group of villagers, and forced to help a woman dig sand to keep the dunes from covering up her hut. If not held back, the sand threatens the entire village. In the heart of this abyss, the drama and the tensions of human life are played out. In response to the pull of disorder and death, visualized in the atemporal homogeneity of the sand, the man finds purpose and learns to live. The novel uses the complex and changing features of sand to create a reality that is by definition inexpressible in language. It uses the man's confrontation with the sand to evoke a confrontation with the most primal depths of a man's being.

When the protagonist embarks on his holiday, he desires a retreat from the world of the city and of human affairs, where his identity has been reduced to the contents of his wallet and a ticking watch. He is out of touch with life, out of touch with himself, and out of touch with time. Subconsciously he is looking for a means of escape from the artificiality that he and his life have become. Instead of escaping, however, he is plunged into a never-ending struggle for survival against the sand. His descent into the sand pit becomes a descent into himself and a descent through more and more primitive levels of temporal experience. Caught in this tension between life and death, he experiences repressed levels of animal passion, brutality, terror, and confusion. Clock time gives way to the cry and demanding hunger of the immediate present, to the ecstasy and pain of momentary feelings, to the disorientation of fragmented, unformed thoughts and sensations without connection. The desire to escape into timeless oblivion pulls him toward the heart of the abyss; but another urging, this one toward order, self-integration, and life, is just as strong.

Ultimately, his flight from time leads to a return, but not to the old way, not trapped in a meaningless, clockwork routine. He emerges with a newfound freedom, based on acceptance of the conflicts that define his existence, and on a sense of time that projects toward the future with purpose and direction. By acknowledging the primitive temporalities inherent within himself, he has been able to integrate them into a firm sense of selfhood. He can now freely choose to stay in the pit and make a commitment to the woman and the future welfare of the village. Here he discovers a new way to extract water, the source of life, from the barren, lifeless sand. Thus the nightmarish world of sand assumes an order and a temporality of its own that is more in keeping with the natural rhythms of human life. In contrast, the temporality of the world outside, which is based on the artificial, pressured time of clocks and regimentation, is seen to threaten human existence. The protagonist, in escaping the rigid order of that artificial temporality, is able to finally reestablish the creative conflicts of genuine human existence.

If we compare his situation to that of Kafka's protagonist, we see that the confrontation with disorder has given rise to a higher integrative level of order where chaos yields creative conflict. If the conflict ceases, as it does in Kafka's story, there is no freedom or growth. We may recall the wager that Goethe's Faust makes with Mephisto. If ever he desires to hold on to the moment, to ease the tensions of existence, and rest in timeless serenity and peace, *that* already spells death and damnation. Mephisto, that son of chaos and pull toward negation, sustains time. Despite himself, he participates in the process of creation. Human living in its highest, most productive, and freest form arises from the dynamic opposition of these forces.

The temporality of Abé's work must be created by the reader on two simultaneous levels. The protagonist invites us to construct the order of events as a world that is hostile to noetic consciousness and also vital to its survival. Thus we must generate the temporalities of his conscious and his subconsciousness self until, as the novel progresses, the two become integrated. With Kafka's story, the temporal realities of the conscious and subconscious become increasingly disjointed in the reader's mind. The result is a mood of strained confusion and despair.

By looking at literary works as constructs of temporal realities, the main themes and problems of time study acquire a new configuration of which the reader is an active participant. Kundera, in his *The Unbearable Lightness of Being* (1984), poses a problem of time, the paradox of time and eternity, and then compels the reader to live in the problem. Kundera asks how events can have any substance or impact if they occur only once in time. As events move into nonexistence, their reality is reduced to mere words, theories, or interpretations that can change or be forgotten over time. In other words, they become lighter than air. The dilemma of human existence lies in the forgetting of the past and the abnegation of responsibility for the present and future. Thus history as well as individual lives become light, half-real, insignificant. If, on the other hand, events recurred eternally, we would be "nailed to eternity" (p. 5) and unable to escape the weight of unbearable responsibility for our actions. With these reflections, Kundera sets up polarities of eternity and change, weight and lightness, which appear paradoxical in everyday life but in the realm of art create tensions from which the reader constructs a complex order of reality.

The narrator–author is clearly the noetic voice that speaks to us directly about his work and his characters as though he were observing them and attempting to explore the essence of their existential situations. He is involved in a process of discovery, not the discovery of what will happen next, but the discovery of some aspect, some possibility of human existence. Thus he moves easily back and forth along the time continuum and consistently disrupts the narrative by reflecting on the characters or including essaylike commentary. In addition, he frequently relates the same events several times but with a different character as the focal point.

As the narrator shifts the focus of his observation to another character, the events we have already encountered assume a different configuration, a different meaning. We reconstruct the same time line several ways, each time the same yet different, unique and separate but simultaneous in our minds. The polarity of return and change thus creates the dynamics of ordering the text while we read. When an event or a fact is repeated, the old contexts are simultaneously present so that they resonate with the new like a musical composition in the reader's mind. The memory of what we read, then, affects how we perceive the present situation, and the present situation causes us to reevaluate our memory of the past. As a result of this continual process of reordering and maintaining simultaneous levels of meaning, we become aware that reality is not fixed, concrete, or permanent, and that our reading about it again and again does not lead to a clearer, more accurate picture. Instead, it only intensifies the mood of ambiguity and changeability, giving the novel its own lightness of being.

Each character in the novel defines a temporal umwelt and an existential dilemma because each in some way is seeking to escape the burden of existence, the arrow of time, the ceaseless flying through the emptiness of time in a straight line. Let us look specifically at two characters. Thomas flees the burden of choice and commitment by narrowing his temporal window to a biotemporal present. His womanizing is indicative of his flight from the existential tensions of nootemporality. Women appear in his life for a short period of time and are forgotten once he satisfies his curiosity about the one-millionth part that makes each unique or dissimilar from others of her sex. He is driven, not by desire for pleasure or the eternal return of the same, but by an obsessive quest for knowledge and the adventure of collecting bits of information. It makes no difference that the bits of information have no connection or casual consequences. His desire for lightness, however, also restricts his ability to choose and act freely. Without the full range of temporal consciousness, he responds to the demands of immediate necessity and is enslaved by the sense that he cannot do otherwise. The mood emanating from the temporal configuration of his reality is the ache born of psychic stress.

His wife, Tereza, on the other hand, tries to hold on to an assurance of permanence against the flood of time. Thomas's infidelity and the changing political situation in Prague are constant reminders that there is no permanence, no real memory of the past or security in a future. Her escape is a longing for dissolution, chaos, atemporality. She feels vertigo, that heady sensation of falling into emptiness, tempting and luring but also terrifying. External reality mingles with a prototemporal umwelt of dreams where time appears suspended and causes are difficult to distinguish from effects. Even while reading the text, we find that the transition from reality to dream is often so smooth as to take us unawares. She dreams of death and being reduced to an anonymous entity, one woman like all others, naked, indistinguishable, soulless, a victim of her husband's lightness, his forgetting. She dreams of sobbing on a tree that has roots, permanence, continuity in time like a great, great, great grandfather (p. 151). It comes from the depths of time in her dream to offer her solace. But in reality the desire for self-integration and temporal continuity is in constant conflict with a yielding toward annihilation. She, like Thomas, suffers from a lack of freedom to direct her life because of this pull from the lower reaches of her being which house the inarticulate fears and desires of a more primitive self.

In contrast to these painful attempts to escape the ceaseless change of time, is the temporality evoked by Karenin, the dog. His life has a natural regularity, moving in a circle like a clock, always in the same path. He becomes the "timepiece of their lives, a hold against the vertigo" (p. 77). When Thomas and Tereza leave Prague for the country, they too become more in tune with the rhythms of nature and their own selves. Their lives move in a circle, rather than following a straight line to the unknown. The monotony, however, breeds happiness not boredom. Thus thematically, Kundera addresses a fundamental problem: human beings have removed themselves from their own natural, temporal roots and have thereby lost the ability to deal effectively with the conflicts of human existence. The return to paradise is in a sense a return to self-integration, and a cessation of flight to primitive and restricted temporalities.

In this novel, Kundera has created a complex structure that turns the linearity of its form into a pattern of return. Through this artistic manipulation of temporal conflicts, he calls upon the reader to engage in the very processes of change and repetition that form the basis of its content. These processes evoke a range of temporalities, and the act of recreating them and finding an order to their disorder compels us to generate and invent new ways of understanding and knowing the complexity of human existence.

Each of the literary works we have examined reveals something to us about the human experience of time, because each compels us to find new ways of structuring reality and temporality. According to Paulson, this process of invention is the means by which all knowledge is created, and the link that allows us to engage in multidisciplinary dialogue. "We need literary knowledge not so as to know literature but so as to know the world and ourselves" (p. 155).

By using Fraser's model of time in a literary context, we are not imposing a scientific paradigm onto literature nor attempting to gain more knowledge of the text. Instead we are looking to open a dialogue about process and temporality. Through such interaction between the disciplines, we can enhance our knowledge and understanding of reality, human experience and the nature of time.

References

Abé K. (1964), *The Woman in the Dunes*, trans. E. Dale Saunders. New York: Random House, 1972.

Atwood, M. (1985), *The Handmaid's Tale*. New York: Ballantine, 1987.

Fraser, J. T. (1975), *Of Time, Passion and Knowledge: Reflections on the Strategy of Existence*. Princeton, NJ: Princeton University Press, 1990.

——— (1978), *Time as Conflict: A Scientific and Humanistic Study*. Boston: Birkhäuser.

——— (1987), *Time: The Familiar Stranger*. Amherst: University of Massachusetts Press.

Kafka, F. (1946), *The Complete Stories*, ed. N. N. Glatzer. New York: Schocken Books, 1976.

Kundera, M. (1984), *The Unbearable Lightness of Being*, trans. M. H. Heim. New York: Harper & Row, 1985.

Paulson, W. R. (1988), *The Noise of Culture: Literary Texts in a World of Transformation*. Ithaca, NY: Cornell University Press.

Ricoeur, P. (1983), *Time and Narrative*, Vol. 1, trans. K. McLaughlin & D. Pellauer. Chicago: University of Chicago Press, 1984.

——— (1984), *Time and Narrative*, Vol. 2, trans. K. McLaughlin & D. Pellauer. Chicago: University of Chicago Press, 1985.

——— (1985), *Time and Narrative*, Vol. 3, trans. K. Blamey & D. Pellauer. Chicago: University of Chicago Press, 1988.

Thomas, D. M. (1981), *The White Hotel*. New York: Pocket Books, 1982.

Yip, Wai-lim (1972), *Hiding the Universe: Poems by Wang Wei*. New York: Grossman.

Perspectives

The recently popular Theory of Everything (TOE) of physics has a very limited goal: to derive an equation that could combine quantum theory and relativity. Alfred North Whitehead's philosophy is a much more honest attempt to write a Theory—or metaphysics—of Everything. Its power and its difficulties come from this lofty ambition.

At the first conference of this Society in 1969, the Whiteheadian philosopher Wolfe Mays (Mays, 1972) examined several of Whitehead's time-related ideas. Whitehead's philosophy has been referred to in the six volumes of The Study of Time *passim, with the latest mention of it to be found in Clarke's paper in the present volume. Clarke draws on Whitehead's early writings to help him formulate a mathematical system in which the physical dimensions of space and time may be derived from a network of overlapping processes.*

Werth's paper reviews some of Whitehead's ideas about time and process. He builds his exposition around the Whiteheadian notion of concrescence. *The word comes from the Latin (*con = *together and* crescere = *to increase) and means to congeal, grow, or take shape. In biology it means the coalescence of parts, originally separate. Nathaniel Lawrence in his primer on the philosophy of Whitehead (Lawrence, 1974, p. 72) explains the meaning of the term as a jelling of "prehensions" (feelings) into an integral, coherent experience. The idea of concresence is a good focus for the essay because for Whitehead, Werth tells us, even physical events are to be understood as mental acts.*

But there is no way one can sum up Whitehead because his philosophy itself is a concrescence of innumerable prehensions. His philosophy shows a struggle for finding the right words to represent his intellectual hunches. It reminds one of Tsvetaeva's art, using words with the right acoustic properties to represent her partly latent, partly articulate feelings.

From the analysis of temporal concepts we turn to another task of philosophy: that of examining value judgments. We assume that we know what time is—if no one asks us to explain it. Then, turn to the role of time in the collective life of a state made up of free citizens.

What Sherover calls the temporalist view in political philosophy is not a name for a proposed new way of conducting our collective affairs. Rather, he is making us aware of something that has been there all along: temporality as the fundamental dimension of all social practices and values.

The ability to predict future events from the memory of past experiences must have served our earliest ancestors as a powerful weapon in the struggle for survival, because it made possible individual and collective planning. The advantages conferred upon our species by the perception of a temporal reality with broad horizons remained a necessity for maintaining civilizations. Sherover examines that intricate mixture of concerns with future and past, value judgments and pragmatism that constitutes, in his words, modern democratic republicanism.

He reminds us that deliberate, rational activity can offer present choice because it aims toward the future and derives from the past. If a society is to be viable, those choices must remain subject to challenge, debate, and decision. It is through such a process that a free citizenry creates its history. It follows that society itself is neither a structure nor an idea, but a process of continuous choices. Political and social functions, no less than the freedom and responsibility of the individual, derive from this, the temporal character of human communities.

The closing paper of this volume turns to the thought of Ralph Waldo Emerson. To many men and women both at home and abroad, Emerson represents the America of lofty goals, spiritual values, individualism, optimism, pragmatic spiritualism: the dreams of the new land. In his essay, "Circles," Emerson (1903, p. 302) remarked that: "The new continents are built out of the ruins of an old planet. . . . New arts destroy the old. See the investment of capital in aqueducts, made useless by hydraulics; fortifications, by gunpowder; road and canals, by railways; sails by steam; steam by electricity."

Helm's paper explores the philosophical and religious foundations of the thoughts of Emerson. Specifically, he is concerned with what has become the American commitment to a permanent revolution. Emerson redirected the Puritan philosophy of seeing human fate as a pilgrim's progress under the guidance of God, to seeing that progress under God as well as under the power of his reason. As Helm puts it, Emerson rewrote "the upward ascent into a forward progress."

The present volume Time and Process *began with thoughts on cyclic and linear change in the medieval science of computus. Thereafter we visited many departments on the campus of the intellect and arrived at Emerson's insistence on becoming as the true essence of man and nature.*

I would like to close these perspectives by quoting from "Fidelity," a poem by D. H. Lawrence written in 1929. The world is comprised of processes, he tells us, that differ only in their relative rates of passage.

> O flowers they fade because they are moving swiftly. . . .
> O flowers they are all time travelling
> like comets, and they come into our ken
> for a day, for two days. . . .
>
> But a gem is different. It lasts so much longer than we do
> so much much much longer
> that it seems to last forever.
> Yet we know it is flowing away. . . .
> The wonderful slow flowing of the sapphire.
> [1977, p. 476].

J.T.F.

References

Emerson, R. W. (1903), *Essays, First Series*. Boston: Houghton Mifflin.

Lawrence, D. H. (1929), *The Complete Poems*. New York: Penguin, 1977.

Lawrence, N. (1974), *Alfred North Whitehead, a Primer of His Philosophy*. New York: Twayne.

Mays, W. (1972), Whitehead and the philosophy of time. In: *The Study of Time*, Vol. 1, ed. J. T. Fraser. New York: Springer Verlag, pp. 354–369.

12

Clarifying Concrescence in Whitehead's Process Philosophy

Lee F. Werth

Abstract Whitehead's process philosophy involves replacing the concept of God as supreme immutable Being (Plato's Absolute) with a concept of a mutable God who is ultimate Becoming rather than immutable Being. Reality is a process which derives values from God but is not predestined by God. Physical events are explained in terms of mental acts, all of which involve valuation. Motion is ultimately explained in terms of action (mental activity). Even inanimate objects are to be understood as "societies" of "actual occasions," these being extremely low grade minds. Whitehead thereby hoped to clarify free will, God's relation to the universe, and mind's relation to the physical world. Process philosophy reveals how there is a "becoming of continuity" rather than a "continuity of becoming." Time is to be understood in terms of this "becoming of continuity."

> "Concrescence" is the name for the process in which the universe of many things acquires an individual unity in a determinate relegation of each item of the "many" to its subordination in the constitution of the novel "one.". . .
>
> Each instance of concrescence is *itself* the novel individual "thing" in question . . . when we analyse the novel thing we find nothing but the concrescence. . . .
>
> An instance of concrescence is termed an "actual entity"—or, equivalently, an "actual occasion.". . . (italics added)
>
> [Whitehead, 1929, p. 321].

Introduction

The task of one who would attempt to clarify any aspect of Alfred North Whitehead's philosophy to professional colleagues unacquainted with it would be difficult enough. Whitehead, as we shall see, freely invents terminology different from that customarily used by philosophers. Moreover, Whitehead's conception of philosophy is itself not a view accepted by the majority of philosophers; argumentation and formal demonstration are generally preferred to speculation, however ingenious the latter. But the present task requires that difficult concepts be explicated for readers who are not professional philosophers. Yet, one may hope that those with a keen interest in the nature of time but who are not philosophers by profession might bring a new outlook, unburdened by preconceived ideas, and it may well be such readers who will be most open to Whitehead's insights even if the details of his philosophical presentation are not fully assimilated. And indeed, who can claim to understand Whitehead's contribution completely?

In studying Whitehead, philosophers write penetrating analyses which seem to others to have little to do with either the history of philosophy or those problems of concern to philosophers working in other areas, such as ethics, political philosophy, or epistemology. Yet, to understand Whitehead is to recognize that he addresses traditional philosophical problems and the philosophical insights of the past.

The present task requires that we first examine the historical context which gave rise to the problems Whitehead's process philosophy allegedly solves. In understanding this context, Whitehead's presentation will not appear ad hoc and gratuitously anthropomorphic.

Historical Context

Traditional Philosophical Problems

We might remember earlier attempts to summarize the subject matter of philosophy: it is the study of the true, the beautiful, and the good; it is the study of what there is, what we can know, and what we ought to do; it is an investigation into the nature of God, freedom, and immortality. Whitehead is of this tradition. He refuses to settle for mere clarity of expression nor a smug skepticism.

Causation and the Mind–Body Problem

Leibniz, a seventeenth-century philosopher whose metaphysics had many features we find in Whitehead's, asked: "Why is there something rather than nothing?" God, supreme immutable and perfect Being, was claimed the sufficient reason for the existence of the universe (Leibniz, 1714a, pp. 527–528). But perhaps Leibniz's question should have been: why does anything happen, that is, why are there any events? Without involving ourselves in the epistemology of alleged proofs of God's existence, let us notice that the question about events emphasizes process whereas Leibniz's question has "being" as its focus. Whitehead reconstrues the investigation as about "becoming" rather than "being."

We shall see that to understand processes of any sort the concepts of causation, mind, and body must be clarified. During the period of "modern philosophy" (seventeenth and eighteenth centuries) philosophers struggled with the mind–body problem, particularly as it related to religious concepts. Causation was construed either as a sort of puppet-string controlled by God above, a push from behind by a prior event, or a pull from the "future" as we strive to attain what we seek, or some combination of these pushes, pulls, and "strings." Materialism as a metaphysical system was an attempt to solve the mind–body problem, a problem that materialists thought resulted from dividing reality into two substantially different kinds of events: mental and physical. Mental events were not in space, only in time. Since physical events occurred in space, as well, the problem was to explain how a mental event could cause a physical effect and conversely. The implications of a failure of a mental event to have a physical effect are that we are unable to control or in any way affect our bodies. This is contrary to our intuitions, although some thinkers have embraced the result and worked out dualistic metaphysical systems which do not involve mind–body interaction. Malebranche, for example, thought God, in knowing our thoughts and acts of will, created the material world anew each instant so that the appropriate physical events occurred at the correct time. The system seemed inelegant, perhaps demeaning to God; but it did explain how direct mind–body interaction could be avoided while salvaging free will (Malebranche, 1688, pp. 279–304).

George Berkeley thought Malebranche had correctly recognized the relevance of God in causing our perceptions of the physical world, both as we seem to be affected by it and as we affect it. But, if our experiences of physical reality are caused by God, why must matter exist as a separate substance in addition to the spiritual stuff of which mind is made? Berkeley rejected the existence of mind-independent matter as metaphysically superfluous (Berkeley, 1710, p. 153).

Materialists likewise rejected what seemed metaphysically superfluous from their perspective. Spiritual causes seemed ad hoc and deus ex machina, whereas mathematical science was empirical and testable. Accordingly, spiritual substance was rejected. The role it played as the active principle underlying mental events was given to matter, that is, thought was regarded as a secretion of the brain, the latter being a rather splendid endocrine gland. The difficulty with the view (perhaps its advantage) is in its elimination of free will and immortality. The sonnets of Shakespeare, all of our strivings and sense of purpose are the effects of bits of matter in motion. Hence, all teleology is reduced to mechanism. Contemporary materialism is scientifically more sophisticated but metaphysically it too reduces teleology to mass in motion, though we know more about the nature of the mass.

Materialism makes action derive from motion. Motion is understood in terms of space traversed per unit of time. Certainly issues concerning relative time as opposed to Newtonian absolute time are relevant, as is Einstein's contribution. Present purposes require only that we see how action stems from motion, and that what Aristotle called "final causation" (the *for* what) is replaced by efficient causation (the *by* what).

The implications of materialism frightened many into idealist metaphysics in which the physical is derived from the spiritual or mental; nineteenth-century American transcendentalism and continental idealisms claim Absolute Spirit is somehow ultimate, the material world being less real, even illusory. Absolute Being is supreme.

Could the advantages of materialism, its emphasis on the empirically testable, be conjoined with some form of idealism? Idealisms in which absolute Immutable Being is ultimate cannot explain temporality. Why would anything happen? Classical theism and its equating God with Supreme Being is unable to account for evil without extremely strained rationalizations.

The strategy in which motion is derivable from action, rather than the converse, requires that idealism abandon Supreme Being and substitute Ultimate Becoming for Ultimate Being. Moreover, if the order of derivation is from the richer teleological causal concept to the more attenuated concept of efficient causation, perhaps our intuitions of freedom and purpose can be salvaged. Mere mechanical motion would be understood as a very simple form of striving, the sort of striving of which simpler particles are capable.

Objective Idealism

Leibniz understood the universe to be composed of a hierarchy of mindlike entities he called "monads." An inanimate object was the manifestation of an aggregate of these. The lowest grade of monad had appetites and mirrored the entire universe, however unclearly, and strived, although only a high-grade soul monad was claimed to have free will. The motion of matter was ultimately to be explained in terms of the striving of monads (Leibniz, 1714b, pp. 533–552). However, Leibniz's universe was a clockwork one devoid of chance and caprice. His explanation of free will seems difficult to reconcile with a universe, the future of which is a foregone conclusion known atemporally by a God not in time.

Charles Sanders Peirce, a nineteenth-century American philosopher who had carefully studied Immanuel Kant (who had studied Leibniz), created a new form of idealism which greatly influenced William James and subsequently Whitehead. Peirce rarely receives credit for his epistemological and ontological insights.

Peirce offers a universe which is basically mental and develops by chance. Habit creates what appears to us to be physical and lawlike; chance brings about evolution and variety. Matter is said to be mind "hidebound with habits" (Peirce, 1892, p. 350). In a letter to Charles Hartshorne, Whitehead refers to Peirce and James as bringing an American Renaissance in philosophy (see letter in Kline, 1963, p. 198). The manner in which Whitehead explains the uniformity of nature in terms of repetition without novelty, would seem to have been derived from Peirce's concept of habit. Physical regularities are derived from what is fundamentally a simple sort of mental activity (subsequently to be explained). Physical regularity based upon habit allows the formulation of the alleged universal laws of nature while still allowing the freeplay necessary for the creative acts of persons, that is, the flexible activity not hidebound by habit, and thereby avoids a deterministic nature which in a Leibnizian ontology seems incompatible with free will. Moreover, Peirce's concept of chance allows even the material universe to escape total immutable regularity. The laws of nature themselves evolve. The simplest particles might manifest novel or chance-based behavior. Hence, Peirce anticipated current cosmological theories of evolution and quantum mechanical indeterminism, and did so in a scientific era characterized by dogmatic determinism (i.e., the belief in a clockwork universe of universal natural laws) (Peirce, 1890, pp. 358–359).

We might think of Whitehead asking Peirce: Just what is it that is creative or strives? Whitehead assimilates Peirce's indeterministic universe and weds a concept of mind to it which is a generalization of William James's concept of mind. (James, too, accepted Peirce's notion of chance.)

Another departure from earlier views is the concept of a finite God who reacts to the universe. Peirce's contribution to theological epistemology enormously influenced both James and Whitehead. Briefly stated, earlier philosophers attempted to prove that a God must exist; typically this God was identified as supreme and perfect Being, although there were those who took the universe at face value and doubted God's perfection. Peirce's finite God was proposed in speculative terms rather than as a logical demonstration. Peirce approached the issue pragmatically: How would our lives be different if we believed in a God? Grounds for claiming a belief is advantageous are not to be conflated with a proof of the object of the belief. In the absence of such proof, however, it is rational to believe when it is advantageous to our lives. Peirce believed the universe was evolving in the presence of a caring and loving deity (Peirce, 1908, pp. 358–379). Whitehead's God is claimed to be a lure for our feelings; we are drawn to a God who thereby influences our strivings and the evolution of the universe.

Let us return to the question of what it is that strives, that is, the issue of the concept of mind implicit in Peirce, explicit in James, and generalized to everything by Whitehead.

William James's Concept of Mind and Whitehead's Generalization of It

Without James's account of human consciousness, Whitehead would not have had the basis from which to generalize (and anthropomorphize) to his basic unit of reality, the "actual occasion," a unit which has both a mental and physical aspect or "pole." An actual occasion is what results from extending James's account of consciousness to its metaphorical limit (perhaps beyond it). Clarification is clearly in order.

James rejected philosophies which employed spiritual substance as the basis of all mental activity. He also found no need for a transcendental ego, a metaphysical pure subject of experience; for example, he rejected Descartes' soul as substance and Kant's deduction of a transcendental ego. James's empiricism was influenced by Hume, who, however, atomized human experience into a bundle of perceptions, the unifying principle not being explained except by appeal to the association of ideas (Hume, 1739, pp. 10–13). What has the ideas? Who does the associating? James substitutes a stream of consciousness for Hume's disjointed sequence of perceptual atoms.

Exactly what is the subject who experiences on the "stream view" of mind? What provides the unity of the stream and the continuity of experiencing as we are immediately aware of it in all its mental and physical diversity? There seems a sort of specious present, as James calls it, during which we are aware of ourselves as having a body, emotions, and thought content. James undertakes a descriptive phenomenology of consciousness and observes:

> Consciousness, then, does not appear to itself chopped up in bits. Such words as "chain" or "train" do not describe it fitly as it presents itself in the first instance. It is nothing jointed; it flows. A "river" or a "stream" are the metaphors by which it is most naturally described [1890, p. 233].

However, our experience is not of an even flow but of specific thoughts, particular sensations, that is, there are objects of consciousness. Do we not move from thought to thought, as the thought content changes? Are the objects of thought not separate from the thinker? Such deliberations have caused earlier philosophers to plunge into the metaphysics of spiritual substance as the active "stuff" which has the thoughts. James makes his contribution: "The transition between the object of one thought and the thought of another is no more a break in the *thought* than a joint in a bamboo is a break in the wood. It is part of the *consciousness* as much as the joint is a part of the *bamboo*" (James, 1890, pp. 233–234). James presents his view of thought, object of thought, and the thinker. With respect to the object of thought he states:

> The object of every thought, then, is neither more nor less than all that the thought thinks, exactly as the thought thinks it, however complicated the matter and however symbolic the manner of the thinking may be. . . .
>
> The next point to make clear is that, *however complex the object may be, the thought of it is one undivided state of consciousness* [James, 1890, p. 266].

One is puzzled by the locution "the thought thinks" until one understands the nature of the thinking subject according to James: "*the 'Self of selves,' when carefully examined, is found to consist mainly of the collection of these peculiar motions in the head or between the head and throat*" (1890, p. 288). And as to the thinker or the conscious subject who is aware of these motions and other thoughts, James states: "The passing Thought then seems to be the Thinker" (p. 324).

> Each pulse of cognitive consciousness, each Thought, dies away and is replaced by another. The other, among the things it knows, knows its own predecessor, and finding it "warm," in the way we have described, greets it saying: "Thou art *mine*, and part of the same self with me." Each later Thought, knowing and including thus the Thoughts which went before, is the final receptacle—and appropriating them is the final owner—of all that they contain and own. Each Thought

> is thus born an owner, and dies owned, transmitting whatever it realized as its Self to its own later proprietor [p. 322].

James speaks of the Thought "appropriating and disowning"; "it is a vehicle of choice as well as of cognition":

> But the Thought never is an object in its own hands, it never appropriates or disowns itself. It appropriates *to* itself, it is the actual focus of accretion, the hook from which the chain of past selves dangles, planted firmly in the Present, which alone passes for real, and thus keeping the chain from being a purely ideal thing [James, 1890, p. 323].

A Thought knows its own predecessor; it finds in it "intimacy" and "warmth." James explains this on-going personal identity: "*Resemblance among the parts of the continuum of feelings* (especially bodily feelings) experienced along with things widely different in all other regards, *thus constitutes the real and verifiable "personal identity" which we feel.* There is no other identity than this in the 'stream' of subjective consciousness . . ." (p. 319).

Whitehead adopts James's account of human consciousness and personal identity. Translating some of the foregoing themes into Whiteheadian terminology, the bodily aspects of the "continuum of feelings" would be referred to as the "withness of the body." James speaks of the Thought "appropriating and disowning." He carefully states that the Thought "appropriates to itself" and is not "an object in its own hands." Such appropriating and disowning would be "positive and negative prehensions," in Whitehead's vocabulary (the term *prehensions* will be subsequently explicated). James speaks of a "pulse of consciousness": "each Thought dies away and is replaced by another." The pulse will be termed an "actual occasion" by Whitehead.

We must not think that James's account of consciousness applies in an unmodified manner to anything other than human consciousness. Whitehead takes James's philosophy of mind and does an imaginative leap with it by extending the metaphors so as to regard all reality as a process. In so doing he is able to explain not only the continuation of personal identity as it applies to human consciousness, but the continuity of existence of physical objects, what Whitehead calls "enduring objects."

James is aware of the diversity of thought content which includes bodily sensations and the customarily recognized cognitive content. But James also acknowledges the usually orphaned content of consciousness which lies between the thoughts we dignify with names. What James calls "transitive" thoughts, as opposed to "substantive," are also constitutive of consciousness. What is the feeling of something missing to be called? We sometimes remember the number of syllables a name has, but not the name itself (James, 1890, p. 243).

In a similar spirit, Whitehead uses the word *prehension* to cover any sort of feeling or awareness we are capable of having, but in generalizing the concept of feelings, he speaks of nonconscious living forms as also prehending. Moreover, inanimate objects and simple particles can prehend. The difference concerns the sort of prehensions of which we are capable, as opposed to those of an electron, for example. Intensity and variety of prehensions vary with the type of creature at issue.

Henri Bergson, Intuition and Flux

Whitehead was not unfamiliar with Henri Bergson, a kindred process philosopher who stressed the importance of intuition, and who claimed that we can proceed from intuition to analysis, but not from analysis to intuition. The transiency of time, or the creative advance of nature, simply cannot be described literally, only intuited. Analyze time into instants and the set derived from such analysis cannot be reassembled into transient time (Bergson, 1903, p. 42).

Newton construed time flow as being independent of experienced nature and as absolute without significant contrast with anything; it somehow flowed in its absolute equable manner. Bergson saw all reality, "concrete" reality, as flowing, as temporally advancing. Although not abstract and mathematical in the Newtonian sense, one still wonders what "flow" could mean when *all* reality is flowing forward in a creative advance. No doubt we will have to rely upon intuition. In any case, the Newtonian absolute present must be abandoned if Bergson's view is to satisfy postrelativity considerations. But Bergson's creative advance need not require an absolute present, merely the denial of any part of reality temporally stopping for a "rest." Nothing real is static; everything is temporally becoming.

It is not that Bergson's view is rejected by Whitehead. Rather it would seem that Whitehead presents a system which being all-embracing and coherent enables us to see the interrelatedness of all those actualities and possibilities which explicate the process or "flow." The fundamental intuition is the same, but Whitehead explains why we have the intuition of things being in flux. He undertakes to provide the anatomy and physiology of the organism which is the universe in its creative advance. (See Whitehead's expression of indebtedness to Bergson, James, and Dewey: Whitehead, 1929, p. vii.)

Speculative Philosophy, Language, and Intuition

Whitehead presents a system of speculative philosophy which explicates process and temporality perhaps better than any alternative philosophical system or analysis. To appreciate Whitehead's success we must understand the goal of a speculative philosopher. The attempt is not to offer a true description of reality by starting with allegedly self-evident premises, and logically deriving descriptions of everything else from those premises. Nor does the speculative philosopher restrict the inquiry to the issues which are empirically describable in either ordinary language or scientific discourse which includes mathematics.

It is Whitehead who best states the nature of his quest:

> Speculative Philosophy is the endeavor to frame a coherent, logical, necessary system of general ideas in terms of which every element of our experience can be interpreted. By this notion of "interpretation" I mean that everything of which we are conscious, as enjoyed, perceived, willed, or thought, shall have the character of a particular instance of the general scheme. . . .
>
> "Coherence," as here employed, means that the fundamental ideas, in terms of which the scheme is developed, presuppose each other so that in isolation they are meaningless. . . . In other words, it is presupposed that no entity can be conceived in complete abstraction from the system of the universe, and that it is the business of speculative philosophy to exhibit this truth [Whitehead, 1929, pp. 4–5].

A certain optimism is required to believe speculative philosophy is in principle possible. Perhaps human experience is so contextually disjointed and our biologically evolved brains so constituted that our experience as human beings is fundamentally incoherent; that is, the different contexts cannot be integrated into a single system of general ideas. The practical affairs of life and the division of technological and scientific inquiry into separate disciplines do not demand an all-encompassing system. Indeed many still consider speculative philosophy a frivolous enterprise at best, assuming the required coherence is possible; alternatively, if such coherence is not possible, the quest for it is naive, a flagrant waste of time and words, in principle unintelligible.

Even in highly circumscribed disciplines such as physics, it is sometimes thought that a TOE (Theory of Everything) is not possible, and if possible, probably not useful. Whitehead's concept of coherence includes all of physics and everything else as well: values, emotions, possibilities. The most ambitious superstring theorist is not likely to attempt the reduction of aesthetics to physics. Getting all of physical reality to cohere so as to explain all physical behavior in terms of a single relationship in which the laws of nature can be understood is task enough.

Yet we have seen advances in science based upon the pursuit of unity and coherence. Some would argue that monotheistic religion also represents an advance, one motivated by the pursuit of generality and coherence. Despite the disparate aspects of human experience, our lives seem to have a unity as lived, as a stream of consciousness. And it seems odd to regard the search for a single all-embracing reality-description to be necessarily doomed. The very rejection of the possibility suggests that we somehow intuit the possibility some would reject. Who has the burden of proof? Why shouldn't speculative philosophy be possible?

How are we to express such a scheme? Any description must begin with language used in clear and paradigmatic ways. Each discipline has its standards. Any departure from ordinary or standardized usages ultimately leads to unintelligibility. How can speculative philosophy avoid it? Physics at least has empirical measurement to provide the bedrock of meaning to its terms. Particles claimed to have color, charm, or strangeness, only metaphorically are so described. Mathematically the meanings are precise even if highly technical and esoteric.

Whitehead was acutely aware of the problems associated with language. If we begin with our experience-descriptions and progressively generalize them to all of reality, we may drift increasingly into metaphor, but if the original usages are clear, then the metaphors will not be totally without meaning; and perhaps with the right metaphors, reality can be described, not as the scientist would, but as would the poet. In *Modes of Thought* (1938), he states: ''Language halts behind intuition. The difficulty of philosophy is the expression of what is self-evident. Our understanding outruns the ordinary usages of words. Philosophy is akin to poetry. Philosophy is the endeavor to find a conventional phraseology for the vivid suggestiveness of the poet'' (pp. 68–69). And in *Process and Reality* (1929) it's claimed that:

> Philosophers can never hope finally to formulate these metaphysical first principles. Weakness of insight and deficiencies of language stand in the way inexorably. Words and phrases must be stretched towards a generality foreign to their ordinary usage; and however such elements of language be stabilized as technicalities, they remain metaphors mutely appealing for an imaginative leap [p. 6].

In *Adventures of Ideas* (1933) we find the same theme: ''Also the success of language in conveying information is vastly over-rated, especially in learned circles. Not only is language highly elliptical, but also nothing can supply the defect of first-hand experience of types cognate to the things explicitly mentioned'' (p. 287).

Language can never substitute for intuition, an imaginative leap, nor first-hand experience.

Whitehead's Ontology

If the foregoing sections have served their intended purpose, we can now see how Whitehead's actual occasions allow the mind–body problem to be solved without action being motion. Rather, all changes are ultimately the result of the strivings of actual occasions.

As the terminology becomes more complex, it strains the metaphors upon which it rests. But notice how insightfully Whitehead accommodates science, religion, the arts—everything, while yet appealing to a very few basic notions which are mutually explanatory.

Everything is explained ultimately in terms of actual occasions, or ''actual entities,'' as Whitehead sometimes calls them. These are atomic in that they are the constituents of reality. However, it is quite easy to import incorrect meanings into Whitehead's atomic actual occasions. Space and time are logically derived from more fundamental relationships, and time concepts derived from actual occasions must be contrasted with those of scientific materialism. In actual practice (doing science) Whitehead's time concepts are entirely compatible with contemporary science if only because Whitehead intended them to be.

Whitehead's reality is not ultimately composed of bits of matter scattered in space–time. An object of any sort is an event, and events have their earlier and later parts. Noncontemporaneous parts cannot coexist temporally, and the present, strictly speaking, is an instant of null duration; therefore, it would seem that nothing could exist. (''Present'' here is relative to context, no absolute present is being claimed.) However, an actual occasion, the atomic constituent of events is *not* itself *in* time. Enduring objects are repetitious occasions, the same sort of actual occasions repeating themselves, thus creating a series which is the event that is the enduring object. (We can easily visualize any seemingly static object as an event if we use time-stop photography. In principle, we might watch the event of a mountain range rising and crumbling, or the continents drifting. What appears to us to be a static object is relative to human perceptual constraints. Fast and slow motion photography help us to surmount such limitations.)

Is an actual occasion a static or changing entity? If changing, does that not take time, in which case time is presupposed in order for the actual occasion to be in process? If it is not in process, but static, then how does process philosophy derive temporal becoming or the *transiency* of time any better than the usual view of pointlike instants of time, or cross-sections of space–time? Does Whitehead's analysis destroy flux as we might imagine Bergson claiming?

This disingenuous dilemma will serve to clarify Whitehead's contribution in which mechanism or the regularities in nature can be explained in terms of the strivings of actual occasions. There is the aphorism that actual occasions are not in time, time is *in them*; and also Whitehead's claim that there is no continuity of becoming but rather a becoming of continuity (1929, p. 53). Although aphorisms are useful, they are limited, even misleading. Actual occasions are active striving agents. They seek—they have a ''subjective aim''; they are telic. Motion takes time, but does action? Remembering James's *pulse* of consciousness, an actual occasion appropriates *to* itself; high-grade occasions are conscious; low-grade occasions simply feel in some very basic way. Here we have the idealist claim that reality is somehow mental. As actual striving agents, actual occasions are mental, sentient in some way. To say a rock is an enduring object composed of a ''nexus'' of occasions which feel, sounds muddle-minded. But if we generalize our experience of ourselves as having a continuum of personal identity derived in the Jamesian manner, we might see how a grossly attenuated form of consciousness could constitute a continuum of the low-grade awareness which is the personal identity of a rock. Or we might imagine higher forms of consciousness whose prehensions, whose cognitive, emotive, bodily, and God knows what other sort of awareness, vastly exceeds our own.

However, as long as we consider the actual occasion as active, as striving toward its subjective aim, we will not arrive at temporality but only a hierarchy of consciousness which seems to relegate each occasion to a sort of solipsism of its indigenous consciousness which presumably is private. This would be so if Whitehead did not have a doctrine of internal relations. No actual occasion can prehend or feel anything without something serving as the objects or data prehended.

Prior to developing internal relations, it should be mentioned that building all reality from entities which prehend or have a sort of awareness has the virtue of avoiding the problem of the *emergence* of consciousness from nonconscious things, a difficulty materialism faces. If a rock has a form of awareness, this may seem unnecessarily complex, but if only one basic kind of building block of reality exists, we gain ontological simplicity; that is, we have a monism which doesn't render consciousness the enigma it is in materialism.

If each actual occasion were only an active striving, we might indeed set up conditions which would only allow for solipsism, each occasion being a private world onto itself. This is, perhaps, how we might regard a "windowless" Leibnizian monad; Leibniz also tried to construct an ontology of atoms of consciousness. However, actual occasions are aware of other actual occasions. They are both subject and object, knower and known. How can this be? How can subjectivity or consciousness itself be seen or known, or be an object of thought? Here again James provides the clue; thought appropriates other thought to itself.

We must think of actual occasions as subject and object, and either way of characterizing them must be done by contrast, or as an internal relation. That is, an actual occasion as subject is meaningless without an actual occasion as object. Externally related entities can be understood in isolation. My car is a Dodge Omni regardless of where it is parked. But the parts of my car are internally related; together they constitute what is in fact my car as a sort of organism. Admittedly, the battery is a battery whether it is in my car or not. But remove enough parts and you have "removed" my car. Likewise an event too has its "parts" and is that event only if the appropriate sequences occur.

How can actual occasions be objects prehended by other actual occasions, and what is it that gives us the becoming of continuity? More terminology must be introduced if concrescence is to be clarified.

When an actual occasion prehends another actual occasion, the prehending is active, and physical time does not apply to the active attempt to realize the actual occasion's "subjective aim" or goal. The prehended occasion, however, is no longer an actual occasion as actual. It has "passed over" into realization of its subjective aim and has reached "satisfaction"; it is no longer a conscious prehending actual entity but it is data for the striving actual occasions to prehend. It is no longer a subject of experience, it is objective data. As such, it can be referred to in terms of its "physical pole" as opposed to its mental. It is "superject," not subject. As objective, it is able to be a relatum for temporal relations with other occasions which are no longer actual. James's thought-appropriation scheme which gives rise to a continuity of consciousness might be employed here; that is, temporality exists in relation to the thought content, not the thinker or "passing thought" which is never an object for itself. James also speaks of bodily awareness. An actual occasion in prehending dative (objective) occasions can have prehensions of what we would normally call physical feelings. These can include an awareness of spatial locations such that we perceive things as appearing *from* a location, with the perceived thing appearing to be *at* a location. This constitutes the experience of regions. Location prehensions, or more precisely, *region* awareness, are logically primitive in our construction of an "extensive continuum." We experience regions as being next to regions, or regions including regions. The "extensive connection" of regions, as Whitehead calls it, is used in deriving points, lines, and so on. This makes space and time (more will be said about time) derived from experience, not presupposed as absolute. Nor are points the mysterious nihilisms of Euclid. For Whitehead, points are derived from region inclusions and their limits. (A nest of boxes sets up a point as a limit based upon inclusion; Whitehead uses this analogy.) Space–time for Whitehead is a coordinate continuum which provides the possibilities for atomization by actual occasions. That is, things require the possibility of location in space–time if they will ever become actual ([1929], part 4, in particular pp. 442–443).

If the foregoing is at least somewhat intelligible, even if unavoidably obscure, we can introduce the expression "presentational immediacy," the experience we have of things appearing to us—even illusions are included. Our visceral and regional awareness as just discussed create the "withness of the body" which was already mentioned when discussing James. In addition to the perceptual mode of presentational immediacy, there is Whitehead's answer to Hume and Kant, that is, the mode of "causal efficacy" which is the awareness that we are somehow passively presented with data coming from something. If sunlight makes us squint, we feel *caused* to squint. This causal mode is as real as the sense data that phenomenalists so lovingly acknowledge, yet Hume calls casual awareness mere habit, and Kant makes it a category of the human understanding. Causal efficacy is *felt*, according to Whitehead, along with other data.

The data prehended cannot simply be actual occasions abstractly considered; something must provide the *content* of a prehension. An actual occasion is just a way of identifying either the unit which feels (subject), or that which provides a package for data or feelings (superject). Moreover, occasions no longer actually striving are what can be prehended; all occasions simultaneously actual cannot interact or be mutually aware of one another, only the occasions antecedent to those which are actual can be temporally ordered. (Actual occasions cannot act *in* time; "when" fully actualized they cease activity, and then can be data temporally related to other occasions.) The data prehended, the content of an actual occasion, are "eternal objects" or what are sometimes called "Platonic forms." Some of these are what classically have been called "universals," but eternal objects include any sort of quiddity. Anything capable of repetition is an eternal object. A color, a shape, a numerical relation, a feeling—if it can happen again it is an eternal object. Such repetition can provide stability to reality in process. Simple occasions do the same thing their antecedent occasions did. Hence physical regularities can be explained. For example, the behavior of gases is predictable: in a gas we have actual occasions constituting a "society" of occasions, a nexus. That society consists of occasions which prehend eternal objects as their antecedent society members did. This creates stable physical processes without the need for some actual occasion to lead the members of its society or nexus. Some stable processes are those events we call "enduring objects," for example, planets, stones, and snakes. Living things strive more creatively and prehend with more harmony and intensity; their prehensions have greater variety, although negative prehensions exclude eternal objects. Fully developed, this explicates free will; we are not compelled by antecedent occasions in our prehending as simple objects seem to be. Even simple things are not without their limited novel behavior. Quantum indeterminism is thereby dealt with.

A hierarchy exists with the actual occasions of the more complex societies having richer more intense prehensions of greater variety, including both conceptual and physical prehensions. Complex societies are called ''nexūs.'' Communities are nexūs, as are biological organisms. We do not generally think of a community as having a personal identity or mind in the way in which a human constituent of that community does. However, where a multitude of actual occasions pursue similar aims, their behavior or actions (as with human societies) will be coordinated, much as the string section of a leaderless orchestra might be coordinated when playing the same score. Each player is independent although adjusting to the surrounding sound. The analogy suggests how a creative, striving, and integrated community (the universe as organism) develops with its contrasting parts being internally related, and an objective unity being achieved through a striving toward a common ultimate end, which we shall see is God, as Whitehead construes Him.

The striving preserves the past as ''objectively immortal'' in the prehensions of actual occasions. Are we not now discussing the creations of Whitehead who is no longer actual but is objectively immortal? The series of occasions prehending the eternal objects of their antecedents preserves the past in the present. The information is passed on in the Jamesian manner. There is becoming of continuity as actual occasions prehend the data of the occasions which having satisfied their subjective aims are not actual (not active), but exist as information to be passed from occasion to occasion and so on. Therefore, a nonactive satisfied and superject occasion achieves what Whitehead calls ''objective immortality.'' That is, the eternal objects which it uniquely synthesized in its activity to satisfy its subjective aim constitute a novel synthesis which is available as data for prehending by actual occasions. Actual occasions inherit the synthesis but act to create their own novelty until they too reach subjective satisfaction and pass from active occasions into an objective immortality. As objective, they have a place in a time series, or more precisely, a location in the extensive continuum relative to other occasions as prehended from the perspective of an actual occasion.

An analogy might clarify the nature of actual occasions as actively striving and as passive data. Consider an author writing a novel. Prior to completion, he or she has not satisfied the subjective aim or the creative urge which underlies the creation of the novel. The book while being written is not yet part of that body we call literature. It is also not available for others to read, assuming our author wishes first to finish it. From the author's perspective, the novel is most ''alive'' during the creative process. He or she incorporates ideas from other books; the creative activity is shaped by the culture and the author's unique needs and perspective. It is the author's world or viewpoint which is being expressed. Until published and copyrighted, the book being created is not yet in that temporal series which includes Homer, Shakespeare, Melville, and other significant writers. Once the author's feelings and ideas are ''incorporated'' in the published book, they become data for others who subsequently are active in their own way. The novel takes a place in the history of ideas.

Inheritance or genetic divisibility is analysis of the different sort of prehensions of an actual occasion. In being actual, the occasion constitutes a unit (James's thought as a thinker) but the content of the novel concrescence can be analyzed in terms of the prehensions. Once activity is completed and creativity and striving cease, then the satisfied occasion also becomes available for analysis in terms of its location in the extensive continuum. Enduring objects, a nexus of occasions, can be coordinately analyzed.

> Physical time makes its appearance in the "coordinate" analysis of the "satisfaction." The actual entity is the enjoyment of a certain quantum of physical time. But the genetic process is not the temporal succession: such a view is exactly what is denied by the epochal theory of time. Each phase in the genetic process presupposes the entire quantum, and so does each feeling in each phase. The subjective unity dominating the process forbids the division of that extensive quantum which originates with the primary phase of the subjective aim [Whitehead, 1929, p. 434].

The completed activity of an actual occasion gives rise to a new objective entity, thus creating the becoming of continuity. *Experience* is temporal if regarded as *content* inherited from occasions that are no longer actual; but the *experiencing* unitary subject as a creator, the actual occasion in its activity as opposed to *passivity* (my term, not Whitehead's) is not in time. Time is in it *via* its prehending of passive (objective) occasions. In this way we are to understand how process philosophy ultimately derives motion from the activity of creative synthesis, the activity of actual occasions in pursuit of their subjective aim.

Concrescence is the word used for the synthesis of data from antecedent occasions by the actual occasion as actively and creatively striving to obtain satisfaction with respect to its subjective aim. As active, it takes eternal objects and renders them concrete and actual. The totality of activity of actual occasions creates the world in process. Depending upon the choices made in the creative synthesis, that is, with respect to the concrescence of eternal objects by the activity of actual occasions, the universe becomes a better or worse place. If intensity and variety are maximized, harmony and goodness are achieved. But if the activity of actual occasions is at "cross-purposes" so that conflict rather than contrast is the product of concrescence, we have evil. The process ought to be an orchestrated symphony, not a banal tune nor cacophony. The "primordial nature of God" is an ordering of eternal objects, the possibilities for concrescence. The "consequent nature of God" is the result of the concrescences that have occurred as the activity of occasions. God saves all in that He remembers all that ever happens, even if actual occasions do not prehend the data. If a child suffers, that suffering is remembered (prehended) even if the child forgets and no one remembers. God knows all but is not all-powerful. Rather, God is love and the lure for feeling. He is the inspiration that brings about the striving and creativity of actual occasions, but He does not ever pass over into mere data and remains actual. Whitehead calls God an "actual entity" not an actual occasion. Whitehead's account of God is not altogether clear, but any account of God must be impressionistic given our human limitations.

Whitehead's account has the advantage of a finite God, who being finite in power, is not the author of evil, and who only can lure goodness into actuality (1929, p. 522).

Real process is concrescence, it is aesthetic, a striving for values. Temporal process is transition, it is mechanistic, but is logically derivative and the result of the completed activity of actual occasions. From an analysis of the data brought into a concrescence, we have duration; the activity of an actual occasion as action is not temporal but gives rise to the temporal ordering of the data prehended. Nonsense? Perhaps, but if all striving is aesthetic, a search for value, then examples from art might help us to understand Whitehead's intuitions.

Mozart is said to have created his symphonies in a single comprehensive act, not serially. The music is temporally ordered, not its creation. A painting of a boat on a turbulent sea, Winslow Homer's *Fog Warning*, is immediately understood in terms of striving and process. The doryman must be picked up by the schooner before the fog obscures the dory. A storm is coming. The painting is a nontemporal (static) concrescence of temporal content. It portrays motion, but more significantly, a striving for life. Objectively, it is only paint on canvas. I remember a photograph of a boy in the Warsaw ghetto. His eyes say everything about the Holocaust to come. A photograph, static shades of gray, how can one be moved by paper? Why do we metaphorically say ''moved?'' On Whitehead's view, motion is ultimately derived from valuation.

Whitehead thought an old hymn captured Plato's intuitions of the immutable, and also the Heraclitian flux of things:

> Abide with me/Fast falls the eventide
>
> Here, the first line expresses the permanences, ''abide,'' ''me'' and the ''Being'' addressed; and the second line sets these permanences amid the inescapable flux. Here at length we find formulated the complete problem of metaphysics. Those philosophers who start with the first line have given us the metaphysics of ''substance''; and those who start with the second have developed the metaphysics of ''flux''. But, in truth, the two lines cannot be torn apart in this way. . . [1929, p. 318].

Whitehead's system captures the meaning of the hymn with two kinds of process, the genetic process ''internal'' to the actual occasion in its active striving toward its subjective aim, a process which is not itself in time, and which explicates final causation; and the coordinate process of transition which explicates efficient causation and time. We have seen how the latter temporal process is based upon the former; teleology is logically prior to mechanism or any sort of efficient causation. However, neither process could exist ontologically independently of the other; they are internally related.

> [There are] two kinds of fluency required for the description of the fluent world. One kind is the fluency inherent in the constitution of the particular existent. This kind I have called ''concrescence.'' The other kind is the fluency whereby the

> perishing of the process, on completion of the particular existent, constitutes that existent as an original element in the constitutions of other particular existents elicited by repetitions of process. This kind I have called "transition." Concrescence moves toward its final cause, which is its subjective aim; transition is the vehicle of the efficient cause, which is the immortal past [1929, p. 320].
>
> An instance of concrescence is termed an "actual entity"—or, equivalently, an "actual occasion." There is not one completed set of things which are actual occasions. For the fundamental inescapable fact is the creativity in virtue of which there can be no "many things" which are not subordinated in a concrete unity. Thus a set of all actual occasions is by the nature of things a standpoint for another concrescence which elicits a concrete unity from those many actual occasions. Thus we can never survey the actual world except from the standpoint of an immediate concrescence which is falsifying the presupposed completion [pp. 321–322].

The process will not pause; any description will be limited and immediately obsolete (Bergson's point). Continuing the quotation:

> The creativity in virtue of which any relative complete actual world is, by the nature of things, the datum for a new concrescence, is termed "transition." Thus, by reason of transition, "the actual world" is always a relative term, and refers to that basis of presupposed actual occasions which is a datum for the novel concrescence [p. 322].

Einstein said, "God does not roll dice"; Whitehead disagreed, God *loves* the game, and will play at *every* chance.

Remaining Questions

Several areas of investigation have frequently consumed the energies of Whiteheadian scholars. We shall briefly attend to: (1) Whitehead's unique contribution—that which is not already implicit in Bergson, Peirce, James, and others; (2) Whitehead's views on freedom and self-creation—whether serious problems exist; and (3) his account of temporality—whether there can be two kinds of flux. To shed light on one of these areas of concern is also to clarify the remaining two.

The Influence of Peirce, James, and Bergson: Whitehead's Unique Contribution

Whitehead, himself, regarded his metaphysics as the product of personal mediation, although he explicitly acknowledges the influence of Bergson, Peirce, James, and Dewey, among others. Scholars have noted the appropriation made of James's account of mind and some of the theological ideas James received from Peirce. Clearly Whitehead's metaphysical concepts were grown on American soil. Victor

Lowe clarifies this influence and also Bergson's contribution, while revealing that Whitehead's contributions are not mere generalizations of earlier ideas or an eclectic synthesis (1962). Lowe is correct, although he seems overly strident in his attempt to differentiate Whitehead's concept of process from Bergson's (pp. 257–268). It is hoped that the quotations included earlier from James's *Principles of Psychology* demonstrate James's contribution (Lowe, 1962, chapter 13).

Bergson's influence is more difficult to explicate. However, Bergson's account of our intuition of free will must be appreciated if Whitehead's presentation of the stages of concrescence are to explain freedom and creativity successfully. The present paper cannot include more than a brief indication of the relevant issues and the critical literature associated with it.

Briefly, Victor Lowe reminds us of Bergson's statement: "metaphysics is the science which claims to dispense with symbols." Since Whitehead describes an elaborate metaphysics of process in terms of symbolic and geometric relations (coining words for the purpose) Whitehead obviously seemed to disagree with Bergson's thesis (Lowe, 1962, p. 260). But perhaps Lowe and other critics overlook Bergson's claim that: "*To philosophize means to reverse the normal direction of the workings of thought.*"

> But to do that, it (the mind) must do itself violence, reverse the direction of the operation by which it ordinarily thinks, continually upsetting its categories, or rather recasting them. In so doing it will arrive at fluid concepts, capable of following reality in all its windings and of adopting the very movement of the inner life of things. Only in that way will a progressive philosophy be constituted freed from the disputes which arise between schools. . . [Bergson, 1903, p. 51, in Alston and Nakhnikian, 1963, p. 77].[1]

Whitehead's account of actual occasions, concrescence, prehensions, and feelings is an attempt to formulate *fluid concepts* which can express the movement of the "*inner life*"; that is, the intuition of freedom which is the basis of Bergson's philosophy, but which requires Whitehead's explication of how *inner* life (subjectivity) is related to the macroscopic world of material things. Motion (a type of transition), a "macroscopic process," is ultimately based upon subjective acts of synthesis (conscrescence), a "microscopic process." Freedom and self-creative striving is the ultimate reason anything happens at all. Without Whitehead's *two* kinds of flux, Bergson's metaphysics is merely a prose poem, but one which Whitehead understood as he also understood the romantic poets, Heraclitus, and the Hebrew psalms.

Temporal flow, or Bergson's reality in its self-creative evolution, is rendered understandable in terms of actual occasions as subject–superjects. Bergsonian *analysis* would involve the analysis of the data synthesized in the satisfaction of actual

[1]Alston and Nakhnikian (1963) contains Alston's criticism of Bergson's *Introduction to Metaphysics* (1903) which is reprinted therein. After drawing attention to the "fluid concepts" passage (p. 77), Alston claims, "It is regrettable that Bergson does not develop further this distinction between two types of concepts" (p. 54).

occasions considered as *superjects*. This would include physical properties which reveal how such occasions are to fit in an extensive continuum, that is, how they are located in space–time. Bergsonian *intuition* would involve understanding the process of concrescence: how the data get synthesized by actual occasions as active *subjects*. The "fluid concept" of concrescence can express an act of creative synthesis in terms of the relations and stages involved. The concrescence of prehensions gives rise to the subject matter of psychology, geometry, and physics (Sherburne, 1961, pp. 41–71; Lowe, 1962, p. 25).[2]

Freedom and Self-Creation, an Avoidable Paradox

Any attempt to clarify free action or self-creation runs a risk: either the act of self-creation is ultimately a reaction or response *determined* by the motives and goal an allegedly free agent has (teleological determinism); or a sort of freeplay or caprice exists which makes an allegedly free act be the product of chance. In the former case a series of goals can be traced to some originally given goal or instinctive desire; self-creation or genuinely free choice is thus an illusion. In the latter case, chance can hardly be the basis of genuine choice even if it can be the basis for novelty and indeterminism.

Since actual occasions are allegedly free to create novel concrescences out of the data given by passive, satisfied, superject occasions, we may wonder in what sense active actual occasions are free? They evaluate data prehended in a novel manner, but is the evaluation process determined or the result of chance? In either case, genuine freedom seems untenable.

The subjective aim of an actual occasion determines its evaluations and which data are integrated in a novel concrescence or negatively prehended and excluded. Is the subjective aim "self-caused"? Although the subjective aim can be modified in terms of the data which act as a lure for feeling (we can be drawn to evil, for example), the original subjective aim is provided by God who has evaluated the eternal objects or potentialities for experience (Sherburne, 1961, pp. 45–46)[3] If a

[2]Sherburne (1961, pp. 41–71) has a detailed presentation of the stages of concrescence accompanied by ingenious diagrams. The detail of these stages is well beyond the scope of the present paper but reveals the subtlety and significance of concrescence which accommodates simple feelings and also the activity of complex cognitive and emotive consciousness. (Lowe [1962] finds concrescence "amazing"; see p. 25 of *Understanding Whitehead*. Deriving geometry and physics from "feelings" is indeed amazing.)

[3]Sherburne quotes Whitehead as to the origin of subjective aims, and also those passages which might be summarized as: God does not compel; God impels (lures). With respect to subjective aim:

> God is the principle of concretion; namely, he is that actual entity from which each temporal concrescence receives that initial aim from which its self-causation starts. That aim determines the initial gradations of relevance of eternal objects for conceptual feeling; and constitutes the autonomous subject in its primary phase of feelings with its initial conceptual valuations, and with its initial physical purposes [1929, p. 374].

Sherburne says, "Each actual occasion has a certain amount of elbow room in its development." But if it is lured by something other than God, or else spontaneously and noncausally "chooses," how is

choice is what God would wish, we react to the subjective aim "given"; if not are we the passive victims of chance? Is action akin to spasm?

Since eternal objects are not created by God (Whitehead, 1929, p. 392), the situation worsens: Does a Platonism of eternal objects ultimately determine all choices via subjective aims given by God? Clearly this matter goes far beyond the scope of the present paper. See Edward Pols's criticism of Whitehead's concept of self-creative actual occasions. The matter seems a serious threat to the philosophy of organism (Pols, 1967).[4]

If Bergson's and James's contributions are appreciated, the horns of the above dilemma are "slipped." Bergson asks us to intuit freedom by considering our "inner" experience. James argues that what appears chancelike from a third-person perspective will intuitively be experienced as freedom from a first-person perspective. Others might see me as reacting randomly; I *intuit* myself as freely acting (Pols, 1967, pp. 131–132).[5]

The above dilemma is avoided only if we consider our intuitions of a free act as not being an illusion.[6] As intuitive, the act can't be further analyzed except in

---

real free will established?

We might aim a flashlight; the light is scattered and radiates outward. The scattered light is not free nor is unscattered light such as a laser beam. Likewise, God provides actual occasions with an aim and also "elbow room" for scattering. The mixed metaphors suggest that actual occasions might need something more "substantial" for free will. I later contend Bergson and James saved Whitehead from the dilemma.

[4]Pols finds actual occasions deficient as agents. Only eternal objects seem to have "power." Therefore a Platonism of eternal objects supersedes genuine freedom. Self-creativity fails, according to Pols, who sees a pernicious sort of Platonism in *Process and Reality*.

[5]Pols includes an account of James's and Bergson's concepts of freedom in relation to those of Whitehead. Pols rightly claims that Whitehead's debt to James is deeper "than is generally realized." However, James's debt to Peirce for the views which appear in James's essay "The Dilemma of Determinism" are even greater and rarely recognized despite James's explicit reference to Peirce.

Pols understands James's attempt to employ the word *chance*, "to designate the causal efficacy of individual subjects, but the word is there used in a strained sense . . . 'chance' there protects the irreducibility of the subject" (Pols, p. 131). James doesn't put it quite that crudely, but Pols's import is correct, as the following passage from James's essay "The Dilemma of Determinism" (1884, p. 154) reveals: "All that its chance-character asserts about it is that there is something in it of its own, something that is not the unconditional property of the whole." Pols claims that "chance" in Whitehead's philosophy is not irreducible and "can be thought of as reducing what is new in the subject to a superjective 'happening' " (Pols, 1967, p. 131).

Although Pols presents a subtle and insightful analysis of freedom in *Process and Reality*, he is fundamentally unsympathetic to James's and Bergson's intuitions. James doesn't equate chance with choice (not that Pols claims he does) but argues that an objective analysis of freedom sheds no light upon an act of volition which must appear to the observer watching me choose one of two equally attractive alternatives as being chancelike, whereas I am intuitively aware of my freedom. The freedom associated with concrescence is to be felt, intuited, not described. Whitehead's fluid concept of concrescence is a guide to recognizing our intuition of freedom and self-creativity; it is not intended *to describe* volition but *to guide* us to a recognition of its significance. Whitehead avoids positing an ego either as a spiritual substance or as ontologically transcendental. Indeed, his no-self theory of subjectivity, in which no irreducible conceptually identifiable entity constitutes the self, is close to Zen and Taoist notions. The superject is describable as data; the subject is intuitable as actor, as self-creative. Ultimately they are the same and internally related. Subject–superjects create the process and are the process.

[6]Is the fallacy of begging the question committed by assuming free will not an illusion? Peirce showed James that the burden of proof is not as it would appear. If free will is logically possible in an

terms of the objective data synthesized, the novel concrescence. The "actor" is not a soul. Actual occasions *together* create the human mind as a *harmony* resulting from individual free strivings of actual occasions which constitute nexūs. A first-person, an "I," is neither a soul nor transcendental ego; it is a freedom to evaluate prehended data.

Two Kinds of Flux, a Becoming of Continuity

The actual occasion is the activity of bringing about a novel concrescence. Concrescence has its successive stages. The word *concrescence* means growing together. Ivor Leclerc writes: "The actuality in becoming is a 'growing together' of the antecedent data into a novel unity. Whitehead uses the word 'concrescence' to signify this process of becoming which constitutes the new actual entity" (Leclerc, 1965, p. 89).

However, the question naturally arises, if concrescence is a process, and if *successive* stages are involved, isn't duration presupposed, and doesn't the process of concrescence occur "in" time? If duration is required by the successive stages of concrescence, how can Whitehead speak of a "becoming of continuity"? Could action be logically prior to motion as it has been argued in the present paper?

Things that constitute a unity can nevertheless be analyzed in some respects; data synthesized in concrescence require the unity of the subject whose act created that novel synthesis (a unity we can only intuitively apprehend, assuming Bergson and James are correct.)

Whitehead writes: "[I]n every act of becoming there is the becoming of something with temporal extension; but that the act itself is not extensive, in the sense that it is divisible into earlier and later acts of becoming which correspond to the extensive divisibility of what has become" (1929, p. 107).

Sherburne (1961) includes the above quotation and adds:

> This is the doctrine "that the creature is extensive, but that its act of becoming is not extensive" (PR 107) and is avowedly based on William James' insight that "Your acquaintance with reality grows literally by buds or drops of perception. Intellectually and on reflection you can divide these into components, but as immediately given, they come totally or not at all [p. 13].[7]

Perhaps the examples provided earlier concerning Mozart's creativity, or the dramatic success of a static picture, will help make the temporally unextended unity of the creative act more accessible.

---

indeterministic universe, one with chance, and if it cannot be *proven* that determinism is true, then if human experience can more effectively be explained in terms of free will, it is pragmatic to take our intuitions of freedom at face value: we feel as though we could have chosen otherwise; we should believe that we could have. This strategy for defending free will is an instance of Peirce's pragmaticism (James, 1884).

[7]Sherburne indicates that *Process and Reality* (pp. 105–106) includes this passage from chapter 10 of James's *Some Problems of Philosophy* (James, 1911).

If the foregoing focuses attention on the relation of motion to action, this paper will have been a success. If the reader consults Whitehead's works, I shall regard this paper as a triumph.

References

Alston, W. P., & Nakhnikian, G. (1963), *Readings in Twentieth Century Philosophy*. New York: The Free Press.

Bergson, H. (1903), *An Introduction to Metaphysics*. New York: Liberal Arts Press, 1949.

Berkeley, G. (1710), A treatise concerning the principles of human knowledge. In: *A New Theory of Vision and Other Writings*. London: Everyman's Library, 1969.

Buchler, J., ed. (1955), *Philosophical Writings of Peirce*. New York: Dover Publications.

Hume, D. (1739), *A Treatise of Human Nature*. Cambridge, U.K.: Oxford at the Clarendon Press, 1965.

James, W. (1884), The dilemma of determinism. In: *The Will to Believe and Other Essays in Popular Philosophy*. New York: Dover, 1956.

——— (1890), *The Principles of Psychology*. Cambridge, MA: Harvard University Press, 1981.

——— (1911), *Some Problems of Philosophy: A Beginning of an Introduction to Philosophy*. New York: Greenwood Press, 1968.

Kline, G. L., ed. (1963), *Alfred North Whitehead: Essays on His Philosophy*. Englewood Cliffs, NJ: Prentice-Hall.

Lawrence, N. (1956), *Whitehead's Philosophical Development*. Berkeley & Los Angeles: University of California Press.

——— (1974), *Alfred North Whitehead: A Primer of His Philosophy*. New York: Twayne Publishers.

Leclerc, I. (1965), *Whitehead's Metaphysics*. New York: Humanities Press.

Leibniz, G. W. (1714a), The principles of Nature and Grace, based on Reason. In: *Leibniz Selections*, ed. P. P. Weiner. New York: Charles Scribner's Sons, 1951, pp. 522–533.

——— (1714b), The monadology. In: *Leibniz Selections*, ed. P. P. Weiner. New York: Charles Scribner's Sons, 1951, pp. 533–552.

Lowe, V. (1962), *Understanding Whitehead*. Baltimore: Johns Hopkins University Press.

Malebranche, N. (1688), The dialogues on metaphysics and on religion. In: *The Philosophy of the 16th and 17th Centuries*, ed. R. Popkin. New York: The Free Press, 1966.

Peirce, C. S. (1890), *Philosophical Writings of Peirce*, ed. J. Buchler. New York: Dover, 1955.

——— (1892), The law of mind, "*The Monist*." In: *Philosophical Writings of Peirce*, ed. J. Buchler. New York: Dover, 1955.

——— (1908), A neglected argument for the reality of God. *Hibbert J.*, Vol. 7. In: *Charles S. Peirce: Selected Writings*, ed. P. P. Wiener. New York: Dover, 1958.

Pols, E. (1967), *Whitehead's Metaphysics: A Critical Examination of "Process and Reality."* Carbondale: Southern Illinois University Press.

Sherburne, D. W. (1961), *A Whiteheadian Aesthetic*. New Haven, CT: Yale University Press.

[8]I wish to express my gratitude to those members of the ISST with whom I discussed an earlier draft of this paper immediately following its Montana presentation. In particular the comments of Gert Müller, Lewis Rowell, Lawrence Fagg, and Douglas Brash, together with the anonymous referee reports of a subsequent draft, were most helpful. J. T. Fraser provided invaluable guidance to this philosopher who sometimes forgets that not everyone speaks his "language." Nathaniel Lawrence's books on Whitehead (1956, 1974) remain invaluable. They console his surviving friends and colleagues, for, in them, Lawrence's ideas survive because they are, in Whitehead's terms, "objectively immortal."

Whitehead, A. N. (1929), *Process and Reality: An Essay in Cosmology*. New York: Harper Torchbook, 1960.
——— (1933), *Adventures of Ideas*. New York: The Free Press, 1967.
——— (1938), *Modes of Thought*. New York: Capricorn Books, 1958.

13

The Process of Polity

Charles M. Sherover

Abstract The idea of "polity" is that of a state run by and for a free citizenry; it has, since its birth in ancient Athens, been at the heart of the modern developments of democracy and republicanism. Polity is not a "thing" but a process—a way of citizens functioning together in utilizing the heritage wrought out of the past in constructing the future. If democratic republicanism is to prosper, a method for evaluating policies and programs is needed that is faithful to the inherent temporality of free citizenship and its enabling conditions.

Free peoples in the contemporary world live under what they generally describe as *democratic governments*. Strictly speaking, this general term is a misnomer. In point of fact these governments, whether nominally headed by a ceremonial monarch or an elected president, embody the principles of classical republicanism. Their citizens pride themselves on their freedoms as they go about their lives—even as they frequently or only occasionally utilize their time to discuss policies, debate programs, become politically active, or merely vote in elections. Yet this essentially temporal freedom from continual political involvement and control is rarely noted. And only rarely is it acknowledged that one prime argument for the representative republic is precisely grounded on this temporal freedom, a pervasive temporal concern that appears in several ways in any examination of our sociopolitical lives.

Five closely interrelated themes that explicate the essentially temporal nature of a free citizenry need to be sketched out. Each of these themes deserves a detailed examination on its own; but it is, I think, important to see them together. The practical outcome of these considerations would be to suggest often overlooked priorities which merit our attention when we attend to our public problems.

Time is as intrinsic to our political processes as are the evaluations and judgments we make about political issues—yet neither political theory nor contemporary political debate pays much attention to the import of the temporal nature of this political activity and this concern. Let us then turn to face these themes in order: (I) the nature of republican government and its historic foundation in a temporalist concern; (II) the future-oriented cast of political intelligence; (III) the temporalist distinction that places a premium on republican government; (IV) the unbreakable interconnection between time and freedom; and (V) a clarification of the nature of pragmatic thinking which insists on facing the alternatives among which we must decide in authentically temporal terms.

I

As we all know, the concept of free government has developed from the ancient Greek invention of civil government, which claims to seek the welfare of its citizens. Aristotle, who wrote the classic text on the Greek political experience, delineated the types of government known in his time. Having lived in the free city or polis of Athens, he tooks its innovation of citizen-government as setting forth a standard for true civil government, a standard which he termed the "idea of polity itself." And his incisive examination, let me suggest, is still pertinent to the more complex situations of our own time.

Two central questions, he urged, need to be faced in evaluating any governmental system. First, how many persons select its officials; second, for whose benefit is political power used? A true *polity* is an organized constitutional government that seeks to function for the benefit of the entire citizenry. When its power is rather used for the sole benefit of the ruler(s)—whether a dictatorial individual, a dictatorial group, or a dictatorial majority—it is a perverse, or we would now say, an *illegitimate* government. Only when governmental power, however selected and distributed, is used for the generalized good of the whole, is it to be recognized as a *true* or *legitimate* government.

Pure democracy is not applauded; rather it is grudgingly accepted as the least perverse of these deficient or illegitimate forms of government: least objectionable because, built on majority rule, its benefit includes most, not merely a few, into its concern. But it is still unacceptable because, incarnating unmitigated majority power, it deliberately rules for its own benefit while excluding all minorities from its concern.

In contrast, a true polity exemplifies a civically organized political structure that seeks to serve its entire constituency as an organized whole; it provides a constitution, which sets out procedures and limitations on authority that bind any majority within it; as such a constitution is a *pre*scriptive standard by which any polis may be judged: its effective dedication is to the common good of the society as a whole in which *all citizens* are to share. How may such a conceptual possibility be institutionalized as an actualized reality?

Aristotle's inherent pragmatism showed here brilliantly. In any general group of citizens, he argued, one finds two somewhat opposed camps, both of which are necessary for the social health: one prizes wealth, the other prizes freedom. For the good of the whole, *both* are needed to function together for the social health. Bring them to cooperate in one unified government and that government will necessarily serve the interests of all, the interests of the entire organized community, just because all opposed interests are represented in its governing structure. Bring oligarchy, the rule of the few who have wealth, into union with the many who prize their individual freedom, and the compromises they must make among themselves for the sake of social harmony will encompass virtually all citizens into the common concern. The move from pure democracy to what he termed polity or *constitutional government* is what, since its Roman development, has been designated as a *republic*; it is the move from *most* to *all*, from the imposition of an unhindered majority will to the compromises inherent in a universal inclusion.

Such a governmental system spells out in advance a set of temporally prescriptive procedures for adjudicating disputes in the light of a common interest, for choosing officials to handle specific delegations of authority—and thereby it also serves to relieve most citizens of the need to bind their own time to the minutiae of civic affairs.

The two key contemporary terms—*democracy* and *republic*—which are commonly used to denote the idea of polity, come from that heritage. We often use them today as virtually interchangeable, but when we attend to their origins, they clearly carry radically different meanings bound around some notion of equality. The term *democracy* has come to have diverse meanings: equality of the right to vote, equal subjection to the law, and especially a social meaning pertaining to the absence of rigid class divisions in the society and a high degree of social mobility. It is these meanings that permit us to distinguish democratic and undemocratic republics. But the other distinctive, and more historically sanctioned meaning refers to political organization, and is completely at variance with the word *republic*. As *political concepts* these two terms point in two opposed directions, two different ways of handling a central temporal concern, and bear on the nature of freedom.

If government is to function *for* the people, the Athenians quickly understood, it must be under their control. Government *by* direct popular participation, by the *demos*, was the unprecedented Greek innovation. And this was clearly understood by them to mean government by participation of the people in the general assembly—what we today call direct or participatory democracy. As all citizens were regarded as fully equal and thereby qualified to serve in virtually all public offices, they saw no need for elections and so they filled most offices by lot, a practice still followed in the United States for the selection of juries.

Two facts quickly became apparent. First, was the general desire for general leisure. As all citizens were required to spend the bulk of their time in the public assembly or other public business, they were not free to control their own time; they certainly did not have the time to furnish their own sustenance. The Greek free citizenry, then, necessarily constituted a prosperous leisure class whose time-expenditures were controlled by the public assembly, and who depended on a slave economy for economic support. It is paradoxical but nevertheless true that their invention of direct democracy or participatory popular rule not only foreclosed individual control over personal time by free citizens but also necessitated slavery.

Direct democracy, let us note, thus entailed a double restriction on individual freedom: the slave's time was constrained by his master, and the time of his master, a free citizen, was constrained by the majority of the popular assembly.

Second, it did not require much political experience to see that some citizens were better qualified than others to handle delicate or involved tasks, such as commanding the army in defense of the city, and so the idea of delegated authority was invented. Particular individuals were elected to represent the interest of the citizenry for a stated period of time in pursuing a specific kind of task. These individuals were given the responsibility to act for, and even to direct, the citizenry in accord with a specific instruction and the ensuing requirement to report back on how the office entrusted to their care had been exercised. And so the idea of representative government was also invented—although it was not until the Romans coined a new term for this: *res publica* or *public property* became the name for a state governed by representatives chosen by the people to handle specific delegations of public authority for the general benefit. It was the Romans who spread the institution of slavery through their conquests; yet, paradoxically, it was the Romans who developed the idea of republican government, which, in contrast to a pure democracy, does *not* require a servile population just because it leaves the citizenry free to control their own individual time without yielding popular control.

It was the Romans who first developed Aristotle's perception that a government directed for the benefit of the entire citizenry had to bring all elements of that citizenry into the governmental structure. This idea of republican government, that it mixes into its structure the varied elements of its citizens, indeed, was the practical definition of polity or constitutional government. It was symbolically proclaimed by the Romans on the standards their legions carried as they carved out the core of the western world—''SPQR'': the Senate *and* the People of Rome.

Indeed, it was the inspiration of the Roman Republic that paved the way for the democratization that began with the onset of the modern age. Their experience, *the only experience of extended republican government* before the American and French Revolutions, was explicitly and judiciously examined by Machiavelli in his *Discourses*, by Montesquieu in *The Spirit of the Laws* (1748), and by Rousseau in *Of the Social Contract* (1762), as they started to work out the texture of modern republicanism. The Roman Republic provided the institutional concepts which enabled first Montesquieu (XV, 1–3, pp. 257–260) and then Rousseau (I, iv, pp. 8–12) to condemn slavery as absolutely illegitimate: Conceiving the modern republic as taking up Aristotle's description of free citizenship they set out an unprecedented standard by universalizing the idea of citizenship as a rightful claim by all society members. Proclaimed as the one fundamental social right due to any person, citizenship as full membership in the polis became the foundation of all specific civic rights and obligations.

Specific rights of citizens then are social rights that are dependent upon the obligations ensuing upon full civic membership; these obligations while exacted from each citizen must be minimal because the popular will is to leave each citizen, as a free being, free to devote his own time not to civic affairs but to the securing of sustenance and enjoying of leisure.

These core concepts of modern societal life incorporate that temporal process of time-bound and time-formed development which *is* history. The historical experience, which produced the core concepts we use today, has often been a painful one. What this historical experience shows is that a polity, or a properly constituted civil government, is neither a *thing* that endures unchanged through time nor a somehow timeless idea; a polity is a *process* of be-ing together in a mutually conditioned temporal experience; the *process of being members of a polity* is a way of be-ing, a process that is certainly time-bound; as lived by each citizen it is also pervasively temporal. And one of its prime, if rarely voiced, concerns is how the control of the individual time of its citizens is to be allocated.

Indeed, the cyclical notion of time, which dominated all Classical thought, led Aristotle to set out a theme which most subsequent political thinkers picked up—the notion that an organized community can be expected to develop through a natural cycle of generation, growth, corruption, and dissolution. It was only with the rise of the modern world that the cyclical presumption of eventual dissolution disappeared from the tracts of political theory—even as the kingdoms in which these tracts were written had already begun their process of dissolution. In a way, this is ironic, for, to the best of my knowledge, the last two serious discussions of this classical theme were presented by Montesquieu and Rousseau—who, for the first time, by foreclosing the legitimacy of slavery, bequeathed to us the essentials of the modern democratic republic.[1]

Modern free societies have taken up the heritage bequeathed by Aristotle's examination of the Athenian innovation of free government. When we abjure their acceptance of slavery, and focus our attention on his descriptions of government that were constituted for the benefit of free citizens, we find that all the rudiments of a contemporary theory of republican government are already in play. We find the insistence that it derives from, and is intended to serve, the entire body of free citizens; the delineation of the three types of governmental activity—executive, legislative, and judicial; the principle that they should be in different hands coupled to the principle that all elements of society should be represented in the institutions of government, yielding the full notion of mixed government. We find the warning against any tyranny—whether by one ruler, a ruling minority, or the ruling majority itself—tied into the crucial distinction between those policies that favor the ruler (individual, group, or majority), and those policies that are to benefit the entire society as one organic whole. We find the thesis that a society, for its own health, cannot be divided into two warring classes of the rich and the poor, and the remedy—a strong middle class that imperceptibly reaches into each of the other two and thus fosters social cohesion so that a mutuality of a common loyalty may unite disparate internal tendencies. These principles still guide us; they are

[1]As the Italian political theorist Gaetano Mosca noted, it was Rousseau who is "the real parent of the doctrine of popular sovereignty and hence of modern representative democracy" (p. 254). With regard to the American experience, there are passages in *The Federalist* that suggest Rousseau's influence; but more to the point, perhaps, is that Montesquieu's principles, which provided the structure of the American Constitution, were compromised on the issue of slavery, a compromise which led de Tocqueville in the 1830s to predict the inevitability of the American Civil War.

commonplaces of theories of republican government. In terms of specific enactments, they sometimes occasion controversy, but generally they are taken in free republics as standards by which to gauge governmental policies.

II

Having generally accepted the standards for a government of free citizens which Aristotle had set out, it is strange that political theory has largely ignored the temporalist framework in which he had set them. Yet, at the outset of his work, Aristotle had clearly set out, on two distinct levels, a temporalist standard for political judgment. On the more general level, every state, as he suggested at the beginning of his *Politics*, is oriented to a future accomplishment: "Every state [or/polis] is a community of some kind, and every community is established with a view to some goods; for mankind always act in order to obtain that which they think good" (1252a). This future orientation is not only a social phenomenon; it characterizes every individual intelligence: "Every art and every inquiry, and similarly every action and pursuit, is thought to aim at some good; and for this reason, the good has been rightly declared to be that at which all things aim" (1094).

If we think through the meaning of these two topic sentences, we can only conclude that when human reasoning is directed to practical concerns, it is essentially constituted in purposive terms and is thus temporally organized. A deliberate decision to act undertakes a course of activity in the present and commits a stretch of oncoming time to the achievement it seeks to instantiate. Whether on an individual level, or on the level of social decision, deliberate rational activity aims toward the future; it aims to bring into the actual dynamic state of existence certain values, norms, commitments, or even a pictured outcome.

Rational activity is purposive activity. Action is undertaken with deliberation to accomplish what has not yet been rendered actual; political intelligence seeks to invoke past experiences that are relevant to the course that is set out. But however wisely or blindly one sets out to pursue some plan or direction, one does set out to accomplish something that will be a new fact of the historical chronicle, something that will depend for its being on the efficacious use of planning intelligence, something that will not come to be except that appropriate steps have been instituted to bring it about.

Political reason then depends upon the appropriation of possibilities, of intelligent assessment of conditions, and helps and impediments along the way, of an intelligent use of an imaginative creativity that is as crucial to the practical as it is to the arts. The future, to which we aim, is not some "thing" into which we are to plow. The future is that which we are to create. What will result from our acts will either be a state of being into which we slide without any awareness of deliberate decision; or, we will have created some new state of affairs which could not have come to pass without our own creative ingenuity in bringing all the contextual forces and conditions at our disposal into a purposeful harmony. If we are obsessed with our passivity through change, we will then discourse on the fateful necessities of historical development within which we will see ourselves as but pawns. When we acknowledge the need for decisive action, we will recognize ourselves as fateful creators of history, as innovators upon whose efforts the outcome of historical development depends.

It is, I think, significant that in all democratic polities, in all truly representative republics, leaders are chosen who claim to be able to bend the perceived direction of ongoing development, who claim to be able to redirect the forces of history and create a future instead of merely acquiescing before it. Most of us perceive social time as open to challenge, debate, and decision. Most of us perceive the present situation—which is to say, take up the facts that have been established and acknowledge the possibilities they present—as the temporal site of options for the future, options among which we must decide, options which however limited they may be, yet present alternatives which are dependent upon our present decision for their eventual oblivion or actualization.

Political intelligence then is always future oriented. It orients itself to what yet can be within the constraints of what must be. It functions on the premise that the future is at least somewhat open. It presumes that, within the realm of the possible, it still is able to resolve what will, in due course, become actual. This is to say, by whatever standards of judgment, rational intelligence constructs the future, it does so in order to transform some presently favored possibility into historical rendered factuality.

This future-oriented stance is inherent in the description of polity as constitutional government. For, as already suggested, a constitution, by its intrinsic nature, has set out in advance a set of procedures by which a free citizenry shall determine the limits of allowable actions, the modes of social interaction, the ways in which alternate social policies may be advanced, debated, and how those debates will be resolved. A constitution is then a *pre*scriptive set of rules enunciated or agreed upon to "henceforth" govern the complex lives of free citizens insofar as their activities may be anticipated as impinging on each other. It is designed to set out in advance of any conflicts the ways in which those conflicts are to be resolved, decisions taken, and authority delegated. To the extent that it is well crafted, it will lead to a social harmony in which differences may be encouraged while contained, and each citizen will be left with that degree of personal control over one's own time, which is the essence of one's freedom, consonant with its general enjoyment by all other citizens. As Kant (1781, 1787) summarized the intent of any constitutional arrangement, it is to "allow *the greatest possible human freedom* in accordance with laws by which *the freedom of each is made to be consistent with that of all others* . . ." (A316 = B373, p. 312). Madison had defined *law* to mean a *rule of action* (*The Federalist*, No. 62, p. 421). A constitution may then be described as that set of fundamental laws which a society accepts as the prescribed procedure to follow in formulating specific rules for specific types of actions that accord with its general animating principles and purposes.

Political intelligence is then prudential or pragmatic intelligence. It seeks to incorporate its values, principles, or goals into a course of activity. And it judges the efficacy of its acts by its success in incarnating those values, principles, or goals into what effectively comes to pass. Political intelligence, in Aristotle's language, is then inherently prudential—thereby entailing the need for close attention to practical sequences in its activities as the virtue of citizenship. To the extent that its guiding principles are moral principles, it is in Kant's language, inherently moral—thereby entailing a deep sense of reponsibility for its animating norms. However framed, the test of an action or a program is its efficacy in accomplishing what it set out to do without disrupting other value-loyalties, virtues or norms along the way.[2]

To acknowledge the priority of our orientation to the future in every deliberate decision is not to suggest that a society can start without acknowledging the legacy from its past. As Paul Ricoeur has noted, "We can perhaps 'transvaluate' [our] values, but we can never create them beginning from zero" (1974, p. 268). In the living present that looks ahead, the presence of the past is always prominent: in the language we speak, the cultural history it represents, the folklore we accept, the values we cherish, the customs we follow, the presumptions within which we have learned to work. Without awareness of the continuity from the past, we would be hardly prepared to face the task of creating our future.

And, indeed, we form the deliberate ongoing continuity of our society by teaching our children about its past, bringing them to share a veneration for its heroes, and finding the roots of a common community not in its supposedly unanchored aspirations but in its past record of successes and failures out of which these aspirations have grown. Without a strong sense of shared historic roots, there can be little sense of present community and a shared vision of what we are striving together to achieve.

Society is indeed a process, a process of historic development. Any particular nation, Edmund Burke (1782) pointed out, "*is not an idea only of local extent and individual momentary aggregation, but it is an idea of continuity which extends in time* as well as in numbers and in space . . ." The ways in which it understands itself and the situations in which it finds itself come out of a development of "the peculiar circumstances, occasions, tempers, dispositions, and moral, civil, and social habitudes of the people, *which disclose themselves only in a long space of time*" (p. 330; italics mine). A society *is* a history, a history not merely of accomplished fact, but more so of a common "we" within whose scope each "I" finds one's own motivating aspirations.

Significantly, as well as symbolically for the civilization that emerged out of the Athenian experience, Pericles had begun his call to the Athenian future, with these words: "I shall begin by speaking about our ancestors . . . [who] by their courage and their virtures, have handed [this land] on to us as a free country" (quoted in Thucydides, p. 144). Acknowledging the heritage of the past carries with

[2]For a fuller presentation of the historical development of this concept of polity, its crucial concept of a *temporalized* constitutionalism, and the ensuing repercussions and implications for all social theory as well as correlation with other themes touched on in this essay, see my *Time, Freedom, and the Common Good: An Essay in Public Philosophy* (1989).

it a responsibility for continually refining and developing what it has bequeathed; by doing so we enrich ourselves by using it as opening up opportunity for what may yet be wrought.

III

As citizens of free republics, we are fortunate in that most of the decisions we make, most of our daily concerns, are not political. Most of our decisions, most of the concerns which preoccupy our living time, are private concerns and decisions about personal matters. Most of our lived time is devoted to questions of livelihood or the leisure that it sustains, to family and friends, to hobbies and entertainments. One of the unspoken glories of a free republic which entrusts its affairs to freely chosen representatives, is that its citizens are freed from preoccupation with public dangers and political excitements, as well as from the minutiae of governmental management; it is precisely because free societies are generally stable, because their procedures are dependably constant, that they serve so effectively as the enabling context for the personal.

As suggested, one of the rarely noted virtues of a free society is that it frees the time of its citizens from preoccupation with the business of administering the polis. Cheerfully, this is left to duly selected employees—who are expected to do what is required of them within the confines of their delegated authority with probity and responsibility and without requiring the time-consuming inefficiency of constant public supervision. Indeed, public anger usually arises only when officials have not been true to their charge, when public corruption or malfeasance is exposed, and personal attention has to be turned from the priority of the private to the intrusive responsibility of public concerns.

To minimize the possibility of such public disruptions while making sure that such potential exposures are pursued when called for, a self-policing system, often called a system of "checks and balances," was worked out since Montesquieu first devised it. That its proper functioning would not be compromised, free republican governments have insisted on the principle of being supplemented by a nongovernmental agency, a free press, whose legally protected carping is expected to announce the alarm when that self-policing system of checks and balances fails to function properly and in accord with the general will.

A free polity, indeed, usually prides itself on providing a stable public backdrop for the private concerns of its members. Because of a continuity of shared freedoms and liberties, we need not be preoccupied with their defense, but generally utilize them without much thought until we find them being attacked or subverted. Because of a continuity of duly constituted procedures for making public decisions, we are for the most part content to periodically take time out from more private concerns to debate and decide whether to reemploy or replace our representatives. The rules and procedures which govern a free people are deemed successful when they are not obtrusive, when they provide a stable public life which does not require constant attention, just so we each can be so much freer to devote more personal time and attention to our private lives.

Indeed, aside from the transformation of the city–state into the nation–state, the greatest departure from the governmental practice of ancient Athens that marks free societies has been the substitution of the republican for the democratic principle in its political life. Strictly speaking, a "democracy" as the Athenians invented it, and Madison explained, precisely depends on the full-time participation of the citizenry in the administration of the political structure (Madison, *The Federalist*, No. 14, p. 84; Sherover, 1989, pp. 156ff). Only a leisure class can conceivably afford the temporal luxury of active governmental participation.

When Montesquieu devised the principles of modern republicanism, he saw that no members of a truly democratic society, dependent upon the efficacious use of their time for their livelihood, could afford the time for direct democratic government, that no free people would *voluntarily elect* to spend their time in that way. If *the people* are to be the ground of their government, if government is to be, indeed, a *res publica*, a public affair and a public property dependent upon the will of its citizens, it needs to honor the public will that the members of the public be left free to pursue their own concerns instead of being tied to operations which were better left to hired hands or representatives.[3] And it must be so constituted that the whole people and not merely a leisure class are regarded as the ultimate authority from which that government derives its authority and to whom it must periodically report.

Rousseau, the most "populist" of all republican theorists, urged that a "pure democracy" or direct self-government, is appropriate only for the gods (1762, III, iv, par. 197, p. 64). Condemning rule by both monarchs and landed aristocrats as unable to govern in the public interest, he joined Montesquieu in resurrecting the principle of classic republicanism by calling for rule that derives its authority from the people as a whole. If the people are to control their government, he argued, their only recourse is to what he termed an "elective aristocracy," the public selection of particular officials who are employed to "administer the constitution" (1762, III, v, par. 201–204, p. 65). Such employees, like any employees, have designated areas of authority, specifically charged responsibilities, and employers (i.e., their constituencies), to whom they must periodically report.

This "republican" system of government, as Madison and Hamilton repeatedly argued, is the only feasible form of government for a free people.[4] This is so for several reasons, two of which depend upon the acknowledged import of the consideration of time. The first has already been outlined: the need of a free citizenry to devote their time to private rather than civic affairs. To this and to the famed principle of checks and balances, which they learned from Montesquieu, they added a new temporal consideration—a strict time limit on the delegated authority of all elected officials, a creative use of chronology to control the utilization of delegated

[3]As one student of mine phrased the problem in very contemporary terms, who has the time, much less the dedication, to do nothing else but watch C-Span all day long?

[4]See *The Federalist*, especially Nos. 1, 9, 10, 14, 31, 37, 39, 51, 63, 71.

political power. But more crucially, they recognized that individual freedoms depend upon limiting the power of governmental officials to intrude on the time of the citizenry. As Hamilton had pointed out, "[Government's legislative power] not only commands the purse, but prescribes the rules by which the duties and rights of every citizen are to be Regulated" (*The Federalist*, No. 78, p. 523), which is to say that governmental power has the capacity to rule and control the time of its citizens. Republican government, therefore, is necessary in order that a free people may control the social impositions placed on their individual control of their own personal time.

This central temporalist argument for a representative republic rather than a participatory democracy speaks directly to some radical proposals being offered today: if people are to be free to spend their own time as they see fit, *their* preference, especially when they are of moderate circumstances, is to devote their time to their livelihood and leisure, rather than to resolving the minutiae of civic affairs. To compel them into participatory schemes is to infringe their freedom to use their time as they see fit. The first freedom anyone desires is the control of individual time. To free a people to control their own time, while still maintaining their oversight over their common good, provides a test of any government that claims to be the government of a free people.

A free people is a people who prize their freedom to control their own time; such people have historically abjured rigid class structures and welcomed high degrees of social mobility, as inherently expressive of a free society. Just because of the great value which they place on personal time, they cannot afford and have therefore abjured the time-consuming luxury of participatory government. Their democratization is their generalization of individual time-control Their republican principle of governance came into being just because the freedom of the people requires freedom *from* having to be preoccupied with details of civic management. Uniquely, republican or representative constitutional government facilitates popular control while leaving the populace free for other time-consuming pursuits.

Republican or representative government thus supports the democratization of a society by maximizing free time for the citizenry; it does this by leaving the nongovernmental activities of its citizens to the voluntary initiatives of its citizens—and on these voluntary citizen undertakings it not only depends but bases its aspirations for prosperity and progress. The expectation is that by limiting governmental agencies for the most part to procedural directives rather than direct intervention in the subsocieties—corporations, unions, churches, clubs—the authority of governmental agencies to interfere with the free time of the citizenry in their own community concerns is minimized.

And, indeed, the degree of personal control over one's own time is a prime mark of a free society, one that distinguishes free political and economic systems from those which are not. John Locke (1689), who comes out of a philosophic genealogy somewhat different from the one outlined here, declared three individual rights as primary and intrinsic to any others—the rights to "life, liberty, and property" (although he showed no qualms in subordinating these individual rights to unmitigated majority rule). But he, and those who join him in doctrines of inherent rights, should have realized that the prime ''property right'' which anyone can conceivably claim to have is not that of any particular permission or possession but the right to use one's own time, as long as it does not harm the public health, as one sees fit. The degree of individual time-control is the first test of individual freedom: only a society that enhances individual control of individual time, in contrast to all systems of dictation and regimentation, is generally regarded as a free society.

IV

In the organization of one's personal life as in the life of one's society, two themes come together: time and freedom. Indeed, if our use of time is in those pursuits that we consider good, we must be able to evaluate the prospects and possibilities before us, discriminate the desirable from the undesirable, the good from the bad. If we are able to pursue our values and normative standards by the use of intelligence, we must also be able to discriminate the important from the unimportant, the useful from the useless, the instrumental value of one pursuit to another and the relative value of alternate claims upon our time, our temporal efforts, attentions, and concerns.

If we find ourselves actually doing this in everyday life, we must be capable of doing so. This capacity to make evaluative judgments and to guide our actions by them is what we mean by freedom. No matter how one explains or explains away the experience of making free evaluative decisions, one cannot avoid continuing to do so. Even the sophist who proclaims that freedom is an illusion finds himself necessarily continuing to make the free evaluative decisions he claims to be a sham. However logical the argument against the reality of freedom may be, it fails just because it is incoherent with our lived experiences, just because it can explain neither the *experience* of freedom nor the fact that its exercise cannot be abjured.

The experience of freedom, of making evaluative decisions and acting upon them, constitutes the time of daily living. These evaluations may be habitual or deliberate; they may concern questions trivial or profound, but without them life would come to a halt. The continuity of these evaluative decisions we feel called upon to make cannot be aborted, just because freedom is itself not free. Freedom is forced upon us because of the temporal constitution of lived experience: not to decide is to decide and the dynamicity of temporal flow demands that it be ceaselessly filled out by evaluative decisions.

The ability to exercise this capacity of making evaluative decisions is the ability to control one's own time; it is what is meant by the power of self-determination. But this freedom of self-determination is socially bound. It is limited and enhanced by the constraints and opportunities which a society constitutes for its members. The traditional distinction between personal or metaphysical freedom and social freedom that characterizes much of philosophic history is existentially meaningless. For I can only exercise my capacity to make evaluative decisions to the extent that my society permits me to do so. The greater the degree of bureaucratic control, the fewer options for time expenditure are allocated to me. To the extent that controversial information and debate is inhibited, that education has failed to stimulate my intellectual capabilities, or that adequate sustenance cannot be earned, my ability to exercise and develop my capacity for making free decisions has been severely frustrated. What I can do, as a social being, presumes the opportunities that my society opens to me and the constraints that it imposes. The freedom to use my time is then first of all a social freedom.

This is why the claim to freedom epitomizes most social demands. For the control of time is control of the self; it is crucial to the entire experience of self-determination. It is for this reason that freedom is so highly valued; this is why the first mark of freedom is the freedom to control one's time. In this claim to-be-free is wrapt up that complex of rights and privileges and obligations and demands which even the freest society necessarily makes finite by organizing the freedoms of its citizens.

Freedom requires time for its exercise. And freedom is always freedom-for-the-future; as Peirce noted, "future conduct is the only conduct that is subject to self-control" (Peirce, 1905, Vol. 5, p. 284). Our concrete freedoms are found in the ability to channel or delineate that time which has not yet been reduced to unalterable fact, to determine which potentialities shall be developed, which possibilities shall be actualized rather than being consigned to oblivion. Freedom is then essentially creative for only in its exercise do we create what shall be. And freedom can only function if oncoming temporality hovers before it as a field that demands that it be marked out and designated.

If freedom requires time in order to function, then time is the condition which makes freedom possible. The reality of freedom, then, depends upon the reality of time. And if freedom is to be effective in a social order, that social order must steadfastly honor its own temporality. A free social order cannot forget that its own being is that of a process, one that brings a legacy out of the past for discriminative development into its future. It cannot forget that its future can only be the creative product of its citizens whose contributions will be limited by the opportunities they find and the restraints they share. To say that a society is not a "thing" but a "process" is to say that history is not merely the chronicle of the past; it is, in its broadest ramifications, the developmental career of the social body that is presently engaged in transforming the heritage it has received into the legacy it will bequeath to those who come after.

A free society calls its citizens into the ongoing building of its history by equipping them to develop their own contributions to the legacy it is constructing. To the extent that it has educated them well, to the extent that it invites their individualized contributions to the social dynamic, it will bequeath a legacy richer than it received. But to take living history as a living bequeathment is to insist that freedom entails responsibility. For the activity of freedom, on the part of the individual as of the social body, always changes the social milieu within which others will have to live. If freedom is inherently social, then all of our acts impinge upon others. And however one may regard his responsibility for what he does to himself, what one does to the time and freedom of others is from the outset a social responsibility.

V

As social beings we must live together. Whether we broaden or narrow the degree of personal autonomy we delineate for our society, our freedom-for-the-future always takes us into the lives of others just as the acts of others come into our own. We are social beings whose temporalities and freedoms tie us together.

Most of our individual decisions and actions have social consequences. Most of our decisions, even private ones, have an impact on others. We live in closely knit societies which provide degrees of autonomy but never isolation. Whatever we do will enter into the lives of others as their decisions and action will enter into ours. We cannot afford utopian gambits that ignore the social costs they might incur. The intelligent exercise of freedom, as the responsibility of a mutuality of citizenship, requires a discipline of evaluation, judgment, and decision that takes the principle of temporal continuity seriously and recognizes that it is inherently incremental.

Even our most rudimentary social concepts come out of a long history of practical application as well as intellectual development. The ideas we invoke and the concepts we use are not merely labels casually pulled out of a dictionary. And none of them was invented "from scratch." Concepts, like the people and the societies that use them, have a history that is crucial to their present meaning. They represent a development of experiments that failed as well as those that succeeded, and often transactions and events that were painful to those involved.

When honored for what it is, a serious concept presents a fund of experience that is often predictive of the ways in which it relates to other concepts and to the practices associated with them. This is to say, again, that a concept is not merely an idle set of words; as representing a historic strain in human experience, it points to practices that honor or defile it. When we think this through, four points, the criteria of a temporally responsible mode of social judgment, will emerge:

1. A concept refers to actual time-bound practices within a context stretching beyond it. To an educated understanding, it reports a history of what has been done in its name and commends and condemns future kinds of activities. It calls us to make evaluative decisions that can be tested in action.

The first mark of a system of socially responsible reasoning is to recognize that the concepts we use are inherently *pre*scriptive; they commend (or foreclose) certain kinds of practices to be undertaken in order to advance certain ideals, values, or norms.

2. Again, our orientation to the future needs some clarification. Too many ethical admonitions reduce this temporal concept to a spatial one, and too many recommendations and proposals and programs follow this spatial language by conceiving our temporally constituted lives as a journey into a "place" called the

"future." Our temporally constituted situation is then misconstrued as the spatial problem of how to get from "here" to "there." The prescription for a journey is necessarily segmented into three independent considerations: first, the starting place; second, the destination; and third, all the various roads or routes between the two—to be clearly examined on a map. There is no intrinsic connection between them.

And we thus find ourselves ensnared in countless debates about whether the end justifies the means, whether it really matters which road we take as long as it promises to lead us, even if by a circuitous route, to the promised land. All roads may, indeed, lead to Rome, as we are told; but Rome will still be Rome once we get there. If we conceive our problem as a journey, then we are justified in separating "means" and "ends." But too much human misery has resulted from the postulation of a desirable end independent of the steps needed to attain it. The more utopian its pretensions, the less scruple seems needed to get us there.

But, time is not space, as William James never tired of reminding us. And time cannot be segmented the way space can be. The future is not a *place* that is merely awaiting our entrance. The future *now* is but a nest of possibilities which suggest a condition of being *yet to be created*. And until it has been rendered actual it cannot be definitively delineated. What will emerge will conform to no preordained pattern but will be the result from *all* that has fed into its creation—including countless other streams of concurrent development that will only then emerge into view.

Rather than the image of a journey across the land, temporal development might better be conceived as the steps an artist takes in reducing an inchoate vision to a tangible work of art which can only be comprehended after its creation—or a *procedure* which a conscientious doctor *prescribes*.

Consider what this suggests. The doctor starts by discerning a problem as having emerged from the presence of a specific historic development which is intrinsic to his diagnosis, he then proceeds to effect a cure by trying out a sequence of actions. The resultant state of the patient's health will emerge from the congruence of these actions together with whatever other unknown factors may intrude. The future state of the patient's health is *not now* known just because it is *now being created*. And the conscientious physician takes the prescribed procedure, not as a means to a predetermined end, but as a process to be tenderly nurtured and continuously reexamined along the way; he watches to see the responses to each step and quickly alters the prescribed course if an unexpected reaction ensues. There is here no divorce of ends-and-means. In a serious commitment to the reality of temporal continuity, a flexible prescriptive procedure is conceived to ameliorate the problem without creating a new one while doing so.

So, the second mark of a disciplined method of socially responsible temporal reasoning is to take the time of action in its intrinsic continuity of development with utter seriousness, abjure any distinction between means and ends, and rather recognize that what we do actually *produces* what will actually ensue. The future is not a place that will be arrived at; like a work of art, it is produced—and until the process of creation has been completed, like the artist, the doctor cannot know precisely how it will turn out.

3. But we cannot each do everything. Take any remotely complex task, and most of us are unqualified to attend to it. Society is built on a division of labor. Each ideally has a function to serve and depends upon countless others to serve theirs. Our individual temporalities overlap each other by feeding into the social temporality we share together.

The third mark of a temporally structured responsible mode of social reasoning, we seek then, is an acknowledgment of limitation in a frank recognition of areas of competence. We *need* each other, and a prime mark of social responsibility is continued acknowledgment that responsibility is being inherently shared. We have a vested interest in the efforts of others and we are, if you like, each other's delegates and representatives in the myriad aspects of our common endeavor; the mark of wisdom is to recognize our finite capabilities, our inherent interdependence, and thereby just how circumscribed our particular responsibilities really are.

4. If the function of the present is to develop the future out of what has been provided, then we must start by recognizing the *absolute continuity* of temporal development. There are, as Leibniz continually insisted, no gaps or jumps (see Sherover, 1981). But this also means that all acts and decisions will carry on into effects often unintended. Any particular action, in a sense, initiates a sequence that cannot be aborted. And so responsibility demands that casual experiments with temporal sequences be viewed with suspicion—for no action, no matter how innocent, can be without repercussion; any action creates an irreversible sequence the consequences of which will become part of whatever ensues. The better part of wisdom, then, is prudential caution—and the more encompassing the conceivable outcome, the greater the prudential care to be exercised.

The fact of a decision presumes the reality of our freedom or ability to make it. But this freedom, while real, is intrinsically limited. It is limited by the factuality of the situation in which we find ourselves, the potentialities or capabilities we bring with us, the possibilities that are discerned, and the limited knowledge we have of whatever else is transpiring. Our decisional freedom, in every instance, is thus bound by all the constituents of the temporal context and the requirements of temporal continuity. The responsibility of freedom is to recognize these and to stand ready to alter our procedures as unexpected developments occur.

When we bring these various considerations together, what we find commended is a genuinely pragmatic method of practical reasoning. For what pragmatism insists upon is that decision and action are conducted in community with others while looking to temporally open but historically formulated situations we seek to resolve. Dependent on the condition of a decisional freedom-for-the-future in order to function, it is pledged to keep the future open. And because of our shared responsibility for what happens to others, it continually tests itself by the way in which activities are working out.

The essence of the pragmatic method was voiced in Charles Peirce's "pragmatic maxim": "Consider what effects that might conceivably have practical bearing you conceive the object of your conception to have. Then your conception of those effects is the whole of your conception of the object'' (1905, Vol. 5, pp. 281–282).

The *practical* meaning of a concept or proposal is not its logical consistency, its internal coherence, or its emotive attraction. Its practical meaning is not even to be seen in its responsiveness to the particular problem that called it forth. For its function is not to address the past but to create, out of what is currently present, a new situation. The meaning of a concept or proposal is *what happens when it is put into practice*—not the goal it seeks to attain but what can actually be anticipated to transpire from its implementation—including the side-effects it may conceivably have on other concurrent events.

In journalistic use, pragmatism is often reduced to a kind of cheap opportunism or to a means–ends utilitarianism. It is neither.

To the reply of opportunism, one need merely point out that without guiding principles of actional maxims, we have no rational means of distinguishing what is important or unimportant in any situation, for evaluating the success or failure of any action. Any new action is itself a difference in a dynamic stream of temporal change. One needs a clear set of valuational loyalties before a rational judgment can say that some conceivable outcome is important or unimportant to what is being attempted, or that some proposed course of action has worked out.

To the charge of utilitarianism one must reply that its dichotomy of means and ends leads to endlessly abstract disputations as to whether the ends justify the means and if so how; such disputations cannot point to a lived concern with concrete reality. If time is itself a developmental continuity, then means and ends cannot be divorced as two separate objects of concern; in time as lived, all that is done forms a continuum; to segment them is as artificial as the segmenting of the continuity of time into minutes or hours or years or centuries, in whatever manner suits one's particular convenience; it violates the principle that every action is ontologically grounded in the nature of the continuity of time. Utilitarianism does this, as suggested, because it conceives of time, not so much as what is to be fashioned, but as a landscape, whose contours can be readily mapped out, just waiting to be traversed.

As Royce clearly pointed out, ''space furnishes indeed the stage and the scenery of the universe, but the world's play occurs in time'' (1904, Vol. 2, p. 125). We are reminded of William James's continuing admonition that time is not space; the temporal is not spatial, even though we may pragmatically utilize spatial metaphors on occasion to mark out time.

The true artist finally realizes just what he was doing only when he finally views his newly completed masterpiece. The physician sees only at the end of his treatment what state of health has been *produced* by the procedures that have been followed from his original diagnosis. Both epitomize the essence of pragmatic thinking; each initiates a temporally ordered ''procedure'' because of a perceived negative condition—a vision still to be captured in its expression, an illness to be cured—because of what Aristotle had termed a ''privation,'' Peirce, ''a failure of habit,'' Heidegger, a ''missing needed tool,'' and Dewey, a discerned valuational ''lack to be fulfilled.''

What is crucial to the physician as to any artist is not merely the final stage of the sequence of acts necessary to attain the final end, but the integrity of the entire procedural sequence itself. As Royce said, in defending the Peircean maxim, the animating purpose does not find its justification merely in its "final expression"; it finds its justification in "its whole embodiment" of the entire sequence of all acts undertaken along the way (Royce, 1904, Vol. 2, p. 136). The responsible moral agent and the responsible social reformer are like the conscientious doctor or the creative artist; they take time seriously as an ordered sequential prescription for the future, as a developmental process whose end-point does *not* already exist; it must first be produced by what we do in *a temporally ordered* way. And what we do—in all its ramifications—is inherently part of the conception or program which provoked its initiation.

The pragmatism of practical reason is then neither opportunism nor utilitarianism. It functions by being responsibly loyal to the temporal continuity of any given situation. It presumes that the ends we project are not predestined stages along a route to be traveled but valuational norms to be incarnated into a developing temporal reality; it invokes those aspects of past experience that seem germane in order to guide it in doing so. It sees the future, not as an ideal state to be attained, but as the increasing incarnation of sought after values—as the patient's maximally possible state of restored health or the artist's completed painting. It proceeds with diagnosis of the conditions offered in the immediate present, conceives itself as carrying a history forward by prescribing curative or creative procedures—modes of activity whose outcome can never be assured—while remaining alert to how particular prescribed procedures "work out" in the particular case at hand.

We are bound, as Dewey continually urged, to a continuum of means and ends, because every step of a process is the goal of an earlier act and every end is but a means to some later state of being. The continuity of time means the continuity of process. What we are *doing* to attain any future instantiation of values becomes part of the entire resultant situation we do in fact achieve. And what we achieve may well improve upon what we have received; but whether it does so or fails, it will produce the point of departure for those who come after.

Our heritage is that of past experiences—always facing problems to be solved, privations to be filled, lacks discerned—always looking to a future instantiation or incarnation of the normative judgmental value that is invoked. Our temporal experience, individually and socially, is structured largely in this way whenever we act, not on impulse, but on the careful assessment and remedial or transformational procedures by which practical reason guides its own activity in the world. The process of our organized societies, the process of any polity, is the social temporality of its citizens.

Its future will be what is actually produced by the corrective and creative procedures invoked because of needs discerned in the situations in which we find ourselves. What will be actually achieved will not be some particular spot already to be pointed out on a map—but rather the product of what is done by all who have some influence on the situation at every step along the way. What they do may well cure the particular social ills and ameliorate justifiable complaints incurred along the way by creating new situations that reach beyond the deficiencies and problems out of which they emerged. Into this continuing becoming, this continuing transformation of heritage into legacy, the process of a responsible polity continually looks toward what practically ensues as it seeks to achieve what reformed institutionalizations of the values its members regard as the good they share together.

VI

"Time," Machiavelli urged, "is the mother of truth" (I, iii, p. 117). It is certainly the mother of the social truths in which we are continually engaged, as Machiavelli, that masterfully dispassionate student of the dynamic of the life of the polis, spelled out in sometimes excruciating detail. Time is the form in which our social engagements find expression; our temporalized responses to its demands and the opportunities we see it as offering frame the varied ways in which we find ourselves carrying the heritage we acknowledge while creating the legacy we will effectively bequeath.

A constitutionally conceived polity conceives itself as a prescription for the future of the society it governs, as a temporal process of working out the problems that it finds given to it. The goods of free citizens are generally seen as degrees of temporal self-control to be attained by education, social contribution, and social desserts. Only in this ongoing process of the social temporality in which we all share are we able to find those responsibilities we are called upon to share, those goods we seek, and the fruits of our varied endeavors, which presume the harmonizing of our individuating temporalities in order to be realized.

"Time is the mother of truth" in every social endeavor in which we may engage. It delimits what has been offered, what potentialities and lessons have been provided, what possibilities can be discerned, and the parameters in which those possibilities may be seized. The "process of polity" is the process of a community that develops its members as individual citizens who, while pursuing their individuated futures, yet find that their time is contributing to the good of the whole. It is the process of encouraging its individual citizens, who are always socially bound, in their ultimate social obligation as individual selves who are continually contributing unique temporal perspectives into the social outlook to which they inherently belong. If each citizen is always engaged in forming the future, while bringing the past into the creation of that future, each free citizen is continually engaged in a temporal time-binding process. It is a continuing process that, at its best, is a common commitment to enhance the socially grounded freedom of its individuated members by using the strength of the whole to nurture themselves and then replenish by creating that future which it will proudly hand over as its legacy to those who come after.

References

Aristotle, Nichomachean Ethics & Politics. In: *The Complete Works of Aristotle*, Vol. 2, ed. J. Burns. Princeton: Princeton University Press, 1984.

Burke, E. (1782), Speech on the Representation of the Common in Parliament. In: *Selected Writings and Speeches*, ed. P. J. Stanlis. Chicago: Henry Regnery, 1963.

Hamilton, A., Madison, J., & Jay, J., *The Federalist*, ed. J. E. Cooke. Middletown: Wesleyan University Press, 1961.

Kant, I. (1781, 1787), *Critique of Pure Reason*, trans. N. K. Smith. New York: St. Martin's Press, 1929.

Locke, J. (1689), *Second Treatise of Civil Government*. New York: Appleton, Century, 1937.

Machiavelli, N., *The Prince & Discourses*. New York: The Modern Library, 1940.

Montesquieu, C. de S. (1748), *The Spirit of the Laws, A Compendium of the First English Edition*, ed. D. W. Carrithers. Berkeley: University of California Press, 1977.

Mosca, G. (1939), *The Ruling Class*, trans. H. D. Kahn. New York: McGraw-Hill.

Peirce, C. S. (1905), What pragmatism is. In: *Collected Papers of Charles Sanders Peirce*, Vol. 5, ed. C. Hartshorne & P. Weiss. Cambridge, MA: Harvard University Press, 1963, pp. 281–282.

Ricoeur, P. (1974), *Political and Social Essays*, ed. D. Stewart & J. Bien. Athens: Ohio University Press.

Rousseau, J.-J. (1762), *Of the Social Contract*, trans. C. M. Sherover. New York: Harper & Row, 1984.

Royce, J. (1904), *The World and the Individual*, Vol. 2. New York: Macmillan.

Sherover, C. M. (1981), Perspectivity and the principle of continuity. In: *The Study of Time IV*, ed. J. T. Fraser et al. New York: Springer-Verlag, pp. 136–146.

——— (1989), *Time, Freedom, and the Common Good: An Essay in Public Philosophy*. Albany: State University of New York Press.

Thucydides, *History of the Peloponnesian War*, trans. R. Warner. London: Penguin Books, 1954.

Tocqueville, A. de, *Democracy in America*, trans. P. Bradley, 2 vols. New York: Alfred A. Knopf, 1945.

14

Emerson and the Sacredness of Becoming

Bertrand P. Helm

Abstract The middle decades of the nineteenth century in the United States witnessed a steady disengagement from the postulates of an earlier, dominant Puritan mind-set, abetted by the Enlightenment philosophy of the Unitarians, and a developing attachment to ideas about the autonomy of persons and a new assertiveness in the American mood.

The New England Trancendentalists were important brokers in these fundamental realignments within the American intellect. Chief among them was Ralph Waldo Emerson. His contributions to philosophy and literature were so salient that, to leading figures both at home and abroad, he was the dominant American mind.

Emerson's earlier religious views and the Neoplatonic constructions which displaced them were not able to withstand the pressure of his studies in the sciences, especially in geology and in the burgeoning evolutionary theory that was to culminate in the work of Darwin. In the later stages of his philosophy, Emerson allows that temporal passage is the vehicle for emergent properties, for essential novelty. Each new fact, he held, is a prophetic sign of one yet to come in an incessant, expanding series of events in this world. We are projecting our future. On such a view, all permanence is illusory.

Emerson's views of flux, process, and time provide some of the background against which American philosophers like William James, Charles Sanders Peirce, and John Dewey developed their own ideas about time and its place in reality. Each of these philosophers attached high value to his work.

Walt Whitman in literature and Charles W. Eliot in education were deeply influenced by Emerson and commended his work to others. Through Whitman, Emerson led the nation to sing a song of itself, and to exalt its own experience. If there is a distinctive American culture, then Emerson is enormously responsible, in the Spenglerian sense, for articulating the morphology of that culture, one which is always divesting itself of its past.

In the middle decades of the eighteenth century, the dominant cultural ideal in the American colonies was tied to a supernatural end. It was animated by hope for a new life in the presence of God. But one hundred and fifty years later, at the turn into the twentieth century, this ultimate cultural goal had been given a historical interpretation. The idea of a supernatural destiny for man had been effectively decommissioned. No longer was it that all things are possible with God. In its new form, it read that all things are possible in an endlessly open future. What happened? What brought about such a change?

The argument here is that the person mainly responsible for rethinking the journey toward the transcendent as immanent history, and who was then able to get his imaginative sketch accepted as a fundamental cultural ideal, was Ralph Waldo Emerson. We examine the stages of the argument under the following headings: (1) antecedents; (2) opposing dualities; (3) providence as mediation between extremes; (4) Neoplatonic hierarchy as mediation; (5) change as ultimate: The sacredness of becoming; (6) the occasions of becoming; (7) the reception of Emerson's process philosophy.

This sequence of stages suggests that real novelty emerges in Emerson's views. The novelty attaches less to the emergence of unique elements in his thought than it does to the emergence of unexpected combinatorial relations between mostly preexisting elements. That is, what was mainly peripheral in his early development becomes salient a decade or two later. Nevertheless, as we shall see, process, becoming, and time, the more Heraclitean features, were never very far from the dominant center of his speculations.

As seen arising out of the context of Enlightenment thought and the progressively more abstract deism of the Unitarian movement, the philosophical development of Emerson shows a series of positions. He reaches back behind Unitarian thought to try to recapture some of the Puritan spirituality and its sense of inwardness. But he also reaches forward to a new kind of practice—the art of creating the authentic life under the conditions of time. The intellectual development of Emerson recounted here, then, is that of a long struggle, a series of balancing acts. Each tentative period of calm actualization seems externally shattered by events such as tragic deaths in his family, or explicit rejection of his views, or social events like the panic of 1837, or the coming conflict over slavery. These eruptions or crises seem to act as catalysts which drive Emerson to some new combination of his basic ideas. Let us examine the stages of his struggle for being.

Antecedents

By the time Emerson graduated from Harvard College in 1821, a highly philosophical branch of Protestantism called the Unitarian movement was hastening the erosion of Puritan orthodoxy, especially as it was construed by William Ames in his *Medulla* (1623). The Unitarian ministers were replacing the more orthodox Calvinists in the administrative and social apparatus behind church pulpits and college classrooms in New England. Emerson's father had been one of these more liberal Unitarians. But the earlier Puritan beliefs were not being completely discarded. It was primarily the content of the Puritan postulates that came increasingly under pressure. Those postulates were being stripped of their prophetic content and thus were losing their provocative emotional impact and their power to discipline through fear. But the form of those beliefs was left intact and lingered on persuasively as an animating, background myth. The Puritan gospel depicts the fallen, dying soul as regenerated by the gift of God's grace. The soul then nurtures and closely disciplines its new life. It moves in a continuing pilgrim's progress, without the need or intervention of any sacerdotal mediation, toward an other-worldly destiny. The Unitarians and, following them, Emerson, retained much of this story line, but reoriented it and stocked it with new players. In the form in which Emerson rewrote it, humankind shall overcome its initial deprived or diminished condition by turning to the powers that are present in reason and in nature. Newly invigorated by the enlargement of energies in soul and spirit obtained from such sources, individuals can advance toward their future destiny in this world.

Rewriting an upward ascent into a forward progress could not become a dominant cultural ideal without somehow changing the conception of the material base of nature. That is, whenever we seek to change one member of a bi-polar pair of terms like *spirit* and *matter*, the adjustment must be fortified by suitable changes in the other member; otherwise, the new view cannot penetrate and become lasting. Thus, the variations that Emerson introduced concerning the self-realization of finite souls in time need to be related to an earlier New England idealism, and its way of explaining matter, space, and time.

As shown by Herbert Schneider (1946) and Perry Miller (1940, 1973), the idealistic philosophies of man and nature, as set out by George Berkeley (1685–1753, the Irish philosopher and bishop), Samuel Johnson (1696–1772, American philosopher, disciple of Berkeley, and convert to Anglicanism), and Jonathan Edwards (1703–1758, American philosopher and theologian, and student of Johnson), realigned ideas about the material world and spiritual selves. Their notions helped supply grounds for Emerson's own more radical changes. Samuel Johnson, partly under the tutelage of George Berkeley, completely dematerialized the surrounding world. Such a cosmology leaves the world lawlike and symbolically intertwined at all levels. For both Berkeley and Johnson, the world was still physical and Newtonian, but for all that, not material. Emerson accepted such an interpretation. Indeed, in its Berkeleyan version, it seemed to free him. In one of his letters (Rusk, 1939), Emerson told Margaret Fuller in March 1841 of the joy he experienced at the first hint of Berkeley's philosophy, and that he never lost sight of it thereafter (pp. 384–385). The idea of matter is simply redundant in the roster of basic ideas used by Berkeley and his American disciple, Johnson. Matter was taken by them as the after-image of a Hellenic, pagan way of reflecting about the world. The reality is God's omnipotent providential spirit. In the face of that terrifying omnipotence, matter, space, and time are just apparitions or figments, mere symbolic devices for suggesting the divine omnipresence and providence.

For Jonathan Edwards, God is the ultimate cause of the world. In the face of God's primary causal efficacy, secondary causes like matter are real, but still not ultimate. As in Berkeley's idealism, physical reality is entirely composed of, and guided by, spirit stuff; it is a function of the divine will. More importantly, for Edwards, the free will we claim to have is not free at all, but is part of the divine necessitation of all things. So our will is not a secondary cause that can separate the believer from God. Human will, taken as a part of the psyche, cannot then be the source of a rebellion from God. Wisely or not, Edwards left human alienation as a mystery surrounding the entire self and its continuity through time. One of Emerson's famous 1841 essays, "Self Reliance," makes enormously more sense when it is seen against this background of Edwards' denial of free will in his important 1754 essay, "Freedom of the Will." After all, Emerson's self-reliance, in its early versions, is grounded in God-reliance.

As a group, the New England transcendentalists, along with Emerson, made a great deal of the tradition-dissolving work of the English Romantics and, to a lesser degree, the German idealists, as R. Wellek (1943) and G. Santayana (1957) show. These transcendentalists did not have to go too far afield to frame the concept of nature as apparition, as a coded system of symbolic laws, within whose shadowy vistas we finite humans work out our destiny. Most of the gain in such a set of ideas was already present in the idealism of these Anglo-American divines, Berkeley, Johnson, and Edwards.

Many of Emerson's philosophical problems arise in this milieu of the dissolution of material nature and its resolution into the view that each part of the material world has mental or spiritual aspects. With the departure of matter, certainly space and time cannot be construed as the containing and sequencing media for properties of matter; nor can they merely comprise God's channels for keeping track of the world. Emerson's problem, then, as reconstructed, goes like this: What can we trust in gaining our bearings if matter, space, time, and a personal God are assigned new roles in, if not dismissed altogether from, the traditional cluster of fundamental concepts? Emerson's later answer of an almost Stoic and aesthetic reliance upon oneself in a shifting space–time field ends up being what is most typical of his mature work. Let us then turn to the developing stages of that answer.

Opposing Dualities

In his earliest notebooks, Emerson began to track the ideas of chance, impermanence, and passage. The journals show that his thought is already laden with a profusion of images about transience and becoming. There are dozens of passages in the early journals in which the concept of temporality is dominant. He gives the impression of a person who is cautiously circling an enormity. In alternating moods of scornful anger and debilitating anxiety, he hurls names and epithets at this gigantic destroyer, time, in order to gain power over it.

Many of these early passages illustrate two dualities. The first duality is the contrast between temporality and eternity. Transience is described in 1820 by Emerson (1960a) in terms of aversive experiences where, in pain and grief, we seek to distance ourselves from a shocking, destructive presence (pp. 29, 174, 200, 236). Time seems to be such a presence to Emerson. Time's powers paralyze us. There is no fairness or justice in time. Especially is this seen in the experience of death.[1] In death, he thought that we ask for and gain nothing from time, but gain all in eternity.

The second duality is immanent in the world, a duality between means and ends. Somehow, he will find a means to ride out the incessant passage of nature in order to gain the end of a tranquil soul. He is even willing to invoke time to help in this passage toward quiescence. He opens three of his early journals (1960a,b), in a sequence called the *Wide World* journals, by trying to press time into his service. In dedicating *Wide World 4*, in 1822, he calls upon time and its trailing stream of hours for aid, but to no avail. In opening *Wide World 9*, in 1822 to 1823, his dedication shows him turning away from a rotten, degenerate past toward a hopeful, consummatory future. In dedicating *Wide World 10*, in 1823, he petitions time (Emerson, 1960b) to slow down, even stop, to allow him a brief, enjoyable moment, but gets back the retort, ''Fool! you implore the deaf torrent to relax the speed of its cataract'' (pp. 105–106). The mood of this second duality of means and ends is that of a prophetic complaint, with its likely model the Old Testament book of Lamentations, while the first duality is more Graeco-Roman with some Christian overtones. Its model is rooted in those classical myths which stress a remorseless fate.

Although these two dualities introduce some order into the wide variety of Emerson's references to time, they do not cover all of his approaches. Two other facts, I think, bear upon the profusion of his references to time. First, he is a poet: He is developing a dependable repertoire of images and phrasings that serve as his initial verbal sketches of the later, more polished verses. In that poetic capacity, it is to be expected that he will explore the relation between man's finitude in time and his symbolic victories over time. Second, Emerson quite early acquired that sort of moralizing philosophy that comes to those who read deeply in classical literature. As such, Emerson had almost from the first picked up the dominant categories that in Western culture provide the chief marks for characterizing reality. Such a roster of basic ideas would include the kinds of being that go by the names: God, self, substance, matter, properties, times, spaces, and events. It is, then, almost routine that assertions about time, process, and becoming would show up in Emerson's reflections. The unusual thing, however, is the frequency with which they are set out in the alternating moods of anger, anxiety, and hope.

Providence as Mediation Between Extremes

In the years following 1824, when Emerson began to read theology under the great Unitarian minister, William Ellery Channing, in order to prepare for his own entry

[1]Ralph Waldo Emerson's father, William, a Unitarian minister, had died unexpectedly in 1811, when Ralph Waldo was eight years old. This early death in the family left him distant and quieted. Henry James, Jr. (1905) believed that Emerson's spirit always lived in the halls of his youth (p. 7). Eric Cheyfitz (1981) says of Emerson that he was the child in *Nature* looking for his father (p. 6). Such an early numbing experience would be particularly troubling since, as Woodberry judged (1907), Emerson was then brought up to be the child of his father, with the duty to uphold the family name (p. 3).

into the ministry, he used many traditional theological concepts to bring his ideas into some semblance of order, especially the idea of the benevolence of God. God's sustaining will reduces the world to manageable order for believers, and allows them to plan on at least some small satisfactions and accomplishments during their lifetimes. But, Emerson believed, it is required that the believer not waste time and that he use time wisely. Some of Emerson's early sermons (McGiffert, 1968) are almost "how to" orations on managing time so as to make good God's beneficient investment in our welfare. Such sermons have the word *time* in their titles, and Emerson repeated them for different churches in the brief span of years after he was ordained and before he left the ministry. His religious philosophy, as set out in his journals and in these early sermons, is strung across well-worn, somber lamentations worthy of many of his Puritan ancestors who were also ministers. His religious view has it that once there was nothing but God, space, duration, and a shoreless abyss. The creation arose from those factors and became separate from its divine ground. Sin, limitation, and time then entered into creation to give us man and nature in their present form.

We should note that Emerson sometimes used the word *nature* to stand for the creation as fallen. Not able to see its Creator, the dull, unseeing eye, with its limiting prejudices, sees the "creation" as nature pure and simple. Under these circumstances, in 1824 Emerson says (1960b) that "nature is the bright ruins of an original creation" (p. 236). The damage began, he believed, following the appearance of relations between those primordial factors, God, space, and duration. For the unregenerate eye, nature is the creation in ruins just like man was a god in ruins. The cause of the ruination would be identical in the two cases. Each is seen as embedded in change and becoming. Man, in so far as he has sinned and separated himself from God, is a fallen being. He has fallen from being into becoming. Thus is Emerson led in 1824 to define man's present nature. "Man," he says (1960b), "is an animal that looks before & after" (p. 237). We are astride the present, looking at the past in fear and the future with hope. But importantly, the hope which Emerson has *for* the future is not *in* the future, but is *after* the future. God will annihilate all the progressive degeneracy of these times and will come again in glory. For Emerson in 1822 (1960b), God's benevolence oversees both the destruction and the subsequent, surrounding cloud of glory (pp. 6, 45, 75–76). In this condition alone does true hope reside.

For such an account, the past is the locus of sin. He says in 1824–1825 (1960b) it is adamantine and fixed forever (pp. 241–242, 324). While the past lies outside our human powers, the present is within our powers. Present time is the power of change, and since the power of change is in us, time is in us. It is not external to us, but flows through us. We, in turn, contribute to its decline toward the disasterous climactic event, the end of time. Because time is in us, almost twinned with us, it is very personal. In 1824 Emerson's image (1960b) is that man is a seneschal of time, one of its stewards (p. 249). Time is personified as master of the feast. It sets the tables and calls the guests. I believe Emerson means that this feast will be like the last supper of the condemned. We are aging inhabitants in an aging world. Decay is everywhere. Emerson (1960b) exults in 1825 that he lives when time is running out: "I rejoice," he says, "that I live when the world is old" (p. 340). In this twilight of decay, he thought, we are driven to the gospel.

In point of fact, however, Emerson was not driven to the Christian gospel by decay and death, but was perhaps hastened out of it by a devastating encounter with death. The editors of his journals (1960c) imply as much (p. xi). In February 1831, his young wife Ellen died of tuberculosis. Emerson was shattered. He almost could not believe it; in some sense, he never accepted it.[2] Her death, however, seems less a cause of his leaving the ministry than it was a catalyst. It intensified his mounting criticisms of form and tradition, criticisms undertaken by a consciousness already deeply restless in American Unitarian Christianity. It dramatically heightened his skepticism about God's providential care. It hastened, in his mind, the dissolution of the idea of divine benevolence as the preferred kind of mediation between time and eternity, and also between past and present as means, and the future as end. It prompted him to consider novel configurations in his basic ideas, combinations that breached his allegiance to Unitarian convictions. Firkins (1915) and Miller (1940) show that Emerson revolted against and outgrew those convictions. The premature eclipse of Ellen's life and his subsequent departure from the ministry now focused Emerson's attention on a most urgent problem: If God's benevolence, if indeed the very concept of a personal deity, is a broken reed that cannot be relied upon, then where is the linkage that will bind the seen, temporal world to the unseen, eternal world? If we are not simply to live in the midst of brokenness, then how can we rethink time as a proper part of the perpetual and the eternal?

Neoplatonic Hierarchy as Mediation Between Extremes

Emerson's first book, entitled *Nature*, appeared in 1836. Several commentators (Matthiessen, 1941; Cameron, 1945; Flower and Murphey, 1977; Allen, 1981; Whicher and Spiller, 1959) treat it his major work, because they judge it to contain the premises or major assumptions for all his later work. As we shall see, and consistent with the views of other Emerson scholars (Aaron, 1963; Neufeldt, 1963; Clark, 1967; Anderson and Fisch, 1969), such a view of *Nature* seems mistaken. For the most part, and in the light of his later work, this first book by Emerson is as important for what it denies as for what it affirms. It denies that the processes of becoming are a degradation from better to worse. It denies some dualisms, like that between time and eternity. Consistent with Berkeley, Johnson, and Edwards,

[2]Apparently Ralph Waldo and Ellen Emerson had a partly religious, partly romantic, desire to pursue immortality together, according to Gay Wilson Allen (1981, p. 72). A year or so after her death, and for reasons not very clear, a troubled Emerson had reopened her casket, and gazed upon her remains. Both Allen (p. 310) and Woodberry (1907) see her death as contributing to a diminishing or ebbing of Emerson's personality (p. 92). Deaths in his family seem to have led to a shrinkage of his personal being, possibly supporting the subsequent emphasis upon the more abstract notion of individuality in his thought.

it also denies that matter is the mediator or connector, the objective element in all natural entities. Rather, Emerson affirms in *Nature* that there is an ideal or spiritual continuity between all natural entities. Such continuity is the unity that underlies all diversity, and makes the very idea of diversity meaningful.

Most of the positive philosophy here is already prepared for in principle in the work of Plato and Plotinus. The famous passage in which Emerson exults that he becomes all eye can be seen easily enough as a commentary on Plato's myth of the Sun and the Good in *Republic*, Book Six. On its Judaeo-Christian side, its metaphysics of light is also prepared for in the work of Augustine, Bonaventura, and Berkeley. Let us now examine some of the specific themes which Emerson pursues in *Nature*, concentrating on those of change, becoming, and temporality.

Nature, he holds, is not something external to us. Appearing in our bodies, it is present in us, though not in us as spirit. In order to identify the purpose of this natural flood streaming through us, and of which we partake, Emerson takes the stance of an Adam. From that vantage point, the world is not fallen, and nature does not differ from creation. His metaphorical device seems to be that of Plato's unfallen eye, in which outer sense and inner sense are in true alignment, and perfect, veridical vision is still possible. The spiritual me and the bodily not-me meet in the eye. As unfallen beings, currents of universal being circulate in us, and we are not radically separated from the divine ground.

Because of our continuity with God and nature, we are steered into experiences of satisfaction and joy. In 1836 Emerson says that it is the role of created nature to supply the commodious means to our consummatory experiences (chapter 1). In the context of our unbroken continuity with nature, Emerson thinks that our being is open in all directions. Space, time, and light are infinite in every direction, and they provide the continuous fields in which our experiences of transcendence occur. So taken, nature disciplines us. She molds and constrains us through environment and society. We are also disciplined by means of space and time. Space and time (Emerson, 1836) tell us that everything is not a heap, a single amorphous mass of things (pp. 36–38). Things and properties are spread out and sequenced. Space and time comprise the principle of individuation, and they discipline us to individuality and freedom. Admittedly, as natural laws come to be known in this now translucent cosmos, space and time seem to vanish because the continuities of reason provide a network of interlocking emblems and symbols which depict the universal for us. But this universal whole is still a changing whole, so temporality remains a feature of the perfection of the whole.

Emerson is ultimately led to ask: Is nature real, or is it an apocalypse or epiphenomenon of mind? Each of his earlier dualisms had denied the separate reality and worth of nature, while they affirmed the reliability of its processes. Emerson updates this feature of dependability in answering the question about nature's reality. The laws are real. Temporality and its laws are not the annihilator or destroyer, rather, they make up the relations in what comes and goes. Emerson explicitly holds that time and process are pivotal. He believes (1960d) that moments of time share in the beauty of the whole (p. 137). But to return to the question, Is nature real?, Emerson concludes that it is a matter of perspective. From the viewpoint of our own active powers of reason, the question remains open. Our senses do seem to present us with an external existent. But for the perfect eye, reason and spirit align the sensory view with a deeper view. There, nature seems to be transfigured or exalted. Material nature as a separate existent seems to withdraw. Moreover, sense experience itself contains hints of nature's metamorphosis.

Emerson (1960d) frequently illustrated these natural hints through his use of Galilean relativity (pp. 25, 142, 182); that is, our perceptions vary as our frames of reference vary. From the vantage point of a moving coach or of a speeding railroad car, nature itself seems to be released into streaming motion. In addition, from the viewpoint of a true poet, like Shakespeare, or of a great painter, such as Raphael in his painting *The Transfiguration*, the world around us becomes fluid and ductile. Art and science meet in one in dematerializing nature here, confirming the results of the earlier Berkeleyan idealism. To achieve such an Adamic viewpoint is to possess a transcendental science which discloses the thoughts of God through nature's transparencies. The inspired reason of the true thinker can grasp God's instant eternities; thus does man achieve immortality. We see that if time and space are relations of matter, then they tend to fade in significance as matter fades into a mere shadow of mind. The fluid durations and the changing forms of nature are the reality, and are sacred. Space and time are perspectival froth riding on that torrent of natural change. They evanesce in the face of the more imperious durations disclosed in natural becoming.

By way of evaluation, I judge that Emerson wrote *Nature* as if he were writing a new ending to the story of the Garden of Eden. That is, how would that religious account go if those universalized humans, Adam and Eve, not only ate of the Tree of Knowledge, but had also eaten of the Tree of Life? They would then be like gods, and not gods in ruins, because they too would have immortality. As immortal, they could see an unfallen creation. This creation would supply their needs, grant gifts of pure joy, and allow transparent insights. Time and space as separators, as principles of individuation, are then transfigured. They disappear in the face of the radiant symbols that comprise the universal laws of the universe, only to reappear in the eternal applicability of those laws.

Change as Ultimate: The Sacredness of Becoming

Emerson's freshly coined myth of a transparent nature, whose parts all signified each other, did not get a chance to appear in mint condition. His spiritualized, poetic conception of nature was not able to withstand the pressure of his continuing studies in the sciences, especially in geology and in the burgeoning theories of biological evolution. And yet another personal tragedy, the death of his brother, hastened the intrusion of these realistic motifs into his philosophy. The effects of the tragedy and the inclusion of these more somber hues are seen in the differences between the radiant Neoplatonic idealism of the first six chapters of *Nature* and the gathering dissent of the last two chapters. Careful analysis by Sealts and Ferguson (1969) has shown that the original version of the book was to have had just the six chapters (pp. 38, 43–45). Emerson had begun to work on a companion volume, to be entitled *Spirit*. But his presumptive partner in the life of the intellect, his beloved younger brother, Charles, died unexpectedly of a lingering tubercular condition. The shock of this death, not unlike the catastrophe of of Ellen's death six years earlier, interrupted the two-essay or two-book project.[3] When he did complete it a few months later, he simply patched two new chapters, suggestive of different orientations, to the original six of *Nature*. According to Packer (1982), the last two chapters were written *after* the death of Charles (p. 48).

The two new chapters, dealing respectively with spirit and with our prospects in this world, made of the book an ungainly whole. Emerson allowed that the materials did not fit together very well, that he was not able to weld cleanly the seam between them. To the metaphysical optimism about man and nature in its first six chapters was added an overlay of skepticism. He sounded the old caution against our forgetting our degenerating condition and finitude. The shocking affront of his brother's death hastened Emerson's turn away from and rejection of a benevolent, unfallen world. The bruising experience seemed to hasten his movement toward the world being disclosed in the natural sciences of his time, sciences which would tend to force an encounter with this world on its own fateful, finite terms. Emerson was already beginning to depart from the Neoplatonic, otherworldly premises of the main body of his book, *Nature*, before the book appeared. The enduring Emerson, I would argue, is the Emerson who began to appear in embryonic form in 1836, the Emerson who had successively worked through a Christian, benevolent worldview, and was now also to evolve beyond a Neoplatonic, spiritualized worldview. He was never completely committed to transcendentalism, although traces of such views remained in his thinking. According to Henry James (1905), Emerson even told his wife, Lidian, that he was "innocent of" all transcendentalism (pp. 21–25). From their earlier centrality and dominance, those views were relegated, because of his combinatorial inventiveness and his increasing integration of aversive experiences, to peripheral and recessive roles.

The new phase in Emerson's philosophy covers the transition from a set of concepts associated with an upward ascent into a body of ideas associated with a forward progress. This line of development covers some five or six years, from

[3]Ralph Waldo and Charles Emerson were exceedingly close as brothers. Ralph Waldo thought of Charles as his alter ego, a part of his own spirit. Thus, as Woodberry believed (1907), Ralph Waldo felt Charles' death to be a "deflowering" of his own life (p. 42). Jonathan Bishop (1964) judges that the death of Charles was the death of part of Ralph Waldo himself (p. 175). Indeed, Ralph Waldo told his second wife, Lidian, that henceforth she would have to be content with just a part of him. Perhaps in response to such a disclosure, Lidian addressed him the rest of their married life as "Mr. Emerson." These events of death in the family, and his resulting responses to them, made Emerson especially sensitive to psychological forces or pressures and personal responses to them. In this connection, see William Bridges (1969), who treats Emerson as a forerunner of some modern psychotherapists. Particularly sensitive was Emerson to his dream life as a source of his creativity (see his 1839 lecture, "Demonology"). In general, he held (1844) that "the text of our life is accompanied all along by this commentary or gloss of dreams" (p. 73).

1836 to about 1841, and includes several distinctive marks. First, he gives up his unitary account of perception; that is, we can no longer perceive the world as it is in itself. Second, he gives up the related cosmological optimism; that is, we are no longer to expect consummatory experiences in an ultimate ground that discloses itself. Third, he discovers positive, limited values in historical events; that is, real increments to knowledge and value are added through historical processes. Fourth, he simply eliminates ends or finalities; that is, humans simply anticipate new anticipations, with no end in sight. Fifth, he displaces assumed spiritual experiences with tentative experimental probings; that is, he is led to present himself as a Baconian wise man, defining himself as an experimentalist. Sixth, he holds that ultimacy and sacredness lie in becoming, not in being. That is, it is becoming or process that is to be secured against violation.[4] Other candidates for ultimacy, such as God or the Neoplatonic One, were no longer fully consecrated, because they were no longer fully trustworthy. Let us briefly consider each of these six themes. Taken together, they frame the new Emersonian worldview.

First, with regard to our knowledge capabilities, he sees the idealistic approach as limited. Almost directly countermanding his earlier notion that the laws of the body and the laws of the spirit meet in the eye, a conjunction which would allow us an inspired insight into the real and the good, Emerson now accepts a disjunction between inner conception and outer law. That is, it becomes clear that the mind's structures are not automatically congruent with the nature of things. The axis of reality is not directly aligned with the mind's processes. From 1836 through 1838, Emerson simply starts over again in trying to show the possibility of knowledge. Although there is a surface similarity between the developing skepticism of Emerson and the mature skepticism of Kant's theory of knowledge, David Van Leer (1986) shows how tenuous are the connections between the two approaches.

Second, concerning his cosmological optimism, it is now clear to Emerson that permanent being is a cheap pretender. Nature is not merely supportive in allowing us some satisfactions, but seems equally disastrous in denying us other fulfillments. There is no known permanent ground which, by itself, can guide us toward what is best, or that allows us to steer successfully away from aversive experiences. Form as a fixed guide is a fraud. Creative powers associated with subconscious processes are better indicators and guides.

Third, with regard to history, he changes his approach. History is not a chronicle of events, but is a continuous saga of human happenings which is doubly tracked with both redemptive and negative forces. Arguably, the two most telling lectures Emerson ever gave were entitled "The American Scholar," delivered in 1837, and the "Divinity School Address" (1838a). The reception of those two lectures mark them out as a salient departure from a prevalent cultural ideal. The key ideas of

[4]The being/becoming polarity is, for the maturing Emerson, an overlay of the waking/dreaming polarity. There is necessitation in the waking mind, but freedom in the dreaming mind. In our dream life, he felt, is the basis of our powers and the fountain of our creativity. See Vivian Hopkins (1951) on Emerson's romantic aesthetic (pp. 178–183), and his efforts to introduce features of the subconscious into his philosophical speculations.

those two performances had been forged in a little remarked series of lectures entitled "The Philosophy of History" which Emerson had finished in the immediately preceding months. For our purposes, moreover, the pivotal lecture in the philosophy of history series is entitled "The Present Age." It was expanded into its own separate series some time later. He argued that the past and the present are caught in a fateful tension. The old and the new set their seal upon everything. They are woven together in the present to give the present the urgency it has. Our *understanding* allows us to dissect forms and laws into the fixed properties and events which comprise the past. But our *reason* discloses what is fluid and labile in the present. Humankind partakes of both the fixity of the past and the fluidity of the present. Emerson says in "History," the first in his *Essays* of 1841, that human life is nothing but a procession, a flight of facts and events. In this flight of events, time becomes the great consoler. Time resembles death because they both require of us that we give up our being. But time is also the agent of a healing spirit, and takes us into new progressions and processes. Emerson (1851c) seconds this emphasis in his famous 1841 essay "Oversoul," holding that our being descends as a stream into us, and takes us we know not where (p. 189).

The fourth of the characteristics of Emerson's revised position concerns ends and finalities. He drops them altogether. This becomes dramatically clear in the essay entitled "Circles," which was published in the 1841 *Essays* (1841a). Published six years after *Nature*, "Circles" is an upwelling of naturalistic conceptions. Around any circle of experience, a new circle, a new horizon, can be drawn, and beyond that horizon is yet another. There are no ends in nature, only transitions to new vistas. The universe is fluid and volatile; everywhere, there is incipience. In 1841, Emerson (1841a) says that "every ultimate fact is only the first in a new series" (p. 214). He has here reached the point where he believes it is morally wrong to believe in finalities. "The only sin is limitation" (p. 217). We are unfinished beings, destined for endless transitions.

Fifth, concerning his experimentalism, Emerson defines his own expanding nature in terms of his probings. "I am," he confesses (1841a) in his essay, "Circles," "only an experimenter. . . . I unsettle all things" (pp. 224–225). He is an endless seeker, with the past at his back, and with new values still to be created and known in the future. But he allows that such a stance is risky, because everything seems to him to be disjointed and sliding. Everything is composed of overlapping and unfirm surfaces, and the challenge is to rely upon oneself in maintaining a precarious balance at the present moment, poised between past and future. It is in this spirit that Emerson slowly disengages from his transcendentalism. Indeed, it was at about this stage of his development that Emerson told Lidian, his second wife, that he was not a transcendentalist, that he was innocent of such beliefs.

Sixth, with regard to his open embrace of process and becoming, Emerson concedes that everything is not fluid all the time. That is, the concept of process does not by itself suffice for a cosmology or anthropology. In the mood of Augustine and Descartes, he holds in his essay "Circles" that we could never even notice the progression and fluidity of things if there were not, by way of contrast, a firm principle of stability or fixity in the soul. Even while the endless generation of circles proceeds, the eternal generator itself abides. Emerson came to speak of this tension or opposition between the power of generation and the resultant generative process in terms that came from Spinoza: *natura naturans* and *natura naturata*. In his oration "Literary Ethics" (1838b) he discloses his familiarity with Spinoza, and ranks Spinoza as a presence somewhat below Shakespeare and Milton. With regard to *natura naturans*, there is incessant coming to be, and it alone is hallowed. In words from 1841 that supply the title for this study, Emerson (1841a) concludes his essay "Circles" by judging that "the coming only is sacred" (p. 226). Nothing else can bear up under the dedication of ourselves to it, for it is the source of creativity.

The Occasions of Becoming

In the three short years between his 1841 book, *Essays: First Series*, and his 1844 book, *Essays: Second Series*, Emerson continued his active, exacting schedule as professional lecturer. He also grudgingly took on a stint as editor of the transcendentalists' periodical, *The Dial*. He maintained his interest in the natural sciences. Geology or earth science became his preferred natural science, effectively displacing astronomy from that role. Arguments in natural theology had often used the facts and images of astronomy as preferred models in order to prove God's causal activity in creating an orderly world. As a result, the shift from the relative changelessness of heavenly, rational models to the more rapidly changing images of earth set out in the geological sciences led Emerson to give still greater attention to the surging, explosive forces associated with this physical world. For some commentators (Van Wesep, 1960; Beach, 1936), Emerson's reading in geology looms quite large, especially that in the works of Lyell. The youthful Emerson had been entranced by the apparent perpetual stability of the starry universe. But the development of the geological sciences showed nature in a speeded-up mode. Thus, Emerson's expanded understanding of nature, in the sense of its greater rates of change, was an external stimulus to support the internal, experiential changes in Emerson's worldview, and to confirm his growing temporalism.

In addition, during this same interval between the two sets of essays came yet another hammer-blow of tragedy in Emerson's personal life. His son, Waldo, just six years old, healthy on a Monday, was dead of scarlet fever by Friday. For weeks growing into months, Emerson was in shock and stunned incomprehension. Following his time of mourning, subtle changes can be detected in Emerson's views, though not simply caused by that crushing event.[5] Probably due to some

[5]Joel Porte (1973) believes that Emerson's tragically troubled consciousness traces to the loss of his son (pp. 85–92). His son's birth had been something of a new covenant with life for Emerson. From the viewpoint of Jonathan Bishop (1964) the birth of Waldo had been experienced by Emerson as a rebirth of himself (p. 122). In effect, the new child would have helped redeem the partial "deaths" Emerson had suffered with the loss of his father, his wife, Ellen, and two brothers, especially Charles. Following the death of his son, Waldo, Emerson felt impoverished, denuded. On the second anniversary of Waldo's death, Emerson told Margaret Fuller of having made no progress in recovering from that terrible fact. According to Allen (1981), some of Emerson's last words in the hours before he died

combination of (1) his deepening knowledge of the physical processes underlying natural change; (2) his increasingly successful defiance of the dominant Unitarian philosophy and theology; (3) his expanding if not prophetic social concerns, especially with regard to his sharpening criticism of the institution of slavery, and with all of them somehow shaken into new combinations by the catalytic effect of his son's death, Emerson's temporalism underwent a shakedown. In any case, new features as well as new configurations can be detected in it. For him, there was no longer a transcendent world of much significance, nor any important future world. There is mainly this existing, present world (Allen, 1981, p. 397), with its dangerously thin surface barely covering a reality of precipitous arrival, and then, a destruction or loss.

There is an account of temporality in the 1844 essays which makes up the next circle beyond the horizons of the 1841 essay. It is not merely the becoming that is sacred. It is, in addition, an organic, consummatory unit event that is the locus of value. Emerson's 1844 essay (1844a), "Character," concludes with the thought that the moment is all. Moreover, no better indicator exists, showing the distance Emerson's views have traveled, than to compare the 1836 *Nature* with the 1844 essay (1844c), "Nature." It is the present hour, in which we sentient beings are immersed, that allows us joy and satisfaction. In effect, eternity and perpetuity can go fend for themselves. The present moment, the now of brief duration, is the locus of our victories. No matter which side is the victorious side, even where there is tragedy that seems our undoing, we are on *that* side of things. Such is the conclusion of the 1844 "Nature." Whether we seem to win or to lose in terms of our own apparent advantages, and wherever the victory falls, that is our side. This, I believe, is the mood of those who have lost everything and who have survived to come out on the other side. They have died to every propelling aspiration, and thus are free to attend to whatever is around them, here and now. Emerson had, in effect, achieved a subjective confirmation for his version of temporalism, in which discontinuities and moments are as important as continuities across the intervals between events.

In two of his remarkable essays in the Second Series of 1844, "Experience" and "Nature," Emerson has departed sharply from his Neoplatonic account of nature in 1836. His new arguments reflect, he says, a more mature position. In the essay "Experience" (1844b), he claims he is not the novice he was fourteen or even seven years earlier. If we count backward with him, God's eternal providence had still been the guarantor for his world fourteen years earlier. Seven years earlier, the Neoplatonic idealistic hierarchies had provided legitimacy to his worldview.

The tone of the 1844 essays was set in the first of them, "The Poet." There, Emerson says (1844d) that "the fountains whence all this river of Time, and its creatures, floweth, are intrinsically ideal and beautiful" (pp. 262–263). The image

---

would relate to his feelings about his son, Waldo: "Oh, that beautiful boy!" (p. 669). In the last analysis, for Waggoner (1908), Emerson developed his philosophy as a way out of personal suffering (p. 97), and this would be consistent with the view of Whicher and Konvitz (1962) that Emerson insulated himself from calamity by living in the soul (p. 40).

of the fountain here stands for a source of power, and does not signify a changeless object. There is, he held, no doctrine of forms in our philosophy. But his assertions in "Experience" set the benchmark for all that was to follow. There is passage on all sides of us and within us, especially in the subconscious. But if we seek to stay the change, we only accelerate the sense of passage. Objects seem most evanescent just when we seek to make them stable fixtures. The fact is that we are always destined to live on surfaces. The art of life is to skate well on them. What we call a form is only a temporary balance in things. Form is an event in a passing history, a fleeting closure of possibilities. This world is a mixture of power and form, of *fieri* and *esse,* of things proceeding at different rates of change. We should savor our value experiences at the momentary crest of each event before the event begins to wane. "Everything good is on the highway," Emerson avers (1844b, p. 305). Value is more a function of passage than of permanence. The secret of life, the law of life, is to find the journey's end in every step. We should try for the greatest number of good hours: such is the wise man's rule. The only ballast of life, he holds, is that we respect the present hour. Our respect for the hour, moreover, should be informed by the understanding that the hour is importantly prospective. The present hour is a kind of stack or assemblage of climactic experiences that is constantly off-balance into the future. Thus does he complete the argument in the essay "Experience." Human life is a flux, a flux of moods.

The more subjective interpretations given by Emerson in "Experience" are neatly balanced by his more objective accounts in the other key 1844 essay, "Nature" (1844c). He again uses the contrast of *natura naturans* and *natura naturata* as threaded into modern thought by Spinoza. *Natura naturans* is the power source, the agency of all becoming. Contrariwise, *natura naturata* is the face of the universe, the appearance of things, the phenomena whose laws are given in terms of space, time, mass, and event. But that is still the surface of things, and is comprised of shifting plates and widening circles. What we call stability and form is just a passing froth, a transient appearance, a simple turning of the hourglass.

Emerson also pointed out in "Nature" (1844c) that most of the changes which touch us pass on by without violence. This is because of the indefinite or boundless space and time in which they occur. Space and time are masters of the feast that tell each celebrant where to be seated for the occasion. They decree the locus of motion and rest. The first and second secrets of nature, Emerson summarizes, are motion and rest (1844c, p. 389). To be is to be placed under the countervailing laws of motion and rest. Change and stability are complementary participants in every event. Thus, for Emerson, a natural existent is a momentary exaggeration that arises out of the tension or warfare between those polar opposites of constancy and change.

There are, in Emerson's later writings, other more incidental developments in his version of temporalism. There is some evidence that he may have started to adjust his view of evolution to account for Darwinian ideas, but the evidence is not striking. There is better evidence that he tried to make his process philosophy operational in wider social circles, in his books on human conduct, and in his social criticism. For the most part, though, his important ideas, with all of their changing ramifications, arose across the years from about 1830 to about 1850. The recognition of the significance of those ideas lagged far behind their initial formulations. That he did not attach any finality to his own speculations about temporality can be seen in the statement he made in his Phi Beta Kappa address in 1867 after the Civil War. ''And time and space—what are they? Our first problems, which we ponder all our lives through, and leave them where we found them; whose outrunning immensity, the old Greeks believed, astonished the gods themselves; of whose dizzy vastitudes all the worlds of God are a mere dot on the margin; impossible to deny, impossible to believe'' (p. 225). Leaving aside his valedictory awe in the face of the problems of time and space, if we are to appraise Emerson's central findings, we would stress the main transition, namely, that the transcendent, eternal goal he first sought to attain in some suprahistorical condition came to be progressively disclosed to him as immanent in the historical procession itself.

The new configuration of basic ideas Emerson created during the 1840s makes temporal experience fundamental. Every moment of experience is bipolar, he thinks, and each of the poles, mental and physical, is caught up in its own kind of change. The mental pole itself is a conflict between conscious and subconscious forces. So each moment reflects struggle, achievement, and then relapse or decline. Our consummations or moments of quiescent being are not attainments of some ultimate ground so much as they are fleeting balances of success and failure. There is value, albeit limited value, in historical events and psychological processes. The upshot is that there are no ends, no finalities. All humans can do is experiment with fleeting achievements as they mold the future. It is this configuration of emphases that was passed on as a legacy to Emerson's ''followers.'' They responded to his accounts of process, passage, struggle, and limited kinds of achievement as his most authentic voice. It is this spirit that touches Walt Whitman, Charles Eliot, Wendell Holmes, and also the pragmatists, Peirce, James, and Dewey. He was able to draw out the ''newness'' of New England, generalize it in terms of change and becoming, and help make it the dominant configuration of the new-model America.

The Reception of Emerson's Process Philosophy

Emerson's process philosophy[6] endured beyond his own lifetime and worked its powerful effects in the shaping of American culture. Let us indicate some of the channels through which such influences proceeded. Aside from his own, direct personal impact upon the thousands who heard him lecture and who studied his essays, his influence worked strongly, even though more indirectly, through his setting of the horizons of opinion which constitute the culture of an era. He drafted

[6]Charles Hartshorne (1973), in his ''Ideas and Theses of Process Philosophers,'' sets out some fifteen theses that most or all process philosophers subscribe to or do not reject (pp. 100–101). Eleven of tthe fifteen can clearly be found in Emerson's work, in my judgment, and at least one more by implication.

the important baseline dimensions of American thought by his influence upon the men who elaborated his imperatives in their own special fields. Particularly should we mention these men: Walt Whitman in literature, Charles Eliot in higher education, O. Wendell Holmes in law, and Charles Sanders Peirce, William James, and John Dewey in philosophy.

Emerson and Walt Whitman knew each other personally. Whitman was effusive in his praise of Emerson's effects upon him (Stovall, 1964). He said (Carpenter, 1967) that when he had been at a simmer, Emerson brought him to a boil (pp. 226–227). Whitman thought that there was something almost titanic about Emerson. He spoke of Emerson (Kazin, 1984) as the divine *liberatus,* one whose words almost had a saving power (p. 30). The relationship between the two men was not always untroubled. But both were aboil with the idea of individuals as power centers who find value in the passionately experienced Now, in the present hour. Post–Civil War Americans learned their incessant, if not flagrant, individualism from Whitman, learned from Whitman to sing a song of themselves as independent, artesian centers of feeling and passion. We human beings are to be constantly in transition toward open, uncharted futures. In charming a nation with such lyrics, Whitman was fundamentally on an Emersonian errand.

Charles W. Eliot is salient in higher education in the United States. It was, of course, Eliot who made Harvard College over into Harvard University. He became president of Harvard in 1869, and began those reforms that have ever since recast American baccalaureate and graduate education. According to Eliot, a university is to be open and inventive, where scholars are to trust their own findings, and not authoritative traditions. Eliot institutionalized Emerson's famous 1837 address, "The American Scholar." For Eliot, we are preparing ourselves and society to move into the future. These are Emersonian themes. Eliot and Emerson knew each other personally. The young Eliot (James, 1930) could not fathom the transcendentalist, Emerson (p. 43), but the maturing Eliot became absorbed with Emerson's ideals (p. 217). Emerson was on Harvard's Board of Overseers, voting with the majority, when it decided to bring Eliot in as President. Eliot knew Emerson's work closely, valued it highly, indeed had a set of Emerson's works as the centerpiece of the library at his summer cottage. Through Eliot, the spirit of the Sage of Concord, Emerson, became the spirit of Harvard's mission in higher education.

With regard to O. Wendell Holmes, there had been the closest connection between Holmes, Sr., his father, and Emerson. Wendell Holmes's father had been something of a cofounder, with Emerson, of the famous Saturday Club, which came into being in Boston in the 1850s, and which focused around Emerson. Wendell Holmes, the son, had learned his Plato from Emerson, and had been imbued with the doctrines of Emerson's essays from an early age. A brief overview of the Emerson–Holmes connection is summarized in Carpenter (1967). Suffice it to say that Holmes's experiential approach to the law and to the Constitution dovetails neatly with Emerson's essays, especially "Experience."

In American philosophy, C. S. Peirce, William James, and John Dewey all figure prominently in that cone of light that is often called "the golden age" of

American philosophy. These three pragmatists, at the turn into the twentieth century, stamped indelibly into the American mind the ideas of time, change, progress, freedom and openness as we move into the future. Extended studies (Carpenter, 1929, 1941) in the history of American philosophy have detailed the influence of Emersonian affirmations on James and Peirce. Again, both men knew him personally, and moved in some of the same intellectual circles with him. The motto of Emerson's 1835 journal, B, *To Think Is To Act,* can easily enough be seen (1960d) as one of the planks of philosophical pragmatism and instrumentalism (p. 4).

Both James and Peirce (Carpenter, 1929, 1941) recognize Emerson's importance for their work. In his own picturesque account of fundamental ideas, James's image of time as master of the feast is almost certainly derived from Emerson's temporalistic, celebratory metaphor that we mentioned earlier. John Dewey, perhaps never having met Emerson, conceded to the Sage of Concord more than these others. Dewey claimed that Emerson was the philosopher of American democracy. Arguably, one of the very best introductions to Dewey's temporalist and experimental logic and metaphysics would be comprised of Emerson's essays from his last phase, "Experience," "Circles," and "Nature." They lead us directly into Dewey's progressive, future-oriented experimentalism.

Emerson, I would conclude, is enormously responsible for creating, in the Spenglerian sense of the term, the *morphology* of American culture, a shaping that is always divesting itself of its past. It is a morphology of endless metamorphoses, an exploration that adds new circles or domains of experience to the one that is presently enjoyed but already is too time-bound.

References

Aaron, D., ed., (1963), *Paul Elmer More's Shelbourne Essays On American Literature*. New York: Harcourt, Brace & World.

Allen, G. W. (1981), *Waldo Emerson*. New York: Viking Press.

Ames, W. (1623), *Medulla theologiae*. Translated as *The Marrow of Divinity*. London, 1643.

Anderson, P. R., & Fisch, M. H. (1969), *Philosophy in America*. New York: Octagon.

Beach, J. W. (1936), *The Concept of Nature in Nineteenth-Century English Poetry*. New York: Macmillan.

Bishop, J. (1964), *Emerson on the Soul*. Cambridge, MA: Harvard University Press.

Bridges, W. E. (1969), Transcendentalism and psychotherapy: Another look at Emerson. *Amer. Lit.*, 51:157–177.

Cameron, K. W. (1945), *Emerson the Essayist*. Hartford, CT: Transcendental Books, 1972.

Carpenter, F. I. (1929), Points of comparison between Emerson and William James. *N. Engl. Quart.*, 2:458–474.

——— (1941), Charles Sanders Peirce, pragmatic transcendentalist. *N. Engl. Quart.*, 14:34–38.

——— (1967), *Emerson Handbook*. New York: Hendricks House.

Cheyfitz, E. (1981), *The Trans-Parent: Sexual Politics in the Language of Emerson*. Baltimore, London: Johns Hopkins University Press.

Clark, H. (1967), *Transitions in American Literary History*. New York: Octagon.

Edwards, J. (1754), *Freedom of the Will*. Indianapolis, New York: Bobbs, Merrill, 1969.

Emerson, R. W. (1836), *Nature*. In: *The Complete Works of Ralph Emerson,* Vol. 1. Boston & New York: Houghton, Mifflin, 1903, pp. 1–97.

——— (1837), The American scholar. In: *The Complete Works of Ralph Emerson*, Vol 1. Boston & New York: Houghton, Mifflin, 1903, pp. 81–115.

——— (1838a), Divinity School address. In: *The Complete Works of Ralph Emerson*, Vol. 1. Boston & New York: Houghton, Mifflin, 1903, pp. 119–151.

——— (1838b), Literary ethics. In: *The Complete Works of Ralph Emerson*, Vol. 1. Boston & New York: Houghton, Mifflin, 1903, pp. 155–187.

——— (1839), Demonology. In: *The Complete Works of Ralph Emerson*, Vol. 10. Boston & New York: Houghton, Mifflin, 1903, pp. 1–28.

——— (1841a), Circles. In: *The Complete Works of Ralph Emerson*, Vol. 2. Boston & New York: Houghton, Mifflin, 1903, pp. 299–322.

——— (1841b), History. In: *The Complete Works of Ralph Emerson*, Vol. 2. Boston & New York: Houghton, Mifflin, 1903, pp. 1–41.

——— (1841c), Oversoul. In: *The Complete Works of Ralph Emerson*, Vol. 2. Boston & New York: Houghton, Mifflin, 1903, pp. 265–297.

——— (1841d), Self reliance. In: *The Complete Works of Ralph Emerson*, Vol. 2. Boston & New York: Houghton, Mifflin, 1903, pp. 43–90.

——— (1844a), Character. In: *Essays by Ralph Waldo Emerson*. New York: Thomas Y. Crowell and Sons, 1951, pp. 324–344.

——— (1844b), Experience. In: *Essays* by *Ralph Waldo Emerson*. New York: Thomas Y. Crowell and Sons, 1951, pp. 292–323.

——— (1844c), Nature. In: *Essays by Ralph Waldo Emerson*. New York: Thomas Y. Crowell and Sons, 1951, pp. 380–401.

——— (1844d), The poet. In: *Essays by Ralph Waldo Emerson*. New York: Thomas Y. Crowell and Sons, 1951, pp. 261–291.

——— (1867), Progress of culture. In: *The Complete Works of Ralph Emerson*, Vol. 8. Boston & New York: Houghton, Mifflin, 1903, pp. 205–234.

——— (1960a), *The Journals and Miscellaneous Notebooks of Ralph Waldo Emerson*, Vol. 1, ed. W. H. Gilman, A. R. Ferguson, G. P. Clark, & M. Davis. Cambridge, MA: Belknap Press.

——— (1960b), *The Journals and Miscellaneous Notebooks of Ralph Waldo Emerson*, Vol. 2, ed. W. H. Gilman, A. R. Ferguson, G. P. Clark, & M. Davis. Cambridge, MA: Belknap Press, pp. 104–143.

——— (1960c), *The Journals and Miscellaneous Notebooks of Ralph Waldo Emerson*, Vol. 3, ed. W. H. Gilman, A. R. Ferguson, G. P. Clark, & M. Davis. Cambridge, MA: Belknap Press, pp. ix–xvii.

——— (1960d), *The Journals and Miscellaneous Notebooks of Ralph Waldo Emerson*, Vol. 5, ed. W. H. Gilman, A. R. Ferguson, G. P. Clark, & M. Davis. Cambridge, MA. Belknap Press.

Firkins, O. W. (1915), *Ralph Waldo Emerson*. Boston & New York: Houghton, Mifflin.

Flower, E., & Murphey, M. G. (1977), *A History of Philosophy in America*. New York: Capricorn Books of G. P. Putnam's Sons.

Hartshorne, C., (1973), Ideas and theses of process philosophers. In: *Two Process Philosophers*, ed. L. S. Ford. Tallahassee, FL: American Academy of Religion.

Hopkins, V. C. (1951), *Spires of Form*. Cambridge, MA: Harvard University Press.

James, H. (1930), *Charles W. Eliot*, Vol. 1. Boston & New York: Houghton Mifflin.

James, H., Jr. (1905), *Partial Portraits*. London & New York: Macmillan.

Kazin, A. (1984), *An American Procession*. New York: Alfred A. Knopf.

McGiffert, A. C., Jr. (1968), *Young Emerson Speaks*. Port Washington, NY: Kennikat Press.

Matthiessen, F. O. (1941), *The American Renaissance*. New York: Oxford University Press.

Miller, P. (1940), Jonathan Edwards to Emerson. *N. Engl. Quart.*, 13:589–617.

——— (1973), *Jonathan Edwards*. Westport, CT: Greenwood Press.

Neufeldt, L. (1963), *The House of Emerson*. Lincoln & London: University of Nebraska Press.

Packer, B. L. (1982), *Emerson's Fall*. New York: Continuum.

Plato, *The Republic*, trans. F. M. Cornford. New York: Oxford University Press, 1945.

Porte, J. (1973), The problem of Emerson. In: *Uses of Literature*, ed. M. Engel, Harvard English Studies, Vol. 4. Cambridge, MA: Harvard University Press.

Robinson, D. (1982), *Apostle of Culture: Emerson as Preacher and Lecturer*. Philadelphia: University of Pennsylvania Press.

Rusk, R., ed. (1939), *The Letters of Ralph Waldo Emerson*, Vol. 3. New York: Columbia University Press.

Santayana, G. (1957), *Interpretations of Poetry and Religion*. New York: Harper and Brothers.

Schneider, H. W. (1946), Immaterialism. In: *A History of American Philosophy*. New York: Columbia University Press.

Sealts, M. H., Jr., & Ferguson, A. R., eds. (1969), *Emerson's Nature—Origin, Growth, Meaning*. New York: Dodd, Mead, and Company.

Stovall, F., ed. (1964), *Walt Whitman's Prose Works*, Vol. 1. New York: New York University Press.

Van Leer, D. (1986), *Emerson's Epistemology*. Cambridge, U.K.: Cambridge University Press.

Van Wesep, H. B. (1960), *Seven Sages*. New York: D. McKay.

Waggoner, H. H. (1908), *American Poets from the Puritans to the Present*. Boston: Houghton, Mifflin.

Wellek, R. (1943), Emerson and German philosophy. *New Engl. Quart.*, 16:41–61.

Whicher, S. E., & Konvitz, M. R., eds. (1962), *Emerson: A Collection of Critical Essays*. Englewood Cliffs, NJ: Prentice-Hall.

——— Spiller, R. W., eds. (1959), *The Early Lectures of Ralph Waldo Emerson*, Vol. 2. Cambridge, MA: Harvard University Press.

Woodberry, G. E. (1907), *Ralph Waldo Emerson*. New York: Macmillan.

Name Index

Aaron, D., 269
Abé, K., 212–213, 214
Achilles Tatius, 77
Adams, J., 33n
Adams, J. A., 113
Agriustia, 37–38
Akhmatova, A., 190
Akhundov, M. D., 71, 81
Allen, G. W., 269, 276
Alston, W. P., 236
Ambrose, Bishop, 37
Ames, W., 264
Anatolios, 36
Anderson, P. R., 269
Andries, M., 89n
Anscombe, A., 36n
Argyros, A., 22n
Argyrus, I., 35n
Aristotle, 29, 31, 57, 144, 221, 244, 246–248, 250, 259
Arlin, M., 112
Arlow, J. A., viii
Atwood, M., 187, 210–211
St. Augustine, 33, 33n, 34–38, 42, 85, 128, 206, 270
Aujac, G., 29n
Aurelius of Carthage, 37n

Bach, J. S., 136n, 138, 167, 174–176
Bachrach, B., 42
Baczko, B., 89n
Bakhtin, M. M., 78
Baluze, E., 38n
Barnes, T., 37n
Bayless, W. N., 42n
Beach, J. W., 275
Bede, 28, 44
Beethoven, L. van, 135n, 137, 138, 147, 154, 158–161, 163, 167, 178–183
Bent, I., 129
Benveniste, E., 151
Bergson, H., 163, 226, 235, 236–237, 238
Berkeley, G., 221, 265, 266, 270
Berlioz, H., 140
Besson, M., 121
Bierman, K., 112
Bishop, J., 272n, 275n
Block, R. A., 108
Blok, A., 196
Bloor, D., 101
Boas, F., 73, 74
Bonifatius, 41
Bonner, G., 33n
Bonnet, M., 118, 119, 121
Bonus, 41
Borst, A., 28n
Brandon, S. G. F., 14
Bridges, W., 272n
Briusov, V., 200
Brodhead, G. L., viii, 125
Brodsky, J., 190
Bronowski, J., 14n
Brown, P., 33n
Brown, S. G., 90
Bucherius, A., 35n, 40n
Burke, E., 250
Burnet, T., 168

Caillé, A., 97

Caldwell, L. S., 114
Cameron, K. W., 269
Carlton, L. G., 114
Carlton, M. J., 114
Carpenter, F. I., 279, 280
Carter, M. C., 121
Castellan, G., 87–88
Chaitin, G. J., 6
Changeux, J.-P., 121–122
Channing, W. E., 267–268
Chateaubriand, A. De, 93–94
Cheyfitz, E., 267n
Christ, W., 28n
Clark, H., 269
Clarke, C. J. S., 25, 26, 67, 217
Clemens, G., 91
Cone, E. T., 130, 142–143, 170
Cook, N., 129
Cooper, G., 170
Cordier, B., 167, 172
Creelman, C. D., 112
Crumb, G., 167, 170–172
Cummian, 29
Cyril, Bishop, 39

Darwin, C., 17–18, 263
De Rey-Pailhade, M.-J., 89n
Deane, H. A., 33n
d'Eglantine, F., 90n
Dehaene, S., 121–122
Demetrios, Bishop, 36, 36n
Descartes, R., 57, 223, 274
Dewey, J., 226, 235, 259–260, 263, 279–280
Dickens, C., 130n, 186–187
Dionysius Exiguus, 29, 36, 41
Dioscores, 39
Ducasse, G. J., 83
Duhem, P., 31n
Dumezil, G., 80n
Dunsby, J., 129
Durkheim, E., 97, 106

Eastwood, B. S., 30n
Edwards, J., 265–266
Efron, S., 185–186
Einstein, A., 11n, 54, 235
Elchardus, M., viii, 71, 88, 89n, 90, 91, 97
Eliade, M., 80
Eliot, C. W., 263, 279
Elkin, A. P., 78
Ellis, M. J., 114, 115
Elster, J., 96, 97–98
Emerson, C., 272, 275n
Emerson, E., 269, 272
Emerson, L., 272n
Emerson, R. W., 217–218, 263–280
Emerson, W., 275–276n
Emmet, D., 61n
English, A. C., 9–10n
English, H. B., 9–10n
Etyches, 36n
Euctemon, 31
Eusebii, 36n
Eusebios, 36, 38

Fallach, E., 112
Fell, J., 35–36n
Ferguson, A. R., 272
Feuerbach, L., 84–85
Firkins, O. W., 269
Fisch, M. H., 269
Flower, E., 269
Forster, E. M., 194
Forte, A., 129
Fraser, J. T., viii, 5n, 6n, 7, 8n, 9–10n, 12n, 16, 18, 19, 21, 28n, 66, 79n, 91, 172, 176, 205–207, 212, 216
Freud, S., 21, 208–210
Friedlein, G., 28n
Friedman, W. J., 110
Fuller, L., 98n
Fuller, M., 265, 275n

Gadamar, H.-G., 101
Gaes, G., 112
Gale, R. M., 169
Gallistel, C. R., 110
Gallois, L., 88n
Gauchet, M., 99
Geminos, 31n
Gill, F. B., 110
Ginzel, F. K., 31n, 32n
Giordano, 140
Gjerdingen, R. O., 177
Glaser, E. M., 121
Glencross, D., 118
Glicksohn, J., 112
Goethe, W., 213

Goldbacher, A., 33n
Goncharova, N., 194, 198
Goodman, N., 176
Gould, S. J., viii, 167, 168, 178
Greene, D., 176
Gregory the Great, 43
Gregory XII, Pope, 25
Grondin, S., 115, 116f
Guillaume, M.-J., 88n
Gunnell, J. G., 89, 91, 92n, 99
Gurney, E., 143

Habermas, J., 95, 98, 100, 101
Halbert, J. A., 114
Hamilton, A., 252–253
Hardy, T., 4, 13
Harris, E., 202–203
Hartel, G., 36n
Hartle, J. B., 55
Hartshorne, C., 222–223, 278n
Hasty, C. F., 125
Hasty, O. P., 185, 186
Hawking, S. W., 55
Haydn, F. J., 138
Heath, T. C., 31n
Heidegger, M., 259
Helm, B. P., 218
Heraclitus, 236
Herwagen, J., 28n
Hesiod, 75
Hicks, R. E., 111, 112
Hilarus, 40
Hindemith, P., 140
Hipparchos, 29
Hippolytus, 27, 35, 36, 37n
Hoffman, B., 41n
Holmes, O. W., 279
Holst, G., 140
Homer, 75, 84
Homer, W., 234
Hopkins, V., 273n
Horace, 43
Horowitz, G., 55n
Hoshizaki, L. E., 114
Houle, G., 157
Hultkrantz, A., 84
Hume, D., 223–224
Husserl, E., 153

Ianuarius, 42
Innocentius, 37n
Isham, C. J., 63–64

Jackendoff, R., 170
James, H., 272
James, H., Jr., 267n
James, W., 223–226, 229–230, 233, 235–238, 238–239n, 257, 259, 263, 279–280
Jaspers, K., 80
Jaynes, J., 66
Johnson, S., 265, 266
Jones, C. W., 28n, 29, 32n, 33n, 34, 35n, 38, 39n, 41, 43, 50
Julianus of Kios, 39

Kafka, F., 187, 211–212, 213
Kallippos of Cyzicus, 31–32
Kant, I., 223, 249, 250, 273
Kantorowicz, 103
Kaplan, C. A., 108
Karlinsky, S., 186, 187
Kazin, A., 279
Keele, S. W., 117
Kerman, J., 129
Kierkegaard, S., 84
King, V. H., 30n
Kinsbourne, M., 111, 112
Klapp, S. T., 119, 121
Klatzky, R., 177
Knoell, P., 33n
Konvitz, M. R., 276n
Kramer, J. D., viii, 130, 169–170, 172–174, 177
Krusch, B., 35, 36n, 37, 38, 39n, 40n, 41
Kuhn, T., 101n
Kundera, M., 187, 214–216

Landis, R., 116
Langer, S. K., 130
Lawrence, D. H., 218
Lawrence, N., 217
Leach, E., 79
Leclerc, I., 239
Leibniz, G. W., 220, 222, 230, 258
Lejeune, H., 117
Leo, Bishop, 39, 40
Leontyev, A. N., 74

Lepine, D., 118
Lerdahl, F., 170
Levi-Strauss, C., 73, 78
Levy-Bruhl, L., 74, 75, 76
Locke, J., 253
Lotman, Y. M., 79
Lowe, V., 235–236, 237
Luhmann, N., 90
Luria, A. P., 74

Macar, F., 112, 114, 115, 116f, 117, 119, 121
MacCarthy, B., 36n
Macey, S. L., viii-ix
Machiavelli, N., 246, 261
MacKay, W. A., 118
MacLingo, W., 119
Madison, J., 249, 252
Maffesoli, M., 102n
Maiakovskii, V., 190
Malebranche, N., 221
Mandelshtam, O., 190
Marcianus, 39
Markov, V., 185, 186
Marrou, H.-I., 33n, 42–43n
Matsuda, F., 117
Matsuda, M., 117
Matthiessen, F. O., 269
Mays, W., 217
McClain, L., 111, 112
McGalley, M. G., 121
McGiffert, A. C., Jr., 268
McGurk, P., 29n
Meletinsky, Y. M., 78–79
Meton, 31
Meyer, L. B., 128, 136, 144n, 170
Meyerson, E., 74
Michon, J. A., 95, 108, 112
Miller, G. A., 176
Miller, G. W., 111, 112
Miller, P., 265, 269
Mommsen, T., 36n, 40n
Montesquieu, C. de S., 246, 247, 251, 252
Mosca, G., 247n
Mozart, W. A., 131–133, 136n, 138–139, 234, 239
Müller, G. H., viii
Murphey, M. G., 269
Musset, L., 42

Nadal, J.-P., 121–122
Nakhnikian, G., 236
Needham, J., 5
Neufeldt, L., 269
Newell, K. M., 114
Newton, J., 226
Nietzsche, F., 103
Nitzan, D., 112
North, J. D., 29n

O'Daly, G. J. P., 33n
Ogg, M., 36n
Ornstein, R. E., 111, 112
Ozouf, M., 89n

Packer, B. L., 272
Parker, N. K., 114, 115
Parsons, T., 98n
Parsons, W., 33n
Pascasinus, Bishop, 39
Pasternak, B., 190, 192, 199, 201
Patch, H. R., 84n
Pattee, H. H., 5
Paulson, W. R., 206, 216
Pedersen, O., 30n
Peirce, C. S., 222–223, 235–237, 238–239n, 255, 258–259, 263, 279–280
Penrose, R., 13n, 58, 63
Pericles, 250
Petau, D., 35n, 40n
Pherecydes, 75
Philips, D., 101n
Piaget, J., 76, 78
Pihl, M., 30n
Pinch, T., 101
Plato, 6, 11, 12, 13, 17, 22–23, 129, 145, 151, 219, 234, 270, 279
Plinius Secundus, G., 31
Plotinus, 67, 270
Pocock, J. G. A., 92n, 94
Pollock, J., 21, 144
Pols, E., 238
Porte, J., 275n
Posner, M. I., 108
Pouthas, V., 116
Prinz, F., 43n
Protagoras, 145
Proterios, Bishop, 39–40
Proust, M., 93–94, 194

Ptolemy, 29

Quesada, D. C., 115

Radice, M., 178
Ravel, M., 137
Reichardt, R., 88
Remondon, R., 42n
Requin, J., 118
Richard, M., 35, 36n
Richardson, C. C., 30, 34
Riché, P., 43
Richelle, M., 112, 117
Ricoeur, P., 206, 250
Rilke, R. M., 191–192n, 196
Rindler, W., 58
Roger, M., 42–43n
Rolland, R., 21n
Romme, G., 88, 89
Rorty, R., 102
Rosenbaum, D. A., 118
Rossini, G., 140
Rousseau, J. J., 19, 100, 246–247, 252
Rowell, L., 125, 129, 130, 136n, 138n, 139
Royce, J., 259–260
Ruchkin, D. S., 121
Rühl, F., 32n, 36n, 41n
Rybakov, B. A., 81

Samuelson, P., 96
Santayana, G., 266
Scaliger, J. J., 35n
Schenker, H., 129, 136, 174
Schiaparelli, L., 38n
Schiller, J. C. F. von, 103
Schmidt, R. A., 113, 114, 115
Schmitt, E., 88
Schneider, H., 265
Schopenhauer, A., 143
Schwartz, E., 32n, 36n, 40n
Sciulli, D., 98n
Sealts, M. H., Jr., 272
Sen, A., 96
Shapiro, D. C., 121
Sheldrake, R., 64
Sherburne, D. W., 237–238n, 239
Sherover, C. M., viii, 217, 250n, 252, 258
Simon, H. A., 108
Smith, R. P., 114
Snow, C. P., 12
Soboul, A., 88
Socrates, 16
Sorabji, R., 57
Sosigenes, 32
Soulsby, M. P., 187
Sparks, M., 185, 186
Spiller, R. W., 269
Spinoza, B., 275, 277
Steblin-Kamensky, M. I., 74
Stein, N., 116
Stevens, W. M., 25, 28n, 29, 30n, 38n, 39n, 40n, 41, 43–44, 46
Stockhausen, K., 151
Stovall, F., 279
Strabo, W., 30n
Strauss, R., 139–140
Styron, W., 187
Suetonius, 32n

Telemann, G. P., 137
Tertullian, 37n
Thales of Miletus, 82
Theodosius I, 39n
Theophilos, Bishop, 39
Thomas, D., 4
Thomas, D. M., 187, 208–210
Thomas, E. A., 112
Thomson, I., 57n
Thucydides, 250
Tocqueville, A. de, 88, 247n
Toporov, V., 83
Treisman, M., 112
Tsvetaeva, M., 185–187, 189–204, 217
Turner, C. H., 36n

Ussher, J., 39n

Van de Vijver, A., 36n
Van der Meer, F., 33n
Van Leer, D., 273
Van Wesep, H. B., 275
Verdi, G., 140
Victurius of Aquitaine, 28–29, 40
Vidal, F., 119, 121
Vitins, I., 191
Vitton, N., 114, 115, 117, 121
Voltaire, 78
Von Uexküll, J., 9n

Vygotsky, V. S., 74

Wade, M. G., 114, 115
Waggoner, H. H., 276n
Wallis, J., 35–36n
Wallon, H., 74
Walton, W., 140
Wang Wei, 187, 207
Warm, J. S., 114
Weaver, W. B., 112
Weisberg, P., 116
Wellek, R., 266
Werth, L. F., viii, 217
Weybrew, B. B., 113–114
Whicher, S. E., 269, 276n
White, M. J., 196n
Whitehead, A. N., 53, 59–61, 92, 162, 217, 219–240
Whitman, W., 278, 279
Whittall, A., 129
Wilmart, A., 38n
Wilson, E. O., 17
Wilson, H. A., 37n
Woodberry, G. E., 267n, 269n, 272n
Wuthnow, R., 106
Wyatt, E. P., 119

Yip, W.-L., 207
Young, M., 106

Zakay, D., 112
Zederblom, N., 83
Zelaznik, H. N., 121

Subject index

Aborigines, time concept of, 78
Achieved rationality paradigm, 97–100
Acta Synodi Caesareae, 38
Action, rational, 248
Action-guiding factors, 94–95
  invariant human nature and, 96–97
  relatively invariant, 98–99
Actual entity, 233–234, 235
Actual occasions, 228–229, 236
  as actively striving, 229–230, 232
  as agents, 238n
  freedom and, 239
  prehensions of, 233
  subjective aim of, 230
  in time, 231–232
Actuality, status of, 62–63
Adamic viewpoint, 270, 271
Adaptation, improvement of, 7
Adaptive behavior, emotions and, 17–18
Adventures of Ideas, 228
Aesthetic judgment
  emotions and, 18
  nature of, 22
Aesthetic paradigm, 71
  elements of, 101–103
  model of temporality for, 103–106
Aesthetic values, 19
Aesthetics, musical, 141–145
Affective bonding, 71
  group, 103
After Russia, 193–194, 201
Aida, 140
American culture, morphology of, 263–280
American Revolution, 246
"American Scholar, The," 279
Analogical process, 136, 139–140
  in music, 127, 134t
Analogy
  Kafkaesque, 143
  music as, 139–140
Analyticity, 58
L'Ancien Régime et la Révolution, 88
Ancient Voices of Children, 170–172
Anger, as adaptive behavior, 17–18
Anthropic arguments, 58
Anthropomorphosis, 81
Archaic society, 80
Argumenta, 29
Art, root metaphor of, 102–103, 105
"Art in the Light of Conscience," 202
Artistic process, 129
  in medium, 131
Astronomical Almanac, 25
Astronomical reckonings, 27, 33–34
Atemporal mood, 21
Atemporality, 5–6, 207, 208, 212–213, 215
  lack of predictability in, 11
Attention
  enhanced, 112
  focused, 111–112
  to temporal information, 122
Attentional resources, 111
Authority
  of sacred, 99–100
  time in quest for, 71
Autonomy, in society, 256–261

Bach Concerto in D minor, 138
Beatitudes, 13
Beauty

definition of, 17
experience of timelessness in, 22
longevity of, 20
in management of emotions, 17–23
Becoming
occasions of, 275–278
sacredness of, 263, 271–275, 280
Becoming of continuity, 219, 239–240
Beethoven Quartet in C-sharp minor, 137
Beethoven's Symphony No. 7, cyclicity and linearity in, 178–183
Before/after distinction, 91
"Before the Law," 211–212
Being, openness of, 270
Beliefs
guidelines to, 9–13
selection of, 8–9
Biological freedom, 15
Biological memory, 10
Biological rhythms, 112
Biology, selective forces of, 15
Biotemporal mood, 22
Biotemporality, 5, 207, 211, 215
continuity of, 210
hallmark of, 21
Blind rationality, 98–99
Bolero, 137
Brain. *See also* Cerebrum; Mind
activity of in time programming, 119–121
event-related potentials of, 119–120
evolutionary forms of, 21
left and right hemispheres of, 177
lesions of and timing deficits, 120–121
Buddhism, *kalachakra* doctrine of, 81

C-Span, 252n
Calculus, 28
Calendar, 25
Christian development of, 32–42
conflict over, 37
Easter on, 30–32
eight-year cycle of, 35–36
in European middle ages, 28–30
French reform of, 87–106
of French Republic, 89
French Republican Convention debate on, 92–95
Hebrew, 34
Hellenistic, 31–32
nineteen-year cycle of, 46
reckonings of, 27–46
Roman, 32
Calvinists, 264–265
Candide, 78
Cantus firmus, 140
Carmen, 140
Causal efficacy, 231, 265–266
Causality, 64
Causation, 11
final, 221
hierarchy of, 12
level-specific, 6
philosophical problem of, 220–222
probabilistic, 15
Cause-and-effect paradigm, 71, 95
Cerebrum, 21
activity of, 119–120
Chaining hypothesis, 113
Chance, 222–223, 238n
Change
human values and, 1–23
ontological priority of, 12
stability and, 277
as ultimate, 271–275
Chaos, 6, 82. *See also* Disorder
atemporal, 207
with disappearance of temporal distinctions, 104
longing for, 215
versus order, 187
primordial, 206
pull of, 21
transition to cosmos from, 71, 75–76
Checks and balances, 251, 252–253
Children, temporal regulation in, 115–116
"Chopsticks," 132–133
Christian schools, early, 42–43
Christianity, time concept in, 83–85
Chronology, 25
Chukot-Eskimo myth, 77
"Circles," 218, 274–275
Circular process, 67, 68
Closed-loop hypothesis, 113
Coda, 165
Cognitive science, 108
Coherence, 227
Commentarial process, 135, 136, 140, 141
in music, 127, 134t

Compactification, 55n
Completeness, 131
Complexities, 5
  definition of, 6
Composer's Voice, The, 142–143
Computer memory, 10
Computistica, 43–44
Computus, 25, 27, 28
  arguments of, 46
  calculations of, 44–45
  Christian development of, 32–42
  cyclic and linear change in, 218
  definition and origin of, 28–29
Concepts
  history of, 256
  practical meaning of, 259
Concrescence, 233–240
  definition of, 219, 239
  fluid concept of, 236–237
Conduct, good and evil as guidelines to, 13–17
Conflict
  definition of, 7–8
  unresolvable, 7–8, 14
Confucianism, 14
Consciousness
  definition of, 224
  emergence of, 229–230
  hierarchy of, 229
  thought and, 224–225
Consequence explanations, 95–96
Constitutional government, 245
  future-orientation of, 249
  as prescription for future, 261
Constitutionalism, temporalized, 250n
Constitutive tension, 7
Continuity, 260
  of becoming, 219
  becoming of, 219, 239–240
  lack of sense of, 21
Continuum-based space-time manifold, 58
Controllability, 142
Cosmos
  definition of, 75
  transition from chaos to, 75–76
Counterpoint, 131n, 136n
Counting, evolution of, 10–11
Creation
  fallen, 268
  process of, 139
Creative act, 239
Creative advance, 226
Creative tension, 210
  lack of, 210–211
Creativity, 8
Cultural ideal, 264
  American, 263–280
Culture
  early stages of, 76
  variation in, 92
Cumaean Sibyl, 196–197
Cursus paschalis, 28–29
Cyclic time, 73, 79, 80–81, 168–169, 247
  in music, 170–172, 177–183
Cyclicity, 167
  in Bach, 174–177
  in Beethoven's Seventh Symphony, 178–183
  deep, 174
  in music, 170–172, 177–183
Cyclus, 29
Cyclus Paschalis, 40–41

Damascus school, 57
Darwinism, 277–278
Death
  Christianity and, 269
  knowledge of, 3
  life after, 4
Decision, freedom of, 258
Deep time
  in Bach, 174–176
  in Beethoven's Seventh Symphony, 181–183
  cyclic, 172
  definition of, 167, 168
  in music, 167–183
Degrees of freedom, 12, 15
Democracy
  definition of, 245
  direct, 246
  pure, 244
Democratic governments, 243
Democratic principle, 252
Democratic republicanism, 217
Descriptive music, 139
Determinism, 237
Developmental psychology, time concept in, 78–79
Disjointedness, 21
Disjunctive process, 127, 134t

Disorder. *See also* Chaos
in literary experience, 205–216
pull of, 209, 213
"Dissonant" Quartet in C major (K. 465; Mozart), 138
Divided line image, Platonic, 23
Djjanna, 84
Doctor Zhivago, 186
"Don Giovanni," 208–210
Don Giovanni (Mozart), 138
Dualities, 266–267
Duration, 60–61
creation of, 150
definition of, 147–148
formation of in music, 162–163
of impulses, 153–154
internal processing of, 110
measurement of in ancient Mesopotamia, 76
meter and, 157
in music, 147–165, 157–158
periodicity and, 151
prediction in music and, 154–156
retrospective versus prospective judgments of, 111–112
subjective, 111
subjective versus objective, 113–114
Dzugur epoch, 78

Earthly signs, 199
Easter
calculation of date of, 30–32, 33–42
dates of for Roman civitas, 36–37
importance of timing of, 32–33
Easter tables, 35n
of Dionysius Exiguus, 41
to unify celebration of, 40
Echoic memory, 152
Edwin Drood, 130n
Ekstasis, 22
Elaborative process, 135, 136, 137
in music, 127, 134t
Election, voluntary, 252
Emergent process, 135, 138, 140
in music, 127, 134t
Emotions
as adaptive behavior, 17–18
beauty and, 17
bonding through, 71
evolutionary origins of, 19–20
expression of, 18–19
guidelines to management of, 17–23
Enduring objects, 231–232
Enjambment, 197
Entertainment, 150–151
Entropy, 8n
Eotemporal mood, 21
Eotemporality, 5, 207, 209, 210, 211–212
probabilistic causation of, 11
"The Epic and the Lyric in Contemporary Russia," 194, 204
Epic hero, 81–82
Equinoxes, position of, 46
Eroica Symphony, 135n
Eternal objects, 231
creation of, 238
Eternal realm, 89–90
Eternity
dynamic, 67
versus time, 84–85, 266–267
time and, 193
Ethical rules, 14–15
Euclidean geometry, 27
Evaluative decisions, 254–255
Evening Album, 190
Events, 59, 61
articulations of, 163
components of, 149–150
duration of, 148
extending over of, 59–60
extensive nature of, 149
persistence and simultaneity of components of, 149
telepathic, 65
wholeness of, 150–151
Evil, conduct and, 13–17
Evolution, organic, 14–15, 17
Evolutionary theory, 263, 277–278
Existence, realms of, 191
Existential tension, 16
nootemporality and, 8
origins of, 7
Experience, symbolic transformations of, 2–3
On the Experience of Time, 111
Experimentalism, 274
Extending-over, transitivity of, 63
Extensive connection, 230–231
Extensive continuum, 230

Fate, 82
Faustian wager, 213–214
The Federalist, 247n
  No. 14, 252
  No. 78, 253
Fedora, 140
Feelings. *See* Emotions
"Fidelity," 218
Figuration, 144
  in music, 141
  process of, 136n
Figurative process, 135–136, 140
  in music, 127, 134t
Figure/ground phenomenon, 140–141
Final end, 260
Finalities, 274
Flux, 226; *See also* Time, flux of
Fog Warning, 234
Force manipulations, 113–114, 115
Foundations of Metaphysics in Science, The, 202–203
Fragmentation
  overcoming of, 199–201
  psychic pain of, 209
Free/strict archetype, 138n
Free will, 15
Freedom, 15, 244–245
  biological, 15
  decisional, 258
  experience of, 254–255
  paradox of, 237–239
  reality of, 254–256
  shared, 251–252
  temporal, 243
Freedom-for-the-future, 255, 256–257
"Freedom of the Will," 266
French Revolution, 87–88, 246
  problem of process and, 92–95
Future, 260–261
  call to, 250–251
  creation of, 248–249, 257–258
  open, 94–95, 98
  orientation to, 256–257
  past and, 85
  religious view of, 84
Future objects, 62–63
Future-related behavior, 3
Futurity, horizons of, 5
Galilean relativity, 271
Game model, 71
General will, 251
Genetic divisibility, 233
Genetic variation, 14
Geological deep time, 168
Geology, 275
God
  as Becoming vs. Being, 222
  causal efficacy of, 265–266
  causation and, 221
  consequent nature of, 233–234
  continuity with, 270
  finite, 223, 233–234
  immutable and perfect, 220
  immutable vs. mutable, 219–240
  as mediator between extremes, 267–269
  primordial nature of, 233
  subjective aim and, 237–238
Good
  definition of, 14
  as guidelines to conduct, 13–17
Government
  in ancient Greece, 244–246
  illegitimate and legitimate, 244
  mixed, 247
  participatory, 253
  process in, 243–261
  representative, 253
Grand Unified Theory (GUT), 54
Gravitational waves, 55
Gravity, 54
  Einstein's theory of, 54
Greek adventure romance, 77–78
Greek mythology, time concepts in, 82–83
Gregorian chant, 140
  ametric, 151
Group, affective bonding in, 71, 103

Habit, 222–223
Handmaid's Tale, The, 210–211
Hartle-Hawking theory, 55–56
Harvard University, 279
Hatred, as adaptive behavior, 17–18
Hausdorff condition, 60
Heavenly City, 84
Hero, independence of, 81–82
"A Hero of Labor," 198–199, 200
Hierarchy, 64–65

Hindus, earthly time concept of, 77
Historical time, 83–84
History
  continuity of, 273–274
  time and, 93–95
  as unresolvable conflicts, 14
Human consciousness
  development of, 74
  William James's concept of, 223–226
Human freedom, 15; *See also* Freedom
Human values
  change, permanence and, 1–23
  time and, 8–9
Hypermeasure, 170, 174–176
Hyperspace, 55–55
  point in, 56–57

"I," 10–11
Ideal world, 191
  escape into, 193–194
Idealism
  of Berkeley, 265–266
  objective, 222–223
Identity, as paradigm for permanence, 10–11
Imitative process, 127, 134t, 135, 137
Immortality, longing for, 4
Implicative process, 136–137
  in music, 128–129
Impulse sequences, 153–154
  duration of, 156–157
"In-between" realm, 195–198
Incompleteness theorem, 6
Indeterminism, 237
Infinity, points of, 57
Insanity, collapse into, 210–211
Instances, 61
Instant, 60
  concept of, 163, 164–165
Institutional memory, 17
Instrumental rationality paradigm, 95–97
Integrative levels, 5
  distinction among, 6
Intelligence
  levels of, 74
  political, 248–249
  pragmatic, 249–250
  preconceptual, representative, 78
Interactive process, 136n, 138, 140
  in music, 127, 134t, 141

Intercalation cycle, 31
Internal clock theory, 112
Internal process, 127
  in music, 128
Internal relation, 229–230
International Society for the Study of Time (ISST), Seventh Conference of, vii-viii
Interrupted process, 144–145
Intuition, 226, 238
  Bergsonian, 237
  speculative philosophy, language and, 226–228
Intuitive time, 76
Invariant, relatively, 90–91, 96–97
  definition of, 98
  redefinition of, 94–95
Invariant ends, 96
Invariant/variant distinction, 90–91, 104
Istar symphonic variations, 135n

Kala, 84–85
Kalachakra doctrine, 81
Kamlaniye seance, 76–77, 79
Knowledge, 4
  capabilities for, 273
  ways of, 12
Kontakte, 151
Kronos, 73, 82

Labor, division of, 3, 258
Language
  intuition and, 226–228
  level-specific, 6
  problems associated with, 228
  resistance to change in, 19
  values built into, 15
Language game, 101
Laterculus Augustalis, 37–38
Law, definition of, 249
Leap year, 32
Learning process, biological, 10
Legality/legitimacy, 98
Lent, reckoning of date of, 34
Life
  after death, 4
  beginnings and endings of, 3
  first appearance of, 7
Life process

interpretations of, 145
metaphysical definition of, 7–8
Limbic system, 19, 21
Linear sequence, 125
Linear time, 168
in music, 170–174
Linearity
in Beethoven's Seventh Symphony, 181–183
characteristics of, 172–174
in music, 177–178
transcendence of, 193–194
Listening, for process, 137–141
Literature
temporality in, 205–216
time felt and understood in, 185–187
Local process, 135, 136
Long-term memory, 176–177
Lunar calendar, 34
Lunar cycle
calculation of, 45–46
early systematic coordination of, 31–32
reckoning of, 44
uncertainties of, 34
Luni-solar time cycles, 50–51
Lyric verse, time and process in, 189–204

Macroscopic process, 236
Magic, 83
Man, ascent of, 14n
Manifold, four-dimensional, 62
Mappaemundi, 29
Materialism, 221
consciousness in, 229–230
implications of, 221–222
Mathematics
definition of, 11
in quantitative physics, 62
truth in, 11
Mathis der Maler, 140
Matter, definition of, 222–223
Mbira music, 151
Meaning, in musical process, 141–145
Means-and-end duality, 257, 267
Memory
biological, 10
duration and in music, 157–158
echoic, 152
event in, 149
institutional, 17
long-term, 176–177
short-term, 176, 182
Mental functions, 2–3
Mental present, 5
Mental processes, 66. *See also* Mind
events in, 65
intercultural differences in, 73–74
William James's concept of, 223–226
Metaphor, musical, 167–183
Metaphysics
Bergson's definition of, 236
of hierarchical theory of time, 7–8
of mathematics, 62
of science, 12
Meter, 151, 156
deep, 169–170
definition of, 174
in rhythm, 152, 157, 162–163
structure of, 161–162
transcendence of, 197
Metonic cycle, 31–32
Metrics, 150
Microscopic process, 236
Middle class, 247–248
Mind. *See also* Brain
evolutionary origins of, 2–3
James's concept of, 235–236
processes of. *See* Mental processes
stream view of, 224
William James's concept of, 223–226
Mind-body problem, 220–222
Minding, process of, 8
Minimalism, 135
Modes of Thought, 228
Monads, 222
Moods, 21–22
flux of, 277
temporalities, 187
Moral courage, 14
Moral law, existential tension in, 16
Moral values, 15–16
Mortal terror, 3
Motion
action and, 239–240
ontological priority of, 12
qualities of, 144
understanding of in materialism, 221
from valuation, 234
Motor program, 109

Motor programming, 122
  conception of, 117–118
Motor timing, 109
  central and peripheral mechanisms in, 115
Movement velocity, 114–115
Multilayer spatial isomorphism, 79
Multilayer world model, 76–77
Mundane world, 190–191
  hostility to, 193
  perceptual norms of, 198
Music
  as analogy, 139–140
  cyclicity in, 170–172, 177–183
  difficulty in describing, 129
  duration and rhythmic process in, 147–165
  figure/ground phenomenon in, 140–141
  linearity in, 172–174, 177–178
  metaphor and deep time in, 167–183
  new, 141, 143
  openness of beginning and ending in, 163–164
  Plato on, 129
  process in, 125, 127–145
  qualities of motion in, 144
  texture in, 144
  time in, 147–148
Musical "now," 176
Musical phrase, 158
Musical process. *See also* specific types of
  aesthetics of, 130, 141–145
  as analogy, 130
  categories of, 133–137
  completeness in, 131
  components of, 128, 141–145
  compound, 140
  cultural attitudes toward, 145
  definition and illustration of, 131–133
  definition of, 141
  descriptive, 139
  history of, 135
  listening for, 137–141
  local, 135
  in new music, 141
  scale of, 131
  taxonomy of, 127, 134t
  usefulness of investigating, 129–130
"My Pushkin," 201
Mythology
  dynamics of, 75
  religion and, 83–85
  temporal conceptions of, 73–85
  world image of, 74–75

Natura naturans, 275, 277
Natural selection, 10, 15
  in organic evolution, 17
Natural world, 83
Nature, 270, 271–272
Nature
  eternal laws of, 89
  as external, 270
  as fallen creation, 268
  integrative levels of, 5, 6
  laws of, 90
  reality of, 270–271
  transitions in, 274
"Nature," 276, 277
Necessity, 82
  as law, 82–83
Neighborhood, 63
Neoplatonic hierarchy, 269–271
Neural networks, 121
Newtonian universe, 265
  permanence and predictability in, 11
Nexus, 144
  of occasions, 229, 231–232
  process, 64
Nirvana, 84
Nisan, 34
Noetic mind, 209, 211–212, 214
  suppression of, 210
Noetic voice, 214–216
Noise of Culture, The, 206
Nonlinearity, 174
Nootemporal mood, 22
Nootemporality, 5, 207, 209–210
  creative conflict and, 212
  existential tension in, 8, 16
  flight from, 215
Number, nature of, 11

Objective idealism, 222–223
Objective immortality, 232
Objects, 61
  actuality of, 62–63
Observer, subprocess of, 63
Oceanic feeling, 21, 22
Ontology, of Whitehead, 228–235

Opportunism, 259
Oppositions, 144
Optimism
  cosmological, 273
  Emersonian, 272, 273
  metaphysical, 272
Order
  breakdown of, 209
  conflict of with chaos, 187
  creation of, 206
  in literary experience, 205–216
  structuring of, 209–210
Order/disorder dialectic, 187, 206
Organic evolution, 9–10, 14–15
  natural selection in, 17
Organic present, 5
Organic processes, 65
Orpheus myth, 196–197
"Other world," 191, 193, 194–195
Overlapping, 163–164
"Oversoul," 274

Paradise, 84
Parallel process, 136
  in music, 127, 134t
Participatory democracy, 253
Particularity, 149
Passing, feeling of, 93–94
Passion, 94
Past
  archetypical, 80
  break with, 88
  future and, 85
  mythical, 78
Pastness, horizons of, 5
Periodic processes, 67–68
Periodicity, 151
  duration and, 152–153
Peripheral feedback, 115
Peripheral hypothesis, 122
Permanence
  appeal of, 3, 10
  evolution of, 11–12
  good and, 14
  human values and, 1–23
  superiority of, 22–23
Permanence-analogues, 9, 10
  hierarchy of, 11
  scale of, 12
Persona, 127
  in music, 142–144
Phenomenal world, 9–10n
Philosophy
  speculative, 226–228
  traditional problems of, 220
  Whitehead's conception of, 220
Physics
  problems in, 53–58
  process-based, 26, 58–65
  speculative philosophy in, 227
Planetary motions, calculation of, 29–30
Planetary theory, 46
Planisphaeria, 29
Planning, individual and collective, 217
Platonic forms, 231
Poet
  as emigrant, 199
  as mediator, 195–196
  responsibility of, 195
"The Poet and Time," 192, 195, 200
"A Poet on Criticism," 203
Poetic devices, 197
Poetry
  creation of, 195
  goal of, 200–201
  as living organism, 204
  as process, 201–202
  task of, 191, 197–198
  time and process in, 189–204
Points, 60
Political intelligence, 248, 249
Political judgment, temporalist standard for, 248
Polity
  in ancient Greece, 244
  definition of, 243, 247
  process of, 243–261
Populism, 252
Possibilities
  appropriation of, 248–249
  closure of, 277
Power of Sound, The, 143
Pragmatic intelligence, 249–250
Pragmatic maxim, 258–259
Pragmatic reasoning, 258–260
Pragmatism, 244–245, 259–260
Precueing paradigm, 118–119
  brain event-related potentials in, 119–120
Predictability, 9, 11

duration and, 154–156
in music, 152
Prediction, chronological aspect of, 56
Preference formation, blindness to, 99
Prehension, 217, 225–226, 230, 236
analysis of, 233
conceptual and physical, 232
content of, 231
Prescriptive procedure, 256–258
Prescriptive rules, 249
Presentational immediacy, 231
Primal-impressions, 153
Principles of Psychology, 236
Privation, 259
Probabilistic causation, 15
Probabilistic truth, 11
Probability
amplitude, 63
calculation of, 57
laws of, 64
Procedural rules, 101
invariance of, 100
Process, 25. *See also* Musical process
alternative view of, 61–65
circular, 67, 68
concrescence in Whitehead's philosophy of, 217, 219–240
connectivity of, 66
continuity of, 260
definition of, viii, 7–8, 128, 202–203
in definition of lyric verse, 189–204
in early work of Whitehead, 59–61
Emersonian view of, 274–275
in French Revolutionary calendar reform, 92–95
inferiority of, 22
interrupted, 144–145
laws of, 63–64
versus local process, 133–135
modern problem of, 87–106
in music, 125, 127–145
organic, 65
periodic, 67–68
phases of, 165
of polity, 243–261
as primitive physical category, 53–68
real versus temporal, 234
social, 71
society as, 217
space-time and, 61–63
study of, vii-viii
temporal scale of, 130
Process and Reality, 228, 238n
Process philosophy
Emerson's, 278–280
historical context of, 220–226
questions remaining in, 235–240
Process physics, 58–65
absence of substratum in, 61
time and, 66–68
Process values, 141–145
Programming hypothesis, 109
Programming time, 118, 121
proprioceptive cues and, 118–119
Progression, musical, 141
Property rights, 253–254
Prophecy, 197
Proprioceptive cues, 114–115
programming time and, 118–119
in temporal regulation, 122
in timing accuracy in visual-motor games, 116–117
Proprioceptive feedback, 113
Prospective judgment, 111
Prototemporal mood, 21
Prototemporality, 5, 207, 208, 215
probabilistic truth in, 11
Providence
as mediation between extremes, 267–269
as mediator, 264
Psychological present, 176
Psychology, time in, 108
Public property, 246, 252
Puritanism
replacement of, 264–266
spirituality of, 264

Quantitas intrinsica concept, 157–158
Quantitative physics, 62
Quantities, laws of, 11
Quantization process, 54
Quantum cosmology, 55–58
Quantum indeterminism, 232
Quantum-mechanical structure, 64
Quantum theory, 63, 64

Raga, 138
Ratio paschae, 38

Rational activity, 248
Rationality
  achieved, 97–100
  instrumental, 95–97
Reaction times, 118
  brain activity and, 120
  precueing, 119
  programming effects on, 119
  programming time and, 121
Reader, role of, 203–204
Real number continuum, 57
Reality
  as arbiter of truth, 101–102
  description of, 90
  dual realms of, 89–90, 193–198
Reason, Emerson's view of, 274
Reasoning
  versus emotions, 19–20
  pragmatic, 258–259, 260
  qualitative, 12–13
Rebirth, symbolism of, 77
Recurrence, 91
Reductionism, 66–67
Referential process, 136–137
Reincarnation, 77
Relative reality, 65
Relativity
  general, 56
  special, 59
  theory of, 25
Religion
  antiscientific basis of, 83
  authority of, 99–100
  mythology and, 83–85
  rational theory and, 99–100
Repetition
  with a difference, 168, 178
  in music, 141, 164, 177–178
  sectional, 177–178
Repetitive process, 135, 140
  in music, 127, 134t, 137
Replicable process, 130–131
Representative government, 246
Representative republic, 243, 253
Reptilian brain, 21
Republic, 17, 270
Republic, 245
Republican government, 246
  checks and balances in, 251, 252–253
  decisions in, 251–254
  theory of, 247
Republican principle, 252
Republicanism, classical, 243
Res publica. See Public property
Response time, 119
Responsibility, 255–256
  shared, 258
Retrospective judgment, 111
Rhetorical process, 136
  in music, 127, 134t
Rhythm
  deep, 169–170
  duration and, 150, 151
  meter and, 152, 157, 162–163
  in musical process, 147–165
  qualities of, 125
Rhythmic drive, 150
Rhythmic impulse, 150
Rhythmos, 151
Rights
  definition of, 13
  inherent, 253–254
Ritual
  in everyday life, 105–106
  temporal distinctions and, 106
Roman Republic, 246
Root metaphor
  of art, 102–103, 105
  mechanism, 95, 96
Der Rosenkavalier, 139–140
Rossini crescendo, 140
Rules, in republican government, 252–253

Sacredness of becoming, 271–273
Saltus lunae, 32, 44–45
Salvation history, 13–14
  linear process of, 25
Satisfaction, 230
Science
  metaphysics of, 12
  truth and, 101–102
Scientific judgments, 20
Self-creation, paradox of, 237–239
Self-integration
  return to, 215–216
  urge toward, 210, 213
Self-interest, pursuit of, 94
Self-organization, 8

Self-realization, 265
"Self Reliance," 266
Self-world, 9–10n
Selfhood
  development of, 3, 10–11
  emotions associated with beauty in, 18
Sensory feedback, 113
  timing accuracy and, 115
  in timing accuracy in visual-motor games, 116–117
Sequence, 91
Sequential order, 96
  knowledge of, 98
Sequential recurrence, 125
Serious playfulness, 101
Shakuhachi music, 151
Shaman, time and, 76–77
Short-term memory, 176
  codings of, 182
Sibyl, 193
Simultaneity, 149, 154, 164–165
Sin, 274
  locus of, 268–269
Skepticism, 272
Slavery, 245
  legitimacy of, 247
Sneering, 18
Social being, consciousness and, 74
Social bonds, time and, 105–106
Of the Social Contract, 246
Social harmony, 245
Social integration, 103
Social life, 102
Social memory, 28
Social processes, 71
Social reasoning, 258
Social rights, 246
Social theory, temporal dimensions and, 95–103
Social variation, 92
Society
  democratization of, 253
  division of labor in, 258
  of occasions, 231–232
  personal autonomy in, 256–261
  personal life in, 254–256
  as process, 217, 250
Sociotemporality, 6
Solar cycle
  calculation of, 45
  early systematic coordination of, 31–32
Sophie's Choice, 187
Space
  in music, 147–148
  mythological view of, 75–76
  time and, 259
  in visual arts, 147
Space-time, 54
  ambiguity in concept of, 54–55
  conceptual simplicity of, 26
  continuum-based, 58
  definition of, 54–55
  geometries of, 57
  from process, 61–63
  processes in defining, 58
  quantization of, 57
  topology of, 63–64
Species-specific universe, 9–10n
Speculative philosophy, 226–228
Spirit, 272
Spirit of the Laws, The, 246
Stability, illusion of, vii
Stagnation, 80
States of arbitrarily fine structure, 57
Storage size model, 111
String theory, 53–55
Subject-superjects, 236–237
Subjective aim, 237–238
Subprocess, 63
  network of, 63
  reality of, 65
Substratum, 54
Superject, 230, 236–237
Supernatural destiny, 263–264
Supernatural world, 83
Superstring theory, 54
  ambiguities in, 56
Svet, 191
Symbolic causes, 15
Syntactical process, 138–139
  in music, 127, 134t

A Tale of Two Cities, 186–187
Taoism, 14
Teleology, 221, 234–235
Telepathic events, 65
Telos, 150
Temporal continuity, 260

principle of, 256
Temporal dimensions, 90–92
disappearance of, 104
fusion of, 104–105
ritual in awareness of, 106
social theory and, 95–103
Temporal flow, 236–237
Temporal freedom, 243
Temporal information processing, 109–110
Temporal moods, 207
Temporal order, 64
Temporalities, 5–6
in aesthetic paradigm, 103–106
affective dimensions of, 21–22
definition of, 206
dimensions of, 89–92
directed and undirected, 9
Emersonian view of, 276
levels of, 207–208
in literary experience, 205–216
moods of, 187
strange, 66
Temporality/eternity duality, 266–267
Texturing, 144
Theory of Everything (TOE), 53–54, 217
uselessness of, 227
Thought
consciousness and, 224–225
direction of, 236
diversity of, 225
passing, 230
Tidal rhythm, 7
Time. *See also* Temporalities
of archaic society, 80
bases of, 109
canonical forms of, 6
as conflict, 4, 7, 20
continuity of, 260
continuum of, 169f, 259
cultural and social variation in relationship to, 92
cycle of, 27–46, 125
cyclic, 73, 79, 80–81, 167–169, 170–172, 247
deep, in music, 167–183
defining nature of, 108
in definition of lyric verse, 189–204
dimensions of, 89–92
directionless, 21
eternity and, 84–85, 193
evolution of, 6
experiential, 26
felt and understood, 185–187
flux of, 226, 234–235, 236–237, 239–240
freedom and, 255–256
of French Republic, 89
hiatus from, 77–78
hierarchical model of, 4–8, 205, 206–216
history and, 93–95
human, 66–67
human values and, 8–9
instrumental-rational conception of, 105
intuitive, 76
linear, 80, 168, 170–174
models of perception of, 111
in music, 130, 147–148
mythological, 36–85
natural philosophy of, 4
philosophical and political issues of, 88–106
points, durationless, 153
points or instances of, 60–61
in political processes, 243–261
process physics and, 66–68
quantification of, 193
reckoning of, 25, 29
rethinking relation to, 90–91
as river, 77, 85
social theory and dimensions of, 95–103
versus space, 257
spiral of, 81
static model of, 82
transiency of, 229
truth and, 261
as two-dimensional, 104
use of, 105
variations in concept of, 85
victory over, 193
Time, Freedom, and the Common Good, 250n
Time-compact globe, 17
Time coordinate, 56
Time of Music, The, 169–170
Time processor, 112
model of, 113f
Time programming
cerebral activity associated with, 119–121
nature of, 121
Timelessness, experience of, 20–21, 22, 23
Timer model, 111

Time's arrow, 125, 167, 168, 172. *See also* Linear time
in Beethoven's Seventh Symphony, 181–183
in music, 177
Time's Arrow, Time's Cycle, 168
Time's cycle, 172. *See also* Cyclic time
Timewriters, 205
Timing
accuracy of, 113–115, 117
central and peripheral mechanisms in, 109–122
central processing view of, 112–113
deficits in, 120–121
feedback control in, 114–117
importance of, 110
movement velocity and, 114–115
subjective versus objective, 113–114
Tonality, 136, 137
Tonic triad, 174–176
Topology, space-time, 63–64
"Tout par compas suy composes," 172, 173f
Tragedy, 16
Tragic hero, 16
Transcendentalism, 222, 263–280, 272–273
Transcendentalists, 266
Transition, 234–235, 274
Transition points, 71
Trees, symbolic, 196
Troilus and Cressida, 140
Les Troyens, 140
Truth
definition of, 9
reality and, 101–102
in science, 12
in selection of beliefs, 8–13
spectrum of, 13
time and, 261
Il Turco in Italia, 140
Twistor theory, 58
Two on a Tower
on human insignificance, 4
truth in, 13

Ugliness, 17–23
Ultimate Becoming, 222
Umwelt, 9–10n, 206–207
hierarchy of, 205
levels of, 209–212
Unbearable Lightness of Being, The, 214–216
Undeterminancy, level-specific, 6
Unitarian movement, 264–266
Universals, 231
Universe
four-dimensional space-time, 55
hierarchical view of, 64–65
history of, 55–56
instantaneous state of, 56–57
Unpredictability, 10
Unreality, 65
Untruth, 8–13
Utilitarianism, 259

Value
change, permanence and, 1–23
definition of, 97–98
epistemology of, 4, 8
as function of passage, 277
in language, 15
in musical process, 141–145
Platonic theory of, 23
as relatively invariant, 93–94
scales of, 4
time and, 8–9
Value judgments
development of, 19
need for, 2–4
versus reasoning, 19–20
Vanessa, 140
Variant, relatively, 90–91
Variation, axes of, 91–92
Venus de Milo, 20
Vibration, 142
Video games, temporal regulation in, 115–116
Virtues, 97–98
Vision, three-dimensional, 11
Visual arts, space in, 147
Visual feedback, 115
Visual-motor game, response durations in, 115–117

Wants
autonomy of, 98
mechanisms of formation of, 97–98
Wave function, 54, 56
Well-Tempered Clavier, 136n, 174–177
White Hotel, The, 208–210
Wide World journals, 267
Withness of body, 225, 231

Das Wohltemperirte Clavier. See *Well-Tempered Clavier*
Woman in the Dunes, The, 212–213
World models, 95–103
Zeno's arrow, 12n
Zodiac, 29
Zurvan, 84–85